# Enclosure and the Yeoman

*To my parents*
*Richard C. and Barbara T. Allen*

# ACKNOWLEDGEMENTS

MANY people have helped me in many ways to write this book. I am grateful to the following who read earlier drafts and gave me valuable comments: Rick Barichello, J. V. Beckett, Maxine Berg, Chuck Blackorby, Joanna Bourke, Kevin Burley, Bruce Campbell, Jean Michel Chevet, Greg Clark, Ted Collins, Jane Cottis, Nick Crafts, John Cragg, François Crouzet, Paul David, Lance Davis, Erwin Diewert, Terence Gorman, George Grantham, Sir John Habakkuk, John Hartwick, Barbara Harvey, Michael Havinden, Dan Heath, John Helliwell, Paul Hohenberg, Phil Hoffman, Jim Huzel, Joanne Innes, Eric Jones, Esther Kingston-Mann, Ashok Kotwal, David Landes, Frank Lewis, Peter Lindert, Don McCloskey, Gordon Mingay, Joel Mokyr, Jeannette Neeson, Patrick O'Brien, Cormac O Grada, Mark Overton, Mike Percy, Gilles Postel-Vinay, Angela Redish, Jean-Laurent Rosenthal, Henry Rosovsky, Tony Scott, Ken Sokoloff, Keith Snell, Peter Temin, Joan Thirsk, Komas Tsokhas, Michael Turner, Dick Unger, Herman van der Wee, Ann Whiteman, Gavin Wright, Jeff Williamson, Ross Wordie. I am particularly indebted to Chris Archibald, Stan Engerman, Mary MacKinnon, Avner Offer, and Bill Schworm for reading the entire manuscript as well as providing enormously helpful support and discussion.

Various earlier papers were presented to seminars at the following universities: Adelaide, Alberta, Australian National University, British Columbia, California (Berkeley and Los Angeles), Cambridge, Harvard, London School of Economics, New South Wales, Oxford, Reading, Stanford, Warwick, Tasmania, Yale. I also presented papers to the following conferences: Cliometrics Conference, 1979; International Conference on Nineteenth-Century Agrarian Structure and Performance, Montreal, 1984; All-University of California Conference on Peasants; Conference on International Productivity Comparisons, 1750–1939, Bellagio; Conference on Peasant Agriculture, Bellagio; and the International Economic History Congress, Berne. I benefited from the discussion at these presentations.

Some material in this book appeared in papers published in the following journals. Their editors and publishers have kindly allowed me to reproduce it: *Economic Journal*, *Economic History Review*, *Rivista di storia economica*, *Journal of Economic History* (published by Cambridge University Press), *Explorations in Economic History* (published by Academic Press,

Inc., who hold the copyright), and *Research in Economic History* (published by JAI Press).

Nancy South, Don Andrews, Susan Ghan, and Michelle Porter provided exceptional research assistance, and Frank Flynn provided invaluable help with computer programming and data analysis. I am grateful for this support, which was essential to the completion of this project.

The Social Sciences and Humanities Research Council of Canada generously funded this research. I also received support from the University of British Columbia and the British Columbia Ministry of Labour Youth Employment Programme, which was a great help at a critical stage.

The archivists and staff of the Record Offices in which I worked unfailingly offered assistance, which I am happy to acknowledge. I thank the Warden and Fellows of St Antony's College, Oxford, for making me an associate member during 1980, 1981, 1982, and 1986. I also thank the Department of Economic History, Research School for Social Science, Australian National University for appointing me a research fellow in 1987–8, when the first draft of the manuscript was written.

I thank Jacquie Metzler of Culpeppers in Vancouver and the landlords and staff of the White Harts in Wytham and Fyfield, Oxfordshire, for creating a hospitable atmospshere for reading and reflection.

Paula Levine provided great encouragement during the research for this book, and Dianne Frank has helped me integrate it into my life. I thank both of them for their support.

# CONTENTS

# MAP AND FIGURES

# TABLES

# ABBREVIATIONS

| | |
|---|---|
| BPP | British Parliamentary Papers |
| cpi | consumer price index |
| GDP | Gross domestic product |
| PRO | Public Record Office |
| RO | Record Office |
| TFP | Total factor productivity |
| *VCH* | *Victoria County Histories* |

But now the grand object is before me. I want several penetrating political arithmeticians at my elbow to point out the combinations between different, and seemingly distinct circumstances, too many of which will, I fear, escape me.

Arthur Young, *Northern Tour*, 1771, iii. 378
(punctuation modernized)

Ce passage de l'état de nature à l'état civil produit dans l'homme un changement très remarquable, en substitutant dans sa conduite la justice à l'instinct, et donnant à ses actions la moralité qui leur manquoit auparavant. C'est alors seulement que, la voix du devoir succédant à l'impulsion physique et le droit à l'appétit, l'homme, qui jusque-là n'avoit regardé que lui-même, se voit forcé d'agir sur d'autres principes, et de consulter sa raison avant d'écouter ses penchants. Quoiqu'il se prive dans cet état de plusieurs avantages qu'il tient de la nature, il en regagne de si grands, ses facultés s'exercent et se développent, ses idées s'étendent, ses sentiments s'ennoblissent, son âme tout entière s'élève à tel point que, si les abus de cette nouvelle condition ne le dégradoient souvent au-dessous de celle dont il est sorti, il devroit bénir sans cesse l'instant heureux qui l'en arracha pour jamais et qui, d'un animal stupide et borné, fit un être intelligent et un homme.

Jean-Jacques Rousseau, *Du contrat social*,
1762, Livre I, chapitre viii

# 1

# Introduction: Agrarian Fundamentalism and English Agricultural Development

> But, without any improper partiality to our own country, we are fully justified in asserting, that Britain alone exceeds all modern nations in husbandry.
>
> *Encyclopaedia Britannica*, 1797, i 249.

THE boast of the *Encyclopaedia Britannica* was well founded. English farmers led the way in adopting new crops and better breeds of livestock. English corn yields were amongst the highest in the world and had doubled since the middle ages. Output per worker was 50 per cent above the next highest European country. At the end of the eighteenth century, British agriculture was indeed the most productive in the world.

English agriculture differed from the European continent's in other, suggestive ways. The technical revolution in farming had been accompanied by an institutional revolution. The open fields were enclosed, and the small peasant holdings were amalgamated into large farms let to tenants who cultivated them with wage labour. By the nineteenth century, a unique rural society had emerged in England. This new society was characterized by exceptional inequality. English property ownership was unusually concentrated. Rents had risen, while wages stagnated. By the nineteenth century, the landlord's mansion was lavish, the farmer's house modest, the labourer's cottage a hovel.

The revolution in rural life was occurring in an increasingly commercial society. From the sixteenth century, London was one of the most rapidly growing cities in Europe. In the eighteenth century this dynamism extended to the provincial towns. From a rustic backwater at the end of the middle ages, England became Europe's greatest commercial power in the eighteenth century, and the leading industrial nation in the nineteenth.

Was there a connection between these events? Usually the answer is 'yes'. Improved farming, the modern agrarian institutions, the increase in inequality, and the First Industrial Revolution are often linked in a system of thought I call Agrarian Fundamentalism. It involves three claims:

1. The technical revolution in farming was caused by the 'modernization' of England's rural institutions. The 'traditional' or 'feudal' peasant farms and open fields of the middle ages had stifled progress. Enclosures and large farms created private property and capitalism; they extended markets and spread commercial attitudes. The result was a productive agriculture.

2. The growth in agricultural productivity gave a strong boost to England's early industrialization—in some formulations it was an actual prerequisite. The manufacturing cities were built with savings from the agricultural surplus, they were peopled with labour freed from farming, and they were fed with the food produced by improved methods. The First Industrial Revolution was the result of the Agricultural Revolution.

3. The increase in inequality was an inherent feature of the Agricultural Revolution. The growth in farm efficiency and the expansion of manufacturing could not have been achieved in an egalitarian society. The idea that there is a trade-off between growth and equity is one of the most entrenched ideas of Agrarian Fundamentalism.

These ideas have had an enduring impact not only on the interpretation of English history but on that of the whole world. In the eighteenth and nineteenth centuries, peasant farming was seen as a stumbling block to the development of the European states. In the twentieth century, Agrarian Fundamentalism prompted the collectivization of agriculture in the Soviet Union and other communist countries. Analogous ideas have been applied by non-communist governments in many poor countries of Asia, Africa, and South America. For all those who contrast a traditional society with a modern one, for all those who argue that the traditional society must be overturned for development to occur, for all those who see inequality as the necessary price of growth—England is the classic case. For that reason, English history is of enduring importance.

## Enclosure, Large Farms, and Productivity Growth

Few ideas have commanded as much assent amongst historians as the claim that enclosures and large farms were responsible for the growth in productivity. This was the consensus amongst the nineteenth- and early twentieth-century works on English economic and agricultural history. Toynbee (1884: 88–9) accepted that

the destruction of the common-field system of cultivation; the enclosure, on a large scale, of commons and waste lands; and the consolidation of small farms into large

... wrought, without doubt, distinct improvement from an agricultural point of view. They meant the substitution of scientific for unscientific culture.

Mantoux (1905) and Polanyi (1944) repeated these sentiments. Lord Ernle (1912: 351–2) concurred: 'Small yeomen, openfield farmers, and commoners could never have fed a manufacturing population. They could not have initiated and would not have adopted agricultural improvements.'

These views are still standard fare in the textbooks. Thus, Wilson's (1984: 33–4, 262) survey of the period 1603–1763 contends:

The full benefits of drainage and root crops were not possible without enclosure ... The land must be freed from communal restrictions that held back the numbers of livestock and technical improvements. The purpose of enclosure was to do precisely this ... Yields may have been nearly doubled.

Further, 'peasant ownership had often meant stagnation, poverty, ignorance'. Mathias (1983: 55–6) agrees that the

Enclosure of open fields, engrossing of smaller plots and holdings into larger agricultural units (units of production and tenure rather than units of ownership) established the basis of improvement ... The break-up of the peasantry was the price England paid for the increased supplies of corn and meat to feed her growing population.

There is always the worry that general texts are out of touch with the understanding of specialists, but in this case the fear is unfounded. Most leading agricultural historians have insisted on the importance of enclosures and large farms when surveying the causes of productivity growth. Thus, Clay's (1984: 114, 119) recent work dealing with the sixteenth and seventeenth centuries states that 'the form of organization which imposed a drag on productivity, was, of course, the open field system'. Moreover, 'there can be no doubt that the larger and more commercially orientated farmers normally secured a higher output per acre than did most peasants ... the continued aggrandizement of larger farmers at the expense of small must, therefore, have played an increasingly important part in the long term rise in agricultural productivity'. Chambers and Mingay (1966: 52), while admitting some possibility of improvement in the open fields, concluded 'Nevertheless, enclosure was necessary because not all open-field villages showed much progress or efficiency and because even where there was progress there were limits'. Large farms were also necessary for advance since 'small farmers generally lacked the acreage and capital to undertake convertible husbandry or improve the quality of their stock and grasslands, and they tended to be ignorant and opposed to change' (Chambers and Mingay 1966: 45). Even Jones, who is otherwise sceptical about the claims for enclosure, believes 'the estate system was an essential part of the way new crops, livestock breeds and farm practices were diffused in England' (Floud and McCloskey 1981: i. 82). 'The agrarian organization which evolved in England made production more flexible and far more responsive to the market than a peasant system could have been' (Jones 1967: 17).

While there is widespread agreement that enclosures and large farms promoted productivity growth, there is a deep difference of opinion as to how they did so. This difference marks a cleavage that splits Agrarian Fundamentalists into two factions—Tory and Marxist. The Tories believe that large farms and enclosures maintained or increased farm employment while increasing production even more; the result was a rise in both yields and labour productivity. In contrast, the Marxists insist that the new institutions reduced farm employment, thereby raising productivity. These different views about farm operations have important implications for the analyses of the contributions of agrarian change to manufacturing development and of the causes of inequality.

Arthur Young was a principal exponent of the Tory view. He was impressed by Quesnay's (1756, 1757) distinction between the *grande culture* and the *petite culture* and applied it to England. Young argued that more labour-intensive cultivation and more livestock raised corn yields. Such cultivation required large initial outlays of working capital. Young believed that large farmers had better access to finance than small farmers, so the amalgamation of farms led to increased capital intensity, greater employment, and higher yields. This enthusiasm for large-scale farming turned into a denunciation of peasant proprietorship after Young's tour of France.

Before I travelled I conceived that small farms, in property, were very susceptible of good cultivation, and that the occupier of such, having no rent to pay, might be sufficiently at his ease to work improvements and carry on a vigorous husbandry; but what I have seen in France has greatly lessened my good opinion of them. (as quoted by Mingay 1975: 190)

To their credit, however, he noted that peasant proprietors worked hard. 'The industry of the possessors was so conspicuous, and so meritorious, that no commendations would be too great for it. It was sufficient to prove that property in land is, of all others, the most active instigator to severe and incessant labour' (Mingay 1975: 190). None the less, only large farmers practised capital-intensive agriculture. 'The husbandry I met with in a great variety of instances on little properties was as bad as can well be conceived'.

Enclosure also raised employment and yields, according to Young, since it gave large-scale farmers more latitude to deploy their capital and thus allowed the benefits of large size to be fully realized. This critique of open field farming has been elaborated by other writers, in particular, Lord Ernle, who identified several features that reduced its productivity. These included: (1) an overcommitment to corn growing, (2) the slow introduction of new crops due to the necessity for group decision, (3) inadequate drainage and weed control due to the intermixture of property, (4) the spread of livestock diseases and the impossibility of selective breeding due to communal grazing (Ernle 1912: 154–6). This indictment remains the standard fare in most modern writing on the subject.

Marxist theory was prompted by the critics of enclosures who contended that they led to employment declines and depopulation. In 1516, for instance, Sir Thomas More had written: 'Each greedy individual preys on his native land like a malignant growth, absorbing field after field, and enclosing thousands of acres with a single fence. Result—hundreds of farmers evicted' (p. 47). How true these claims were we shall see, but they were undoubtedly common, and Marx based his theory of agricultural transformation on them.

In spite of the smaller number of its cultivators [after the peasantry was eliminated], the soil brought forth as much produce as before, or even more, because the revolution in property relations on the land was accompanied by improved methods of cultivation, greater co-operation, a higher concentration of the means of production and so on, and because the agricultural wage-labourers were made to work at a higher level of intensity, and the field of production on which they worked for themselves shrank more and more. (Marx 1867: i817)

Notice here two things: first, the insistence that large-scale, enclosed agriculture reduced employment per acre, and, second, the vague treatment of crop yields. This is an awkward problem for Marxist scholarship. Capital-intensive farming is still the usual explanation for yield increases. 'Agricultural development was predicated upon significant inputs of capital, involving the introduction of new technologies and a larger scale of operation' (Brenner 1976: 49). However, Marxists need to avoid Young's unwanted conclusion that more capital meant more employment. Gone are the armies of turnip hoers. Instead, farmers buy cattle and sheep to heap the land with manure. 'Animal production had to increase in relation to arable in order to provide manure and ploughing to counter the tendency to declining fertility of the soil' (Brenner 1976: 308).

## Agrarian Change, Labour Release, and Manufacturing Growth

The second tenet of Agrarian Fundamentalism is that the increase in agricultural productivity led to the growth of manufacturing. There are many possible links—the provision of savings, the supply of food, the extension of the home market, the release of labour—and they have all been worked into the analysis at one time or another. But labour release has always been the most central, probably because thinking on the subject was prompted by the concern that enclosure led to depopulation. Later I will show that there are other good reasons for this focus.

In the fifteenth and early sixteenth centuries no connection was seen between the depopulation associated with enclosure and the growth of manufacturing. The displaced farmers were assumed to become permanently unemployed. Wolsey's 1517 Commission on Depopulation reports that eighty people were expelled from Stretton Baskerville and that they 'have remained idle and thus they lead a miserable existence, and indeed they die wretched' (Fisher and Jurica 1977: i. 117). In *Utopia* Sir Thomas More (1516: 46–7) asks of such people 'what can they do but steal—and be very properly hanged?'

It was not until the mid-seventeenth century that a link was suggested between enclosure and manufacturing growth. The context was explaining Dutch commercial ascendancy and devising policies for England to emulate it. The political economists who pursued this question developed a two-sector model, in which the economy was conceptually divided into agricultural and commercial sectors, to explain the growth of trade and manufactures. In 1663 Fortrey applied this model to enclosures. He admitted that they destroyed villages and led to the conversion to pasture— 'one hundred acres of which, will scarce maintain a shepherd and his dog, which now maintains many families, employed in tillage'. But he denied that the displaced farmers remained unemployed: 'Nor surely do any imagine that the people which lived in those towns they call depopulated, were all destroyed, because they lived no longer there'. Instead, 'they were onely removed to other places . . . and employed in the manufacture of the wooll that may arise out of one hundred acres of pasture'. Thus enclosure led to the growth of industry. 'The manufactures and other profitable employments of this nation are increased, by adding thereto such numbers of people, who formerly served only to waste, not to increase the store of the nation'. Enclosure led to weaving—not stealing!

Fortrey's argument, with its assumption that enclosure reduced farm employment, fits comfortably with the Marxist theory of technical change in agriculture, and, indeed, Fortrey's argument was accepted without revision by Marx. 'The expropriation and expulsion of the agricultural population, intermittent but renewed again and again, supplied . . . the town industries with a mass of proletarians' (Marx 1867: i. 817). While this statement is often regarded as an inflammatory attack on enclosures, the intent of its seventeenth-century precursor was to defend them.

Arthur Young was keen to argue that enclosure and farm amalgamation were all for the best. He had argued that these changes increased agricultural employment. Could he square the circle by also arguing (with Fortrey) that enclosure and large farms released labour to industry? He found a way by introducing population growth into the analysis of agrarian change.

In *Political Arithmetic*, Young applied Fortrey's argument to the amalgamation of farms and began with the premise 'if the small farms are thrown into large ones, many of the people will disappear: let us (which we need not do) grant this fact' (Young 1774: 70). The conclusion, of course, was that the freed labour increased the production of non-agricultural goods and services. 'The fewer employed [in agriculture] (consistently with good husbandry) the better; for then the less product is intercepted before it reaches the markets, and you may have so many the more for manufacturers, sailors, and soldiers' (Young 1774: 296). But, as the parenthetical qualification in the premise suggests, Young did not believe that farm amalgamation lowered employment. Instead it increased output. So he added, 'we may suppose more people who eat it'. Thus, farm amalgamation (and enclosure) maintained or increased farm employment but led to greater production of food. Population expanded in consequence, and the increment was employed in manufacturing. The theory that population expands as food production expands is now called Malthusian, but it was emerging in the work of Wallace (1753), Steuart (1767), and Young half a century before Malthus's *First Essay*.

In Tory thinking still, population growth is the source of the industrial labour force. Professor Chambers has been the most forceful proponent. 'If agrarian change, as symbolized by enclosure, cannot be regarded as the chief recruiting agent of the industrial proletarian army, where did the new drafts come from?' The answer: 'The movement of population had taken an upward turn in village and town alike and provided an entirely new supply of human material' (Chambers 1953: 338). According to the Tories, the manufacturing work-force was the result of the 'natural' drive to reproduce rather than of social changes like enclosure or large farms.

## Income Distribution

Tories and Marxists agree on the facts about income distribution during the agricultural revolution: labourers were so wretched in the first half of the nineteenth century that it is hard to believe they had shared in any advance. On the other hand, landlords raised rents as they reorganized their estates, so they prospered from agricultural productivity growth. The question is why inequality increased. The Marxist and Tory answers differ because they are elaborations of their differing explanations of productivity growth and labour release.

Marxists believe that the growth in inequality had two immediate causes. The first was the concentration of property ownership that accompanied enclosures and farm amalgamations. 'The depriving of the peasantry of all landed property has beggared multitudes' (D. Davies 1795: 57). The second cause of rising inequality was the employment effects of large farms and enclosures. These reduced labour demand. Not only were most rural people becoming exclusively dependent on wage income, but the demand for that labour was falling. Enclosures and large farms were the cause of the rise in productivity, but they also caused low wages and unemployment for the majority of the population, and high rents for a rich minority. Inequality and productivity growth were inextricably linked.

The Tories dismiss the Marxist suggestion that the gentry and aristocracy might be blamed for the rise in inequality. They discount the concentration in landownership as a cause either by denying that it occurred or by saying that it cannot be blamed because the increase in concentration was lawful and thus also legitimate. Tories also deny that improved agriculture reduced labour demand; instead, they attribute low wages to population growth. Here Malthus comes into his own. He contended that the population would expand if the wage exceeded the 'subsistence wage', that is, the cost of raising a child and supporting him or her through life. Conversely, the population would fall if the wage were below the subsistence level. This demographic assumption implies that the population will converge to the size that maintains wages at the subsistence level. Suppose, as Tories do, that improved agriculture raises the demand for farm labour. Who gains? Initially, the wage rises; in consequence, the population expands. But the growing population drives the wage back to the subsistence level. The result is more wage earners but no increase in their standard of living. As more labourers work the land more intensively, output rises, but landlords receive all the gain as rent. So the Tory concludes that it was inevitable for the agricultural revolution to increase income inequality. That is why economics was called the dismal science.

## A Counter-Tradition

While Agrarian Fundamentalism has been the dominant interpretation of English agricultural history, there has been dissent. It began in England but has been strengthened by the confrontation of Agrarian Fundamentalism with the facts of agrarian change in Europe and more recently in Asia,

Africa, and Latin America. This book weaves together and extends several strands of doubt and criticism. Some strands emphasize justice, others efficiency.

The English critics of enclosure and farm amalgamation were persistent dissenters from Agrarian Fundamentalism. From the fifteenth century onward they claimed that enclosure and large farms were unjust—they enriched the large landowners at the expense of the poor. Sir Thomas More (1516: 46–7) objected that 'The nobles and gentlemen, not to mention several saintly abbotts, have grown dissatisfied with the income that their predecessors got out of their estates. They're no longer content to lead lazy, comfortable lives, which do no good to society—they must actively do it harm, by enclosing all the land they can for pasture, and leaving none for cultivation'. The same spirit of indignation led Cobbett to curse 'the system that takes the food from those that raise it, and gives it to those that do nothing that is useful to man'. (Cobbett 1948: ii. 42.) The Hammonds (1932: p. viii) assert that 'The main question for the historian is this: Were the poor sacrificed or not in the enclosures as they were carried out?' They answer that the poor 'were sacrificed and needlessly sacrificed'. 'Needlessly', since the more enclosures raised output—an Agrarian Fundamentalist view they accept—the greater was the potential for alleviating mass poverty.

What animated these writers was a sense of injustice and a sense that things might have been different. 'This state of things never can continue many years! By *some means or other* there must be an end to it' (Cobbett 1948: ii. 55). This attitude is very different from that of Marx. While he incorporated the facts marshalled by the critics of enclosures and large farms into his theory of historical development, he thought that those changes were progressive and desirable. They created a rich society that would make socialism possible and, indeed, inevitable.

The Marxist position highlights a weakness in the moral condemnation of enclosure. If the critics could say only that it was unfair to the poor, their criticism was vulnerable to the rejoinder that enclosures and large farms were necessary 'for the economy', that they promoted economic growth, however much they hurt some people. The second strand of thinking on which this book is based affirms that large farms and enclosures were not necessary for the technical revolution in agriculture.

The leading eighteenth-century thinkers in England and France were hostile to small farms, but a counter-tradition emerged by the mid-nineteenth century. In England a pivotal book was W. T. Thornton's *Plea for Peasant Proprietors* (1843), which argued that small-scale owner-occupiers were more efficient than English tenant farmers. On the factual level, Thornton pointed to many examples of productive peasantries. On the theoretical level, he disputed the incentive arguments advanced to explain the superiority of English arrangements. He countered Young's (1771*a*: iv. 343–5) claim that 'high rents are an undoubted spur to industry' with the observation that proprietorship guaranteed the peasant the full return to his exertions and thereby provided a motive for improvement. He inverted Young's belief that capital-intensive agriculture was labour-intensive agriculture by emphasizing its converse—that peasant proprietors typically devoted their 'leisure' to investing in farm improvements. In his *Principles of Political Economy* John Stuart Mill drew on Thornton's critique as well as on the opinions of continental friends like de Tocqueville, who favoured peasant ownership, and travellers' accounts of improving peasants to argue that peasant agriculture could sustain a revolution in farming technique.

We have surely now heard the last of the incompatibility of small properties and small farms with agricultural improvement. The only question which remains open is one of degree; [and] the comparative rapidity of agricultural improvement under the two systems [peasant proprietorship and English capitalism]. (Mill 1848: 154)

The cogency of Mill's argument, however, was not enough to dislodge the fundamentalist consensus from English thought (Dewey 1974).

By the late nineteenth century, debate about the efficiency of peasant agriculture had shifted to central and eastern Europe, where it was a critical issue. Did modernization require the replacement of peasant agriculture by capitalism (as the Russian westernizers believed) or could an advanced society be erected on a peasant base (as the more traditional thinkers contended)? If small-scale peasant agriculture was so inefficient, why did it persist in the face of competition from capitalist farmers? There were several responses. Lenin (1899, 1908) claimed that capitalist farms were indeed more efficient and, in Russia, were driving the peasants out of business. After the Revolution, the collectivization of agriculture represented the fulfilment of this analysis. On the other hand, in Germany and France, peasant agriculture was not succumbing to capitalist competition, and Marxists in those countries questioned the correctness of Agrarian Fundamentalism. Kautsky's (1900) explanation for the persistence of peasant farming harked back to Arthur Young's observation that French peasant proprietors worked harder than labourers. Another response denied that peasants spurned improvement. In 1939 Doreen Warriner published a comprehensive assessment of peasant farming in central and eastern Europe. She showed that peasants in Holland, Denmark, Switzerland, Bohemia, and western Poland realized crop yields like those in the United Kingdom. Slovakian and Hungarian peasants were not far behind (Warriner

1939: 99). Throughout central Europe the traditional open field system had been abandoned. Only in some parts of Transylvania and Slovakia was a three-field course with one field of bare fallow still followed. In most places, new crops had been introduced, fallows eliminated, commons enclosed, and the village rotation abandoned. Consolidation, however, was rarely pursued, so farms were still split into many fragments (Warriner 1939: 10). 'In general, in European conditions, there is good reason to think that peasant farming is efficient in the sense that the productivity of labour is as high as on large farms, and that peasant farming as such offers no hindrance to technical progress' (Warriner 1939: 7). President Gorbachev's decision in the 1980s to break up collective farms in the USSR and replace them with family farms held on long leases is belated vindication of Kautsky and Warriner.

What of peasant farming outside of Europe? When Sir Arthur Lewis surveyed agriculture in *The Theory of Economic Growth* (1955), he began in Arthur Young's footsteps—'There is almost always some difference [in efficiency] in favour of large size [farms]' (p. 129)—but he also noted that peasant proprietors worked harder than others and so had an offsetting advantage. The novel part of Lewis's discussion was to confront the platitudes of Agrarian Fundamentalism with a new set of facts, in this case, those of Japanese history:

The typical farm in Japan is still only between two and three acres in size; nevertheless productivity per acre on these farms is two to three times as great as in other parts of Asia. Productivity per acre in Japan increased by nearly fifty per cent in the thirty years before the first World War, and had doubled by the middle 1930's, without significant changes in the size of farm. (Lewis 1955: 136)

Lewis rejected Agrarian Fundamentalism and concluded that peasant farming was not an impediment to rising agricultural productivity.

Since the 1950s, the performance of peasant farmers has been systematically studied in many parts of the world. A consensus has emerged that is the opposite of Agrarian Fundamentalism, both Tory and Marxist (Berry and Cline 1979; Booth and Sundrum 1985: 98–125, 186–99). First, production *per se* exhibits constant returns to scale—that is, the views of Young and Marx that large scale is more efficient than small is usually rejected in statistical studies. Second, many family members (e.g. grandmothers) cannot find employment off the farm at the going wage, so they work on their farms, which are consequently cultivated more intensively than farms operated with wage labour. Young's observation about the hard-working French peasant has been repeatedly confirmed. Third, some of this hard work is devoted to capital formation (like drainage, terracing, ditching), so small farmers have more capital per acre than large farmers in so far as capital can be made by family labour. Fourth, the greater intensity of cultivation on small farms means that they produce more output per acre than large farms. The higher land productivity involves both a higher yield per sown acre of the main field crops and greater total production per acre overall due to the cultivation of additional, labour-intensive crops. Fifth, as *purchased* capital goods and fertilizers become available, small farms lose some of their advantage *vis-à-vis* large farms since the latter usually have access to cheaper credit. Again this was Young's view. As commercial capital goods become available, large farms use them more intensively than small farms. Under this circumstance, the large farms manage to reap the same yield per sown acre as the small farms; contrary to Young's view, the large farms still do not achieve higher yields. Moreover, small farms continue to produce more total output per acre by using otherwise unemployed family labour to produce vegetables and other high value crops. The Agrarian Fundamentalist strictures about large farms and productivity growth have been repeatedly confuted in the Third World.

The other tenets of Agrarian Fundamentalism have given poor predictions in the developing world. The notion that agriculture contributes to manufacturing growth by releasing labour has been continuously refuted by the failure of industry to absorb the mass of unemployed or marginally employed workers. Instead, attention has shifted to promoting development by increasing the employment of labour in agriculture (Booth and Sundrum 1985).

The other prediction of Agrarian Fundamentalism—that development necessarily raises the incomes of rich landowners without benefiting poor labourers—has had a mixed record. When the concentration of landownership has increased and technical change has displaced labour, then the English experience has been replicated. Some countries have avoided this result by land reforms that ensure egalitarian property ownership and limit tenancy. In such cases, income inequality declines as productivity advances. There are many examples of countries repeating the English experience, but the pattern is avoidable.

The failure of Agrarian Fundamentalism outside of England raises the question of whether the doctrine is really a good description of what happened within England. Indeed, a few English historians have expressed dissatisfaction with the fundamentalist model, particularly the importance attached to enclosures. The path-breaking paper was Havinden's 'Agricultural Progress in Open Field Oxfordshire' (1961*b*). It shook the fundamentalist consensus by presenting several examples of common field villages introducing grasses like sainfoin—making exactly the sort of changes that

Young and Ernle thought impossible. Yelling (1977: 166–7) has expanded the evidence by presenting a few examples drawn from the 1801 crop returns of open field villages growing turnips. Kerridge (1967: 19) asserted that 'whether fields were open or enclosed has little bearing on the agricultural revolution'. While this research has attracted considerable specialist interest, its scope has been limited to enclosure, and it has not dislodged the fundamentalist consensus.

## An Alternative Approach to English Agricultural Development

This book criticizes Agrarian Fundamentalism in both its Tory and Marxist versions. The truth or falsity of these views turns on factual issues like the effect of enclosure and farm size on cropping, employment, and yields, the ease with which surplus agricultural labour shifted into manufacturing, the trends in the distribution of landownership, and the movements of wages, profits, and rents.

One way in which this book differs from others is that I shall investigate these issues statistically. My data sets allow open and enclosed farms, large farms and small farms to be compared. Arthur Young pioneered this approach in the eighteenth century, and I will reanalyse his data. (The results are a surprise since they contradict his views.) But Young's data are not sufficient to answer all of the questions, so many other data sets have been assembled.

Young's data deal with the whole country, but the data I have collected deal mainly with the south midlands—roughly the stretch of country between Oxford, Cambridge, and Leicester. This is prime agricultural land and it has been at the centre of all enclosure controversies. There is enough geographical diversity within the region for different environments to be compared. There is enough uniformity to ensure that geography does not overwhelm social institutions.

The investigations reported here lead to a rejection of Agrarian Fundamentalism. I contrast my conclusions with the received view by summarizing the argument of the book as follows.

PART I. THE RISE OF THE YEOMAN AND THE LANDLORDS'
AGRICULTURAL REVOLUTION

I argue that there were two agricultural revolutions in English history—the yeomen's and the much more famous landlords'. Part I traces the rise of the yeomen and their subsequent elimination during the landlords' revolution.

English rural society in the thirteenth century consisted of lords, free tenants, serfs, and cottagers. The Black Death of 1348/9 ushered in a century of population decline that destabilized this social system. Labour mobility increased, real wages rose, and rents collapsed. Enclosure and conversion to pasture were the most creative seigneurial responses to the new order. Between 1450 and 1525, about one-tenth of the villages in the midlands were destroyed. These enclosures eliminated small-scale agriculture and represented an abrupt transition to capitalist relations.

Paradoxically, these enclosures led to the consolidation of peasant agriculture in the remaining open field villages. The depopulation that followed enclosure so alarmed official opinion that the Crown began protecting peasant farmers. Legislation, investigation, and prosecution were tools, but the most effective response was the extension of property rights to peasants by the Tudor courts. Copyholds and beneficial leases became secure forms of tenure by which many small farmers held their land. This legal revolution ended the worst abuses and created the tenurial underpinning for the yeoman farmers who flourished under Elizabeth and the Stuarts. The yeomen were owner-occupying family farmers—true peasants. Their economic significance cannot be understated, for they were responsible for much of the productivity growth in the early modern period.

Despite their progressive farming, the yeomen were eliminated in the landlords' agricultural revolution, which consisted of enclosure, the concentration of landownership in great estates, and the creation of large, leased farms. Since the sixteenth century, enclosure has been a favourite culprit in the disappearance of the English peasantry, although its significance has been hotly debated. Marxists have usually argued that enclosure involved the expropriation of peasant property by manorial lords, while Tories have replied that enclosure had little effect on the distribution of landownership. In fact, the relationship was complex. Early enclosures, especially those before the mid-sixteenth century, frequently involved the destruction of villages and the expulsion of their inhabitants as lords seized peasant land. In the seventeenth century, evictions were rare (because of the solidification of yeomen's property rights), but enclosure still usually involved a concentration of landownership and a venting of the population surplus to the needs of agriculture. In the eighteenth and nineteenth centuries, enclosure had little effect on landownership or population. The early enclosures are, therefore, good examples for Marxists, the later enclosures exemplify Tory views, and the seventeenth-century enclosures fall between these extremes.

The real collapse of yeoman agriculture occurred in the eighteenth century, in open field villages as well as in enclosed. Many yeomen were freeholders, and they sold their property to great estates. Other yeomen held their land on copyholds for lives or beneficial leases for lives or long terms of years, and they lost their land when large landowners stopped renewing these agreements. These real estate dealings were due to the creation of modern mortgages which increased the propensity of great estates to buy land.

PART II. ENCLOSURE AND PRODUCTIVITY GROWTH

I begin my reassessment of the landlords' agricultural revolution with enclosure: agrarian fundamentalists contend that it was necessary for technical advance since they believe that open fields were inflexible and inhibited the adoption of new crops and the conversion of arable land to pasture. I investigate this charge by dividing the south midlands into three fairly homogeneous natural districts—the heavy arable, the light arable, and the pasture. The same management scheme maximized profits throughout each district, so I assess the performance of open fields and enclosures by comparing the diffusion of modern methods district by district.

The diffusion of new techniques bears out the fundamentalist claim that enclosure led to agricultural improvement. In the heavy arable district, the tenacity of the soil meant that poor drainage was a serious problem. The solution was the installation of hollow drains, and, indeed, enclosed villages were far more likely than open villages to undertake this investment. In the light arable district, adopting the Norfolk rotation (turnips–barley–clover–wheat) and upgrading the breed and management of sheep were the bases of advance. Again, enclosure greatly accelerated the adoption of this system of management. In the pasture district, converting arable to grass was the key to progress, and enclosure led to a substantial increase in pasture. In all three districts, open field villages adopted the new methods to some degree, so Lord Ernle's charge that they were 'impervious' to new techniques is an exaggeration. Nevertheless, the open villages were far less innovative than the enclosed.

It is a far cry from that conclusion, however, to establishing that the enclosure movement made a substantial contribution to productivity growth in early modern England. Crop yields and labour productivity (not crop rotations) were the two critical indicators of advance. Both about doubled between the middle ages and the nineteenth century. Enclosure, however, made only a minor contribution to these increases.

I compare crop yields around 1800 in the three natural districts. In the light arable district, yields in open and enclosed villages were identical. In the pasture district, enclosed yields were perhaps a tenth higher than in open villages. In the heavy arable district, enclosure boosted yields about a quarter. Furthermore, most elements of capital-intensive agriculture made no contribution to the growth in yields. Yields were uncorrelated with the share of land planted with clover and beans, the main nitrogen-fixing crops, or with livestock densities, or the use of exotic or purchased manures. Convertible husbandry was unnecessary for the growth in yields. The only modern technique that raised yields was hollow draining, and its diffusion was indeed responsible for the growth of yields in the heavy arable district. Even there, however, open field farmers had accomplished three-quarters of the growth in yields between the middle ages and the nineteenth century. In the other districts, open field farmers accomplished almost the whole advance. Hence, when yields are used as the yardstick of progress, the enclosure movement played only a minor role.

Next I consider employment and labour productivity. The impact of enclosure on employment is the main battleground between Tory and Marxist fundamentalists. I develop two methods to resolve the dispute. The first involves listing all the tasks in farming and determining their cost; by applying this accounting to the details of open and enclosed farming in each natural district, I can compare the total cost of labour and thus total employment. The second method uses the details of the several hundred 'representative farms' surveyed by Arthur Young in his tours of the 1760s. Since he recorded farm employment, we can compare employment per acre for open and enclosed farms. Despite the radical difference in the procedures, the results are similar. Enclosure had little, if any, effect on employment when it did not affect the balance of tillage and pasture; when it led to the conversion of arable to grass, employment declined. These calculations provide no support for 'optimistic' Tory views. In pastoral areas, the Marxist story is closer to the facts.

This result opens the possibility that enclosure accounted for much of the growth in labour productivity. Its effect, however, was mixed. In some circumstances, output per worker increased considerably, but in most places the gain was only in the order of a tenth. Even though enclosure often led to some decline in employment, labour productivity grew little since output also sagged. This result is not surprising in view of the small boost that enclosure gave to yields, the tendency of enclosure to reduce the share of arable, and the fact that tillage gave more output per acre than pasture. As with yields, the enclosure movement made only a modest contribution to the growth of labour productivity in England.

Early modern discussions about output and employment were usually vague and inconclusive. On two issues, however, there was agreement: enclosure led to changes in cropping and farm methods, and enclosed farmers paid higher rents than open farmers. Both beliefs were true. Tory fundamentalists see these beliefs as connected—the rise in efficiency increased the farm's capacity to pay rent and thus accounted for the rent rise. My finding that enclosure made little contribution to productivity calls this conclusion into question and suggests that the rent increases represented a redistribution of income from farmers to landlords. I investigate the issue by computing the Ricardian surplus for open and enclosed farms; this surplus is the difference between a farm's revenue and the income that its labour and capital could have earned elsewhere. Ricardian surplus indicates both the overall efficiency of the farm (in technical terms, its total factor productivity) and its capacity to pay rent and taxes.

The agricultural revolution in the heavy arable district worked much as Tory fundamentalists suggest—enclosure led to investment in hollow drains, which raised yields. Ricardian surplus rose as well, and its increase matched the growth in rent. In this district there is no evidence of income redistribution at enclosure. In most other districts, however, the Tory view is inconsistent with the facts. In the light arable district, the growth in productivity was very small, and higher rents required some redistribution of income from farmers to landlords. The evidence for the pasture district is more difficult to interpret, but efficiency gains following parliamentary enclosures appear to have been minor, and there is very strong evidence that open field rents in the mid-eighteenth century were less than the value of the land. The balance of probabilities is that rent increases following enclosure in this district in the eighteenth century involved a considerable redistribution of income from farmers to landlords.

The investigations presented in this book do confirm the two reasons usually adduced for the importance of enclosure. It did lead to highly visible changes in land use and farming methods, and it did lead to higher rents. However, when I measure the effect of enclosure on yields and labour productivity, its importance diminishes, especially when the gains are contrasted with the growth in efficiency across the early modern period. Despite the enthusiasm of many landlords for enclosure, its contribution to agricultural productivity growth was small.

PART III. CAPITALIST AGRICULTURE AND PRODUCTIVITY GROWTH

If enclosure was not responsible for the growth in yields and labour productivity in England between 1500 and 1800, perhaps the other features of the landlords' agricultural revolution can take the credit; namely, the concentration of land in great estates, the concomitant elimination of peasant proprietorship, and the amalgamation of small family farms into large farms dependent on wage labour. The alternative, of course, is that the yeomen themselves were responsible for the rise in efficiency. In this section,

I directly compare the contributions of the yeoman and capitalist systems to the growth in productivity.

I begin with yields. Tory thinkers like Arthur Young claimed that large farms secured higher yields than small farms since the former employed more capital per acre than the latter. I investigate this claim with Young's survey of farms. These data show that capital per acre was either independent of size or declined with size for both arable and pastured farms. Young, himself, used his data to correlate size with yield. He believed they showed a positive relationship. In fact, the correlation is tenuous and too weak to explain the growth of yields in early modern England. Moreover, Young's yield data pertain to villages, rather than farms, and so are poorly adapted to investigate the correlation between yield and farm size.

I use probate inventories to construct a more suitable sample for measuring the correlations between farm size, capitalization, and yield. This sample shows capital per acre declining with size. Of greater historical importance, the sample exhibits no correlation between size and yield. Large farms were not necessary for bountiful harvests. Moreover, the rise in yields preceded the eighteenth-century shift to large 'capital' farms. It was small-scale farmers in the open fields—the English yeomen—who accomplished the biological revolution in grain growing.

Next I use Young's farm survey data to measure the effects of increasing farm size on employment and labour productivity. As with enclosure, Tory fundamentalists expect employment to grow with size, while Marxists expect the reverse. Again, the facts support the Marxist view. The employment per acre of men, women, and boys all declined with size. The decreases were greatest for women and boys. Eighteenth-century farm amalgamation rendered most rural women and children redundant in agriculture.

Labour productivity increased for two reasons in the early modern period. Half of its growth was due to the rise in yields achieved by seventeenth-century yeomen. This increase did not account for England's exceptional productivity *vis-à-vis* the continent in 1800, for yields increased generally in northwestern Europe, raising output per worker throughout the region. Instead, England's superiority was mainly due to the declines in farm employment consequent upon the amalgamation of farms in the eighteenth century.

The investigations reported here imply a radical shift in our understanding of productivity growth in early modern England. The seventeenth-century yeomen were the decisive contributors. They accomplished most of the growth in yields and about half of the rise in labour productivity over the period. The main contribution of the landlords' agricultural revolution was a further shedding of labour in the eighteenth century. While this did push labour productivity to record levels, it was of secondary importance compared to the achievement of the yeomen.

PART IV. AGRARIAN CHANGE AND INDUSTRIALIZATION

Productivity growth in agriculture is important since it may lay the basis for industrial advance. In this section I contrast the contributions of the yeomen's and landlords' revolutions in this regard.

One way that agricultural productivity growth can aid industrialization and economic growth is by providing savings to finance manufacturing investment. In England, most of the productivity gains of the agricultural revolution accrued to the large landowners as rent increases. Instead of saving, however, they were net borrowers and spent their fortunes on stately homes and elegant living. Agriculture made little contribution towards promoting industrial capital formation.

The main contribution of the agricultural revolution to England's economic growth was, therefore, its immediate impact on the national income. There were two ways in which the rise in agricultural efficiency increased gross domestic product (GDP). The first was by expanding agricultural output; the second was by releasing labour to other sectors. In the seventeenth century, agrarian change made contributions in both ways. The expansion in agricultural output was the more important, and it was the achievement of the yeomen's agricultural revolution. A reduction in farm employment was a consequence of the early phase of the landlords' agricultural revolution. Most of the people forced out of agriculture left the midlands and swelled the population of the Metropolis where their employment expanded output in commerce and manufacturing. This reallocation of labour did boost GDP, but the rise was smaller than the increase attributable to the yeomen's agricultural revolution.

Enclosures and farm amalgamation were more extensive in the eighteenth century, but they made less of a contribution to the increase in national income. Technical change no longer raised yields, so it did not directly raise agricultural output and GDP. Enclosures and farm amalgamations reduced employment per acre, but the labour released was not successfully re-employed in manufacturing. By the early nineteenth century, the agricultural revolution was producing paupers—not proletarians. Samuel Fortrey's dream of industrious weavers gave way to Thomas More's nightmare of thieves. Some of them, indeed, were hanged; many were transported to Australia. The rise in farm productivity through labour release, therefore, did not translate into a rise in the national income. The overall impact of agricultural change on GDP was very small in the eighteenth and nineteenth centuries. The landlords' agricultural revolution, in its most intense phase, made little contribution to economic growth.

PART V. THE DISTRIBUTION OF THE BENEFITS OF TECHNICAL PROGRESS

Agricultural productivity growth created the potential to raise the incomes of all people in early modern England. In fact, the rich were the main beneficiaries.

Price changes were a principal avenue by which the benefits of productivity growth were distributed. Thus, consumers would have gained had food prices fallen. Indeed, such a development would have reduced the inequality of real incomes since the poor spent a higher share of their income on food than did the rich. However, real agricultural prices rose in England between the late middle ages and the nineteenth century. This trend increased inequality by reducing the real incomes of the poor relative to the rich.

The gains from productivity growth accrued to agriculturalists as rising incomes. Labourers, however, did not gain since the real wage fell, then remained low, as productivity increased. Likewise, farmers (as the owners of working capital) did not gain since the real rental price of livestock and equipment was trendless from the fifteenth to the nineteenth century. All the benefits of rising productivity accrued to landlords—real rents increased about sevenfold between 1450 and 1850.

The tendency to increased inequality was least pronounced during the yeomen's agricultural revolution in the second half of the seventeenth century. The rise in crop yields ensured that the real price of agricultural products did not rise. Likewise, the preservation of farm employment stabilized the real wage. In the eighteenth century, these trends reversed and inequality grew. The cessation of output growth meant that the real prices of farm products increased. The decline in labour demand increased unemployment and lowered agricultural labour income. As a result rents increased sharply during the main phase of the landlords' agricultural revolution.

If the ownership of property had been equally distributed, the fact that productivity growth raised rents would not have increased inequality. But landownership was highly concentrated and became even more so over the early modern period. Most of the income generated by agricultural productivity growth, therefore, accrued to the gentry and aristocracy.

## *Two Agricultural Revolutions*

There were two agricultural revolutions in English history—the yeomen's and the landlords'.

The yeomen's revolution occurred mainly in the seventeenth century, although its legal basis was laid in the sixteenth. This revolution was marked by a doubling of corn yields; it raised England's national income, and the benefits were distributed widely. Small farmers who held their land on copyholds and beneficial leases gained as land values increased. Labourers held their own since employment was maintained.

The landlords' revolution consisted of enclosure and farm amalgamation. This reorganization began in the fifteenth century but occurred mainly in the eighteenth. The early enclosures increased farm output and released labour when the population was low, so they probably contributed to a rise in GDP—a rise that accrued mainly to large landlords through higher rents. Enclosure and the growth in farm size in the eighteenth century did not increase output—they reduced farm employment. The released labourers did not raise the national income since they were not re-employed in manufacturing. The only gainers were large landlords.

The conclusion is unavoidable—most English men and women would have been better off had the landlords' revolution never occurred.

PART I

*The Rise of the Yeoman and the
Landlords' Agricultural Revolution*

# 2

# Enclosure in the South Midlands

This lordship [Little Dalby] is . . . an antient inclosure; and none of the inhabitants know when it took place. I thought at first to have discovered the date of it from the age of the trees in the hedgerows; but, none of them which I have had an opportunity of examining are more than about 120 years old; but if the inclosure went no farther back than this, we should have learnt the date of it from tradition. I then searched the parish register, to find whether any depopulation had taken place since the time of Elizabeth; but could find none, and therefore concluded that the inclosure was at least as early as her reign. That there has been a depopulation I conclude, not only from the natural consequence of inclosing, but from the foundations of buildings which are discovered in the closes near the church.

Mr. Professor Martyn, quoted by Nichols, *History and Antiquities of the County of Leicester*, 1795–1815, ii (1798), Part 1, p. 160

ENCLOSURE was the pivotal event in the development of agrarian relations in early modern England, both because of its direct effects and because of the official responses to them. Enclosure was most disruptive in the midlands, so I concentrate on its history in that region (H. Gray 1915; Baker and Butlin 1973). In particular, I focus my study on 1,568 parishes in the counties of Rutland, Northampton, Huntingdon, Cambridge, Bedford, Buckingham, Oxford, Leicester, Warwick, and Berkshire.[1] Map 2–1 shows the boundaries of the region. Its area is 2,850,866 acres or 9 per cent of the surface of England.[2] The region is prime agricultural land, and has been involved in every wave of enclosures. While it was not highly urbanized, its agriculture was influenced by London and the northern industrial cities. It was the centre of several protoindustries. For all of these reasons, the south midlands is a suitable region for analysing the effects of enclosure.

[1] I excluded villages in these counties that were in the Cotswolds, the Chilterns (and all land south of them), the fens, the Arden part of Warwick, and Charnwood Forest. Hence, my sample villages lie in the 'midland' regions identified by Thirsk (1967*a*; 1987: 24) and Kerridge (1967).

[2] In addition, these parishes contain 4,328 acres of fen and 30,263 acres of land on the Chilterns.

## Open Fields

In the midlands, open field farming was the norm in the middle ages. Most farm land was in tillage producing wheat, rye, barley, oats, and beans. The land growing these crops surrounded the village and was usually divided into two, three, or four large, contiguous fields, which frequently served as cropping units. In a typical three-field village, one of the fields in any year was fallow, another was planted with wheat or, less often, rye (winter crops), and the third was planted with a spring crop. In the middle ages that was probably barley or oats (occasionally mixed), but in the early modern period beans became most common. The cultivation of each field rotated through the sequence. In fact, more flexibility was possible since fields were subdivided into furlongs (groupings of parallel strips), and they could function as cropping units. But those divergences were epicycles on the principal movement.

A distinguishing feature of open fields was that farms were not consolidated but consisted of strips scattered across the fields. The strips were typically one rod (5.5 yards) wide and 20 to 40 rods (one half to one furlong) long, so they comprised one eighth to one quarter of an acre. Often they were 'S' shaped. The boundary between the strips was either a drainage ditch or a grass border. Sets of parallel strips were grouped into the furlongs, which, in turn, were grouped into the fields.

Although grain was the main product of an open field village, some livestock was kept. Horses (at an early date oxen)[3] were maintained as draught animals, and there were also dairy cows and sheep. Animal husbandry involved using four kinds of land. First, the common provided pasture for the stock. Not every resident of a village had the right to put animals on the common: that right resided only in the owners of land in the fields and of some cottages. In order to prevent overgrazing, commons were frequently 'stinted,' that is, restrictions were placed on the number and kinds of animals that could be pastured. Second, farmers might also have enclosed pasture for their exclusive use. Third, hay for the animals was mown on the meadow, which usually adjoined a stream or river. Winter flooding encouraged the growth of grass but also obliterated property boundaries and, on occasion, changed the shape of the meadow itself. Flexible and equitable management was essential. Fourth, the village sheep flock was pastured at night on the fallow field and the other fields when they were not growing crops. The sheep ate weeds on these fields. The aim of folding the flock was not so much to provide forage for the sheep as it was to weed the fields and manure them.

---

[3] Langdon (1986) discusses this issue.

Open or common fields take their name from this last characteristic. No one had the right to exclude the village flock from foraging on his or her strips when the field was fallow. Usually a manorial court set the dates on which flocks could be pastured on particular fields; the court also enforced these rules.

## The Methods of Enclosure

Enclosures were effected in three ways—by agreement amongst the proprietors, by sole proprietorship, or by parliamentary act.[4]

Enclosure by agreement amongst the proprietors was the most obvious way to carry out enclosure. The simplest method was for the owners of all the land to renounce common grazing on the fallow. The strips could then be fenced, or not, as was convenient, but they would be under full private control. This method of enclosing was usual in many continental European countries. A field divided into long, narrow strips growing different crops was the result and is still a common landscape. Some enclosure took place in this way in England outside the midlands;[5] within them, it was rare. Enclosure effected in this way was not particularly disruptive.

In the midlands, enclosure, however it was effected, was usually accompanied by consolidation—the various owners exchanged strips to create large blocks of property. Sometimes only small acreages were involved, but exchanges of hundreds or thousands of acres were more frequent. The new property boundaries were usually demarcated by hedges and ditches, and common grazing was renounced so that exclusive private control over land was established.

Enclosure by agreement of all the proprietors was like an elaborate conveyance: the landowners sold each other strips so they ended up with compact holdings, and then they renounced common grazing. Two problems complicated these agreements. First, if the enclosure led to the conversion of arable to pasture, then tithe owners, who received a tenth of the grain crop, suffered a loss of income since grass usually generated less revenue per acre than arable. They had to be brought into the agreement, and that was usually done by providing them with an annual cash payment (a composition) in place of the tithe in kind. Second, some of the proprietors might not have been in a legal position to sell their property. For instance, if an estate was settled and the land entailed, the owner was only a life tenant and could not commit his land after his death. On coming into his estate, his heir could undo the enclosure by suing for recovery of the strips. In the

[4] Yelling (1977: 1–93) provides an exhaustive discussion of methods of enclosure.
[5] Postgate (1973: 287) presents examples. See Yelling (1977: 71–93) for a survey.

seventeenth century, enclosers tried to forestall such actions by initiating fictitious suits in Chancery to get the Chancellor to endorse the enclosure. Later, private acts of parliament were used since they could amend the family settlements, thus binding the heirs. (English and Saville 1983: 29).

Enclosure by sole proprietorship was a limiting case of enclosure by agreement. If a single individual bought up or otherwise acquired all the land in a village, it was tantamount to enclosure. Subject to any agreements with tenants, the sole owner could separate the intermixed farms into compact holdings, bar common grazing, and suspend collective decision making. Subject to agreement with tithe owners, he could convert arable to pasture.

Before the eighteenth century, all enclosures were effected by agreement or sole proprietorship. After 1750, enclosures by parliamentary act became standard, and 96 per cent of the land enclosed in the south midlands between 1750 and 1849 was reorganized by this means. Parliamentary enclosures were probably an outgrowth of the use of private acts to endorse enclosures by agreement.

Parliamentary enclosures were initiated by a petition to parliament by proprietors in a village. Unanimity was *not* required as was the case with enclosures by agreement. The concurrence of the owners of only 75–80 per cent of the land (the proportion was not rigid) was sufficient for the bill to proceed.[6] Since landownership was highly concentrated, bills could proceed with a majority of the owners opposed. Enclosure acts appointed commissioners to carry out the enclosure. They established who the landowners were and appointed a surveyor to value the property. The commissioners redrew boundaries to create consolidated holdings. All proprietors (including those opposed to the enclosure) received land in proportion to the value of their holdings in the open fields and their grazing rights on the common. Usually tithes were eliminated and tithe owners awarded land in compensation. Common grazing and collective control over cropping were abolished. The new arrangements were set out in the enclosure award.

## When Were the South Midlands Enclosed?

In 1300 most of the midlands was in open field or common pasture and by 1900 almost the entire region was enclosed, but dating the change between those years is a very difficult problem. Enclosures were important events, and they left traces in many public and private documents, but, with the exception of the parliamentary enclosures, there is nothing like a national

---

[6] The concurrence of all manorial lords and tithe owners was also necessary.

register. The chronology of enclosure must be worked out by collating as much information as possible. Historians have been working on this task since the late nineteenth century, and the results presented here depend substantially on that earlier work.

I have tried to reconstruct the enclosure history of the south midlands in the following way. A list of the 1,568 parishes in the region, as they were defined in the nineteenth century (with their areas) was compiled, and a file card with this information was prepared for each. The dates of parliamentary acts and awards and the acreage awarded, as given in Tate's *Domesday* (1978), was entered on each card. Evidence relating to non-parliamentary enclosures was then added. That evidence, of course, is more difficult to find, but many of the sources have been examined and summarized by earlier investigators.[7]

Their results were supplemented by a fresh search for new material. While some work was done on all counties, efforts were concentrated on those for which less research had been done by others. The following sorts of documents were examined: published histories, a very large number of glebe terriers,[8] the summary of the Chancery decree rolls prepared by Beresford (1977), the published details of deserted villages, the files of the Deserted Medieval Village Research Group, the result of the 1517 inquiry on enclosures (Leadam 1897), the unpublished results of the 1607 inquiry (PRO c. 205), and the manuscripts listed in county record office catalogues under 'enclosure'. These sources are not sufficient to date the enclosure of every acre in the south midlands, but they do date most of it.

Two practical difficulties arise in this work. The first relates to dating. While parliamentary enclosures could be precisely dated, other enclosures could often be dated only within limits, as between successive glebe terriers. Provided the span was not broad, a date within the span was arbitrarily assigned to the enclosure. As a result, an annual tabulation of non-parliamentary enclosures by year is inaccurate, so I analyse the data in 25-year periods.

---

[7] The Vale of White Horse in Berks. by Cottis (1985), Oxon. by H. Gray (1915), Havinden (1961a), and the recent *Victoria County History* (*VCH*) volumes for the county, Bucks. by Turner (1973b), and Jenkins (Bucks. RO, D99/70), Beds. in small way by Burgess (1978), and in the unpublished parish history abstracts in the Bedfordshire RO, Cambs. by the post-war *VCH* volumes, Northants to a substantial degree in the unpublished Wake papers (Northants RO, Wake collection), Leics. by L. Parker (1948), Beresford (1949), Thirsk (1954b), and Leicestershire *VCH* v, and Warwicks. to a substantial degree by Tate (1949), Beresford (1950), and Barratt (1955). Shirley (1867) was also helpful.

[8] Glebe terriers were documents prepared periodically which described the property of the parish church. Glebe terriers noted the exchange of church lands in enclosures as well as the negotiation of compositions.

The second practical difficulty relates to acreages enclosed. Again, for parliamentary enclosures, precision is possible, but for other enclosures that is not the case. While the Tudor inquiries often recorded acreages, these are areas for the land converted from arable to pasture, which was usually less than the area enclosed (L. Parker 1948). Consequently, a somewhat larger figure than that recorded in the inquiry was presumed enclosed. The situation is more difficult for many enclosures before 1485 and many seventeenth-century enclosures. We may know that a whole village was enclosed, but the records do not indicate the area. However, in these cases we know the total area of the village.

I dealt with this disparate material by allocating the total area of the village to dates of enclosure roughly in proportion to what I judged was the acreage enclosed at each date. Thus, if a village was enclosed at one time, I counted the full area of the village as enclosed at that date. If part of a village was presented as enclosed by the 1517 inquiry (say in 1502) and the rest was enclosed by a parliamentary award (say in 1767), I divided the area of the village between the two events as seemed reasonable in view of the recorded acreages in the sources. The resulting areas, which sum to the village total, are then assigned to 1502 and 1767. When known, the acreage of 'old enclosures' in the eighteenth century was helpful in making these allocations. In many cases, there was not enough information to date the enclosure of all the area of a village; in that case the residual is recorded as unknown. As with dates, it is evident that the acreages of early enclosures are not known precisely, but enough is known to reveal broad patterns.

There were three main waves of enclosure in the south midlands (1450–1524, 1575–1674, and 1750–1849), and the rate of enclosure increased from wave to wave. (See Figure 2–1.) There were hiatuses in enclosing in 1525–74 and 1675–1749. The sixteenth-century slowdown in enclosure has been debated by historians, but is supported by these figures.[9]

Figure 2–1 excludes 305,473 acres whose enclosure could not be dated within 25-year periods, but the enclosure of 71,074 of those acres could be dated within the main waves and intervening troughs. Table 2–1 shows the results of that tally. Of the dateable acres, 7 per cent were enclosed in the first wave, 18 per cent in the second, and 60 per cent in the third. Including medieval enclosures and those in the troughs, 11 per cent of the south midlands had been enclosed by 1524 and 32 per cent by 1674.

---

[9] See Wordie (1983: 492) for a recent survey of much of the literature. In addition, Bowden (1952; 1962: 4–6; 1967: 368) has repeatedly argued that there was a pause in enclosing—even a reconversion of pastures to corn—at about this time although his reasoning has been disputed by Wright (1955) and J. E. Martin (1988).

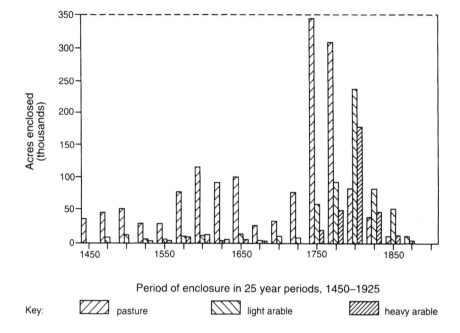

FIG. 2–1. Enclosure in the South Midlands, by Natural Districts

The chronology of enclosure in the south midlands is different from that in the rest of the country. Wordie's (1983) reconstruction shows that Tudor enclosures were negligible in England as a whole, and that more enclosing occurred in the seventeenth century than in any other. In the south midlands, Tudor enclosures were relatively more important, while the parliamentary enclosures involved a far greater acreage than the seventeenth-century enclosures.

*Table 2–1. The chronology of enclosure in the south midlands, 1450–1850*

|  | Acres enclosed |
|---|---|
| pre–1450 | 103,439 |
| 1450–1524 | 182,824 |
| 1525–1574 | 62,044 |
| 1575–1674 | 477,500 |
| 1675–1749 | 143,294 |
| 1750–1849 | 1,562,073 |
| 1850– | 85,293 |
| Undated | 234,399 |
| TOTAL | 2,850,866 |

## Geography and the Chronology of Enclosure

At least since Gonner's *Common Land and Inclosure* (1912), historians have recognized that there is a link between geography and the timing of enclosure. The link is strong in the south midlands. To establish it, I divided the region into three natural districts and worked out the enclosure chronology for each.

The natural districts were defined in terms of their agricultural potential. (See Map 2–1.) The pasture district (1,742,299 acres) included much of Leicestershire and Northamptonshire and large parts of Warwickshire, Rutland, and Buckinghamshire as well as small areas with substantial pasture potential in other counties. The object of enclosure in this district was usually the conversion of arable to grass. The heavy arable district (401,886 acres) was mainly located in Cambridgeshire, Huntingdonshire, and northern Bedfordshire. The soil in this district was very heavy clay. It supported miserable grass, so conversion to pasture was not a motive for enclosure here. Instead, the aim was improvements in drainage. The light arable district (706,681 acres) was scattered across the south midlands on several geological formations. The common feature was that the soil was light enough to grow turnips, and the land was usually cropped under a Norfolk rotation (turnips–barley–clover–wheat) after enclosure. Conversion to pasture was not a motive for enclosure in this district either.

The chronology of enclosure was different in the three natural districts. (See Figure 2–1.) The 1450–1524 and 1575–1674 waves were located mainly in the pasture district. About half of that district was enclosed before 1750. The little early enclosure in the arable districts was mostly small fields rather than whole villages. I suspect that these enclosures were isolated patches of good pasture, so that pre–1700 enclosure was even more orientated towards converting arable to pasture than Figure 2–1 suggests. Enclosure before the eighteenth century was rarely directed to the improvement of arable husbandry.

The premature enclosure of Grantchester, Cambridgeshire, provides a dramatic example of this generalization. This village was in the heavy arable district and was unsuited for grazing. However, in the 1660s John Bing enclosed the village by buying or becoming the beneficial leasee of all the land in the village. He converted 1,575 acres to pasture. Despite extravagant claims about the value of his enclosures, he quickly went bankrupt. That is no wonder in view of the grassland potential of the village. What is important, however, is that when the various lessors recovered the land, they reinstituted open field farming for corn. Neither they (nor Bing) saw any point in enclosing the land if it was to be used as arable. The village

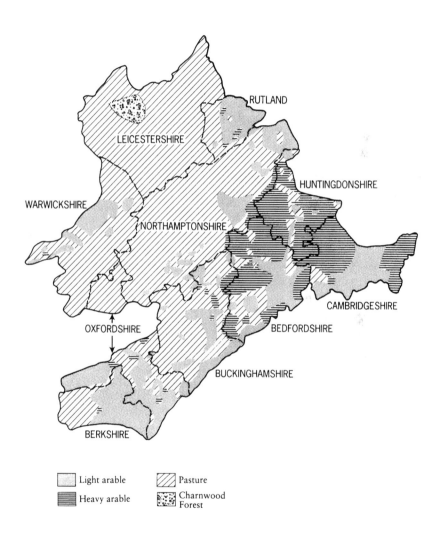

MAP 2–1.  The South Midlands divided into Natural Districts

remained open until it was enclosed by parliamentary act in 1803 along with the rest of the heavy arable district. (*VCH*, Cambridgeshire, v. 207).

Some 96 per cent of the land enclosed in the third wave (1750–1849) was enclosed by parliamentary act. Parliamentary enclosures were much more important in the arable districts than in the pasture district. As Table 2–2 shows, three-quarters of the former were enclosed by parliamentary act, while only 45 per cent of the latter were enclosed by that method. The greater importance of parliamentary enclosures in the arable districts reflects the late date of their enclosure.

*Table 2–2. Enclosure methods*

|  | Non–parliamentary | Parliamentary act confirming agreement | Parliamentary act | Total |
|---|---|---|---|---|
| Heavy arable | 110,348 | 1,160 | 290,378 | 401,886 |
| Light arable | 168,937 | 8,873 | 528,871 | 706,681 |
| Pasture | 941,951 | 20,325 | 780,023 | 1,742,299 |
| TOTAL | 1,221,236 | 30,358 | 1,599,272 | 2,850,866 |

Gonner (1912: 197–237) was the first to notice that the parliamentary enclosures occurred in two bursts—the first in the 1760s and 1770s and the second during the Napoleonic Wars. He believed that geographical factors explained this pattern. Figures 2–2 to 2–5 test this conjecture for the south midlands. Figure 2–2 shows that the south midlands as a whole exhibited the same two-peak pattern that Gonner first observed in the national distribution. Figures 2–3 to 2–5, which separate the enclosure chronologies of each district, show that the odd overall pattern results from concatenating the dissimilar histories of the natural districts. Thus, there was only one principal peak in each district—the pasture district's peak was between 1750 and 1770 while the arable districts' peaks were during the French Wars. Moreover, Figures 2–2 to 2–5 show that enclosure was rarely aimed at improving corn growing until the very end of the eighteenth century. Enclosure for the preceding four centuries was intended to convert arable to pasture.

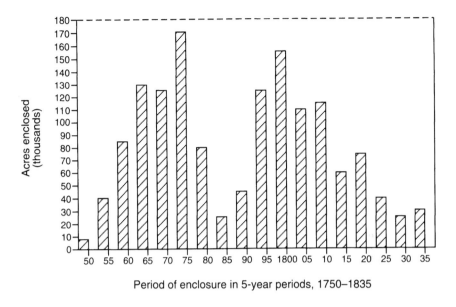

FIG. 2–2. Chronology of Parliamentary Enclosure: all Natural Districts

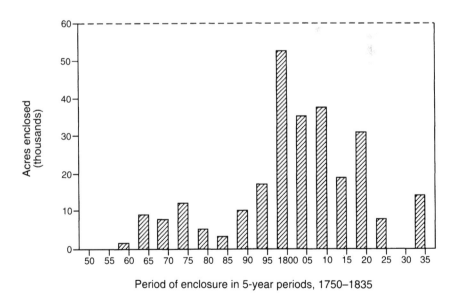

FIG. 2–3 Chronology of Parliamentary Enclosure: Heavy Arable District

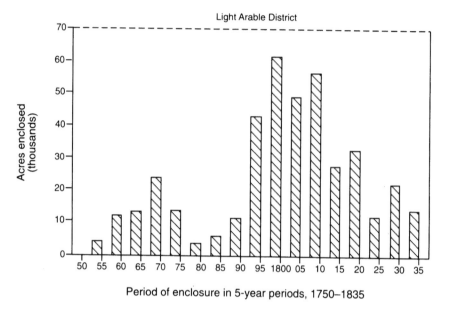

FIG. 2–4  Chronology of Parliamentary Enclosure: Light Arable District

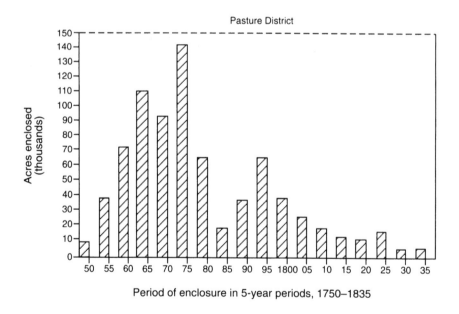

FIG. 2–5  Chronology of Parliamentary Enclosure: Pasture District

# 3

# Enclosure and Depopulation

And the aforesaid jurors say that Henry Smith was recently seised in his demesne as of fee of 12 messuages and 4 cottages, 640 acres of arable land to the annual value of 55*l* with appurtenances in Stretton super Street [Stretton Baskerville] in the aforesaid county [of Warwick], with each of the aforesaide messuages 40 acres of arable land, suitable for and ordinarily in cultivation were accustomed to be let, farmed and occupied from time immemorial. Thus was the same Henry Smith seised on the 6th. December 9 Henry VII [1493]. He enclosed the messuages, cottages and lands with ditches and banks and he wilfully caused the same messuages and cottages to be demolished and laid waste and he converted them from the use of cultivation and arable husbandry into pasture for brute animals. Thus he holds them to this day, on account of which 12 ploughs that were employed in the cultivation of those lands are withdrawn and 80 persons, who similarily were occupied in the same cultivation, and who dwelled in the said messuages and cottages, were compelled to depart tearfully against their will. Since then they have remained idle and thus they lead a miserable existence, and indeed they die wretched.

Commission on Depopulation, 1517 (in Fisher and Jurica,<br>
*Documents in English Economic History*, 1977: i. 117)

ENCLOSURE is the oldest explanation for the destruction of the English peasantry. Henry Smith's enclosure of Stretton Baskerville in 1493 is a paradigm, which defines the eviction model of enclosure. According to this view, enclosures were effected by manorial lords who evicted their tenants (all peasant farmers), converted the land to pasture, and operated it as a large capitalist enterprise. The village was deserted as the population left. In the sixteenth century, the former inhabitants were presumed to 'have remained idle', although later theorists like Fortrey and Marx had them redeployed to manufacturing. According to the eviction model, enclosure was the prime mover that destroyed the peasantry, created the great estate, and released labour for industry.

In the twentieth century, enclosure has fallen out of favour as an explanation for these developments. The case for this revision is a solid one for parliamentary enclosures, but the evidence for earlier enclosures is thin and conflicting. I will argue here that early enclosures often did, indeed, eliminate peasant agriculture and concentrate the ownership of property in the hands of manorial lords.

The case against the importance of enclosure in destroying the peasantry rests on the evidence of the land tax assessments. The tax was first imposed in 1692, but the administrative records become usable for studying the effects of enclosure only in 1780 when the assessment form was changed to one that listed and distinguished the owner, occupier, and tax liability of every property in the village (E. Davies 1927: 271–2). Since a new form was prepared every year, the distribution of landownership before and after enclosures can be compared. Initially, research concentrated on the fate of owner-occupying small farmers. A. Johnson (1909: 128–54), H. Gray (1910), and Davies (1927) showed that enclosure after 1780 had little effect on their number and that they seemed to have disappeared before the late eighteenth century in open as well as in enclosed villages.

More recent studies have concentrated on the effect of enclosure on the overall degree of inequality in property ownership. Studies of Leicester (Hunt 1959), Warwick (J. M. Martin 1967), Bedford (Payne 1946), and Buckingham (Turner 1975) have generally shown two things: first, parliamentary enclosure had no dramatic effect on inequality although villages enclosed in that way were slightly less egalitarian than open field villages. Second, villages enclosed before 1780 had much more unequal distributions of property ownership than villages enclosed after that date. Other than that last finding, virtually nothing is known about enclosure and ownership inequality for earlier periods.

Knowledge of the impact of enclosure on population is even sketchier than knowledge of its impact on landownership despite the endless disputes over alleged depopulations. Again there is much more information about parliamentary enclosures, and they do not appear to have been depopulating. Gonner (1912: 407–11) used early nineteenth-century censuses and time series of baptisms, burials, and marriages from 1750 to 1800 to compare population growth in open and enclosed villages. The difference was negligible. In contrast, the work of Hoskins (1950a) and Beresford (1954) on deserted villages suggests that fifteenth- and sixteenth-century enclosures led to depopulated villages. How frequently, however, is unclear. The consequence of Elizabethan and Stuart enclosures for population is anybody's guess.

The finding that parliamentary enclosures neither depopulated nor increased inequality has given strong support to Tory Agrarian Fundamentalism. In this chapter I will argue that Tories have taken too much comfort from the land tax assessments. The character of enclosure depended on when it occurred. Enclosures in the first wave (1450–1524) were often like the enclosure of Stretton Baskerville—the result was high inequality, depopulation, and often the destruction of the village itself. Enclosures in the second wave (1575–1674) also increased inequality (but to a lesser degree) and forced many people born in the village to move elsewhere. It was only the enclosures in the third wave (1750–1849) that had no effect on population or inequality. Since the studies of the land tax assessments have focused mainly on the latter group, they have created an excessively optimistic impression of enclosure. Early enclosure, at any rate, led to inequality, capitalist agriculture, and labour release.

## Enclosure and Deserted Villages

The first task of this chapter is to measure the effect of enclosure on the destruction of villages, the level of population, and inequality in landowner-ship. The enclosure chronologies of the south midlands villages discussed in Chapter 2 provide the framework. I supplemented this information with data about landownership, population, and deserted villages. By cross-tabulating the characteristics of parishes against their date of enclosure, I can compare the social impact of enclosures undertaken at different times.

The destruction of a village and the expulsion of all its inhabitants was the most extreme result an enclosure could have. Beresford and Hurst (1971: 181–212) have identified 370 deserted villages in the south midlands. In the nineteenth century, about half of these villages continued to exist as distinct parishes administering their own poor rate. Population was certainly low: there was generally a church and often a manor house but frequently little else; however, the streets and foundations of the medieval village were often apparent and attested to a much greater population at an earlier date. In the other half of the cases, the church had gone, and the lands of the medieval village had been attached to a neighbouring village for administrative purposes. Population was lowest in these villages. Again, however, it is often possible to detect the site of the medieval settlement and frequently its name lingers on as that of a hamlet, field, or stately home. By adding the Beresford–Hurst information to the village enclosure histories I have assembled, I can determine the relationship between enclosure and deserted villages.

The connection between enclosure and deserted villages is tighter, the further back one goes. (See Table 3–1.) The period 1450–1524 saw the largest number of village desertions. These involved the enclosure of

Table 3–1. *Enclosure and deserted villages*

| Enclosure period | Number of villages destroyed | Acreage enclosed | Percentage of all land enclosed in the period |
|---|---|---|---|
| pre –1450 | 64 | 45,545 | 44 |
| 1450–1524 | 156 | 123,437 | 68 |
| 1525–1574 | 19 | 14,679 | 24 |
| 1575–1674 | 54 | 50,845 | 11 |
| 1675–1749 | 10 | 10,001 | 7 |
| 1750–1849 | 13 | 17,208 | 1 |
| 1850–1924 | 1 | 1,068 | 1 |
| Undated | 53 | 31,524 | |
| TOTAL | 370 | 294,307 | |

123,437 acres of land—68 per cent of the total enclosed in those years.[1] After 1525 the connection between enclosure and deserted villages gradually weakened. Only nineteen villages were enclosed and depopulated in the enclosure trough of 1525–74, and the land involved amounted to 24 per cent of the total enclosed. In the second wave (1575–1674), fifty-four villages were enclosed and depopulated. The land involved amounted to 11 per cent of the new enclosures.

After 1675, enclosure rarely led to a deserted village. Table 3–1 shows twenty-four villages deserted after 1675, but the associated enclosures amounted to only 1.6 per cent of the total land enclosed. Moreover, these deserted villages are far-fetched examples of depopulation.[2] Nuneham Courteney is a case in point, for it was the prototype of Goldsmith's 'Sweet Auburn' (Beresford and Hurst 1971: 54–6). It was indeed destroyed at enclosure *c.*1759 when the village was emparked, but a new village was constructed to replace the old. The brick houses still flank the road from Oxford to Dorchester, and in 1801 the population was 278. In his Dedicatory Note to 'Deserted Village', Goldsmith wrote:

---

[1] The destructions of 53 villages have not been dated, but most of them probably occurred in 1450–1524.

[2] Moreover, Table 3–1 spuriously overstates the villages destroyed in that period. The table was constructed by tallying as a deserted village *every* village included in the Beresford–Hurst list of deserted villages. The table is consequently the most favourable that could be constructed to support the charge that enclosure led to deserted villages. When one examines the histories of the villages that were enclosed after 1700, it is clear that they were not depopulated at the time of enclosure. Generally, they are listed as deserted villages because they contain some settlement abandoned at an early date. Thus, the deserted village enclosed most recently (in 1858) was Horspath, Oxon. However, it is on the Beresford–Hurst list since it contains 'Old Horspath', not because the main village was destroyed. Most of the entries in Table 3–1 for the years after 1674 can be similarly disregarded.

I have taken all possible pains in my country excursions for these four or five years past to be certain of what I allege. Some of my friends think that the depopulation of villages does not exist, but I am myself satisfied.[3]

Exactly what evidence satisfied Goldsmith is unstated, but the statistics bear out the scepticism of his friends.

The history of deserted villages leads to an important conclusion. During the early Tudor period, enclosure and the destruction of villages did go together. But only 7 per cent of the south midlands were enclosed at that time. If the enclosure of the rest resulted in depopulation, that result was effected in a less overt way.

## Enclosure and Depopulation

To explore the effects of enclosure on population at other times, it is necessary to follow the population of a sample of villages. The 1801 census provides official counts for all villages.[4] For 779 villages, I estimated the population in 1676 with the Compton Census and in 1563 with the ecclesiastical census of that date, or in 1524/5 with the returns for the Lay Subsidy.[5] These are convenient dates since they come at the end of the first and second enclosure waves and thus allow their effects to be compared.

---

[3] As quoted by Beresford (1954: 96).

[4] I have used the summaries of the census in the population tables in the *Victoria County History* except for ambiguous cases and for Leics. where the summary table does not use a consistent definition of villages. In those cases I consulted the original returns, i.e. *Abstract of Answers and Returns Pursuant to Act 41 Geo. III, for taking an Account of the Population of Great Britain in 1801*, BPP 1801–2: vi, vii.

[5] These returns as well as those for the 1377 poll tax were extracted from *VCH*, Leics., iii. 163–74; *VCH*, Oxon., v. 318, vi. 358–61, vii. 231, viii. 274–5, ix. 189; Money (1899), Cornwall (1959), Chibnall and Woodman (1950), Burgess (1978), Palmer and Saunders (n.d.), Allison, Beresford, and Hurst (1965 and 1966), British Museum Harleian MSS 595 and 618, and a print of the microfilm of the Compton Census supplied by the William Salt Library, Stafford.

To estimate population from these enumerations, it is necessary to inflate the number of taxpayers, communicants, etc. by a 'multiplier'. These multipliers have been the subject of much debate. I proceeded as follows: the 1524/5 Lay Subsidy taxed the adult male population. I assumed that the number of taxpayers equalled the number of households and estimated population by multiplying by 4.25. The 1563 ecclesiastical census recorded an estimate of the number of families or households in each village (Hollingsworth 1969: 80). I estimated population by multiplying by 4.25. The Compton Census contains an estimate of the population of each village in the Diocese of Peterborough (Northants and Rutland counties); for the other counties in the south midlands, the adult population is recorded. I estimated the total population as 50 per cent more than the adult population. (I am grateful to Dr A. Whiteman for suggesting this interpretation of the Compton Census.) On average, the 1524/5 Lay Subsidy and the 1563 ecclesiastical census gave similar population estimates as did the 1676 Compton Census and the hearth tax returns for 1662–74. See Allen (1986) for the pertinent regressions.

42 *The Rise of the Yeoman*

Since villages with greater area had larger populations, I compare population density.

The earlier an enclosure, the greater was its impact on population. The demographic histories of villages still open in 1800 were very similar to the histories of those enclosed between 1676 and 1800. (Compare rows 1 and 2 in Table 3–2.) Hence, eighteenth-century enclosures had a negligible effect on population. In contrast, villages enclosed between the mid-sixteenth century and 1676 had a much slower rate of population growth *c*.1550–1676 than did villages enclosed later (row 3 in Table 3–2). Enclosures between the mid-sixteenth century and 1676 may not have lowered population, but they did keep it from growing—most likely by forcing the natural increase to emigrate. Finally, enclosures before the mid-sixteenth century had a dramatic—negative—impact on population. The average population density *c*.1550 of villages already enclosed was only half the density of the ones that remained open (row 4 in Table 3–2).

*Table 3–2. Enclosure and population density, c.1550–1801*

| Enclosure history | Population density | | |
| --- | --- | --- | --- |
| | *c*.1550 | 1676 | 1801 |
| 1. Open in 1801 | 0.0854 | 0.1403 | 0.1856 |
| 2. Enclosed 1676–1801 | 0.0779 | 0.1443 | 0.1968 |
| 3. Enclosed *c*.1550–1676 | 0.0709 | 0.0883 | 0.1279 |
| 4. Enclosed before *c*.1550 | 0.0468 | 0.0781 | 0.0807 |
| Overall | 0.0756 | 0.1224 | 0.1720 |

*Notes:*
Population density is measured in people per acre.

In parentheses, the number of villages in each row are: 1 (148), 2 (283), 3 (133), 4 (23), overall (781).

Villages were classified as open or enclosed at the end of each time period. A village was classified as 'enclosed' if more than 75% of its land was enclosed and 'open' if less than 25% of its land was enclosed. Some villages went through a phase of partial enclosure in which 25–75% of their land was enclosed. This table excludes any village that was partly enclosed at any point between the mid-sixteenth century and 1801. The population histories of partly enclosed villages were not inconsistent with the histories reported here. See Allen (1986) for some tabulations of the population histories of partly enclosed villages.

For counties other than Rutland and Northants, 'mid-sixteenth century' (*c*.1550) means 1563, and the population estimate was derived from the ecclesiastical census of that year. No such returns survive for Rutland and Northants, so 'mid-sixteenth century' for them means 1524, and the lay subsidy of that year is the basis of the population estimate.

To measure the population decline in the pre-*c*.1550 enclosures, it is necessary to estimate the village populations before enclosure. The only source that allows population estimates for a large sample of villages between the Black Death and the middle of the fifteenth century is the Poll Tax of 1377.[6] While this is considerably earlier than the first wave of

[6] I estimated population from the poll tax returns by increasing the number of people taxed by 50 per cent to allow for children and some adult underenumeration. (Hollingsworth 1969: 122–3).

enclosures—and so may not accurately represent population immediately prior to enclosure—it is a full generation after the Black Death and so should reflect the immediate population readjustments following that catastrophe.

The population fell 20 per cent in all of the villages for which I can estimate population in 1377 and in the middle of the sixteenth century (Table 3–3). This decline is consistent with Campbell's (1981: 154) recent estimate that the English population as a whole fell 20 per cent between 1377 and 1524. The decline was greatest in the villages that were enclosed —their average population density was halved. Furthermore, the villages that were enclosed *c.*1550 had lower population densities in 1377 than the villages that remained open, but the differential was smaller. This result raises the possibility that population decline preceded enclosure. Nevertheless, Table 3–3 confirms that enclosure before the middle of the sixteenth century was associated with a considerable decline in population.

*Table 3–3. Enclosure and population density, 1377–c.1550*

| Enclosure history | Population density | |
|---|---|---|
| | 1377 | *c.*1550 |
| 1. Enclosed in 1801 | 0.1116 | 0.0734 |
| 2. Enclosed 1676–1801 | 0.1002 | 0.0881 |
| 3. Enclosed *c.*1550–1676 | 0.0897 | 0.0707 |
| 4. Enclosed before *c.*1550 | 0.0738 | 0.0375 |
| Overall | 0.0936 | 0.0751 |

*Notes:*
Population density is measured in people per acre.
   The overall population densities include villages that went through a partly enclosed stage.

## Enclosure and Inequality

The eviction model also implies that enclosures increased the inequality in landownership when the lord seized the property of his tenants. Ideally, one would test this claim in the same way I tested the analogous claims about population—by comparing indicators of inequality in landownership in the mid-sixteenth century, 1676, and 1801 for villages enclosed at different times. Unfortunately such a research strategy is impossible since there are no sources with which one can compute overall measures of inequality before the late eighteenth century. What I can do is use the land tax assessments to compare inequality across villages *c.*1790 and correlate those measures with the date of enclosure. These comparisons are, of course, vulnerable to the objection that other factors affected the observed patterns, but those patterns are so consistent with the history of population that I believe they do reveal the differential impact of the various waves of enclosure.

I measured inequality of landownership in the following way. For 1,013 south midlands villages whose enclosure chronology I had established, I obtained a copy of the land tax assessment for 1790 (or the nearest suitable year).[7] I estimated the acreage of land belonging to each owner by dividing the total acreage of the parish amongst the owners in proportion to their tax assessments. I also divided the villages into categories depending on whether they were open or enclosed in the mid-sixteenth century, 1976, and 1801.

How the effects of enclosure varied over time is clear from Table 3–4, which presents five indicators of inequality in the distribution of the land ownership—the average number of owners, the share of the land owned by the largest owner, the minimum number of proprietors owning 75 per cent of the total land, the average size of a property, and the percentage of owners who were cottagers.[8] The characteristics of villages open *c.*1790 and villages enclosed since 1676 were very similar. Inequality was much greater, however, in villages enclosed before 1676. Indeed, the ownership of land in villages enclosed before the middle of the sixteenth century was even more unequally distributed than in those enclosed between 1563 and 1676. Whatever indicator one chooses, the results are the same.[9]

The usual conclusion drawn from the land tax evidence is that enclosure had no impact on inequality in landownership. The evidence reviewed here confirms that finding for eighteenth- and nineteenth-century enclosures, but shows that earlier enclosing was related to much more inequality in landownership. Indeed, the earlier the enclosure, the greater the inequality.

The concentration of landownership *c.*1790 is entirely congruent with the earlier results on the history of deserted villages and population. At one extreme, enclosures in the period 1450–1524 were associated with the greatest inequality *c.*1790, frequently with the destruction of the village, and a substantial drop in population. At the other extreme, enclosures after

[7] The sample does not include any villages in Rutland or Berks. Villages in other south midlands counties are included only if it was possible to date their enclosures within the principal waves.

[8] In Table 3–4, any property of ten computed acres or less was tallied as a cottage.

[9] Inferences about intertemporal changes based on cross-sectional comparisons would be misleading if the distribution of property ownership changed for reasons besides enclosure. However, the change that we know did occur during the late 17th and 18th centuries was an increase in inequality in open field villages. (See Ch. 5.) That change means that comparisons between open field villages and ancient enclosures *c.*1790 understates the increase in inequality that occurred during the early enclosures.

*Table 3–4. Enclosure and inequality in property ownership, c.1790*

|  | Open in 1801 | Enclosed 1676–1801 | Enclosed c.1550–1676 | Enclosed before c.1550 |
|---|---|---|---|---|
| Average number of owners in village | 30 | 27 | 14 | 5 |
| Percentage of land owned by largest owner | 38 | 45 | 61 | 75 |
| Minimum number of owners for 75% | 6 | 5 | 4 | 2 |
| Average size property (acres) | 57 | 68 | 118 | 248 |
| Percentage of owners cottagers | 46 | 45 | 26 | 18 |

*Note*: In parentheses, the number of villages in each column are: 230, 380, 138, 83.

1676 had a negligible effect on inequality in landownership or population. Elizabethan and Stuart enclosures exhibit a pattern between the two extremes—they resulted in an intermediate level of inequality in landownership, and they led to depopulation in the restricted sense that the natural increase in the population was forced to emigrate. Thus, the fifteenth-century enclosures provide good examples for Marxists; the eighteenth-century enclosures do the same for Tories. The seventeenth-century enclosures do not fit either model very closely.

## The Early Tudor Enclosures

Why did enclosure between 1450 and 1525 result in deserted villages, extreme inequality, and depopulation? The traditional answer is that those enclosures were effected by manorial lords evicting their tenants and converting the open fields to grass—the history of Stretton Baskerville, in other words, was the norm. However, an opposing explanation of the association of enclosure and low population has been advanced by Hilton (1957) and Dyer (1980: 244–63) following Postan (1950). That theory contends that depopulation preceded enclosure: as a result of the mid-fourteenth-century population decline, settlement in marginal grain-growing areas was abandoned as their inhabitants drifted to more productive farming districts and took up vacant tenements. The abandoned farm land reverted to rough pasture, passed into the hands of the manorial lord, and became *ipso facto* enclosed.

Examining the historical record is the only way to decide the relative importance of these two kinds of enclosures. That record reveals still more complexity, for there were, in fact, three kinds of enclosures in the fifteenth and early sixteenth centuries: (1) eviction enclosures in which lords expelled their tenants, (2) abandonment enclosures such as Hilton and Dyer described, and (3) enclosures that did not result in depopulation at all.

A large number of enclosures *not* causing depopulation may come as a surprise, but the phenomenon is implicit in the numbers previously presented. In particular, according to Table 3–1, one-third of the enclosing in the south midlands did *not* result in deserted villages. These enclosures are candidates for non-depopulating enclosures. Indeed, the five villages enclosed but not deserted for which I can estimate population in 1377 and at some time in the sixteenth or seventeenth century had an average population density of 0.1130 in 1377 and 0.1320 later.[10] They were not depopulated. It was a different story for the deserted villages on the Beresford-Hurst list. For thirty-four villages that were enclosed and deserted, the average population density in 1377 was 0.0687. After enclosure, it fell to 0.0157.

The villages that were not destroyed also differed from the deserted villages in the concentration of property ownership. This conclusion is clear from a comparison of twenty-four villages enclosed before the middle of the sixteenth century that were *not* deserted with forty-seven villages enclosed at the same time that were. The former had an average of ten owners *c.*1790 while the latter had an average of two. The other inequality measures show similarly impressive differences. Landownership was much more concentrated in villages that were deserted than in villages that survived.

What happened in villages that were enclosed without being destroyed? Probably these villages had a large number of freeholders and enclosure was accomplished by agreement amongst them. Many owners probably meant many residents who were not evicted, so the village survived and the population did not drop. These enclosures, in other words, probably anticipated the seventeenth-century pattern of enclosure upon agreement of all of the proprietors.

In the remaining two-thirds of the cases, enclosure was accompanied by low population levels and the destruction of the village itself. But which came first, enclosure or depopulation? The most obvious test is to examine population levels in 1377. Table 3–5 shows a cross-tabulation of the pre- and post-enclosure populations of the thirty-four deserted villages mentioned previously. Most ended up with population densities below 0.02 and all but one less than 0.03. Five of the villages had population densities less than 0.02 in 1377 and so were clearly depopulated before enclosure. On the other hand, many villages had population densities higher than the *average* population densities of some classes of open field villages shown in Table 3–3. There is no sharp line to divide the villages into those depopulated before

[10] In this calculation, the 1801 population for one village was used in preference to an estimate derived from the Compton Census. The latter estimate was considerably and implausibly larger.

*Table 3–5. Pre- and post-enclosure population densities of thirty-four deserted villages*

| Population density in 1377 | Post-enclosure population density | |
|---|---|---|
| | 0.00–0.03 | 0.03–0.05 |
| 0.00–0.03 | 4 | 1 |
| 0.03–0.06 | 9 | 1 |
| 0.06–0.09 | 8 | 1 |
| 0.09–0.12 | 5 | 1 |
| 0.12–0.15 | 4 | 0 |

enclosure and those depopulated after. There are some of both, but the latter sort predominated.

The history of rents also distinguishes the two types of enclosure, eviction and abandonment. This evidence is particularly important since it highlights the class conflict that underlay eviction enclosures.

Both types of enclosure were a response to the population decline between 1348 and *c*.1450. Labour became scarce, the real wage rose, and the real value of land fell in all uses.[11] The fall was greatest in labour-intensive farming (i.e. corn) and least in land-intensive farming (i.e. sheep). In some villages where the land was especially infertile, the value of the corn was less than the income that the farmers could now earn elsewhere; in those cases, farmers left the village and moved to more fertile areas. The result was an abandonment enclosure. In other villages, however, the value of the corn still exceeded the (now augmented) value of the farmers' labour. They had no incentive to move, and there was a surplus with which to pay rent. However, in the pasture district, that rent was less than the land would yield if it were devoted to grazing. In these villages, therefore, there was a fundamental conflict between the interests of the landlord, which were served by conversion to pasture, and those of the farmers, which were fully met by the continuance of grain growing. Eviction of the farmers by the landlord was a frequent resolution of the conflict.

Eviction and abandonment enclosures differed in an important way which helps identify them. In abandonment enclosures the land was valueless as open field arable since the value of the corn was less than the value of the labour needed to cultivate it. In eviction enclosures, however,

---

[11] The theory advanced here is opposed to the traditional theory that it was the rise in the price of wool that prompted the first wave of enclosures. That theory is untenable since the price of wool did not rise with respect to the price of corn. (Lloyd 1973: 50–1, 68–9.) The relative price that did change was the wage rate, which rose with respect to all agricultural product prices. The fall in the population after 1348 must have been the ultimate cause of the rising wage.

the land did generate a positive rent in the open field state, although the rent was higher after enclosure. Therefore, a positive pre-enclosure rent signals an eviction enclosure.

The evidence on pre-enclosure rents points to many eviction enclosures. Thus, Hoskins (1950a: 76) reports that the open fields in Ingarsby, Leicestershire, were worth 8*d.* an acre when it was enclosed in 1469. This was a high value for open fields, which usually rented for about 6*d.* per acre, and, indeed, was high compared to enclosed pasture, which typically rented for 12*d.* per acre. (See Table 9–1.) The 1517 Commission on Enclosures reported positive pre-enclosure rents for nine villages (Leadam 1897: 66). By the early sixteenth century, it had become common opinion that enclosure raised land values *and that they were usually positive before enclosure.* Fitzherbert's 1523 discussion 'Howe to make a Township that is worthe xx Marke a Yere, worth xx li. a Yere' values the land in the village to be enclosed at 6*d.* per acre (Tawney and Power 1924: iii. 22–3). Enclosures of open fields that were this valuable were not abandonment enclosures. Fitzherbert's observation suggests that the abandonment enclosures had ceased by the end of the fifteenth century, if not earlier. Thereafter, either eviction enclosures or non-depopulating enclosures were the norm.

The enclosure movement between 1450 and 1525 was complex because it involved three kinds of enclosures. The first were the abandonment enclosures that resulted from the drift of population away from infertile land. The second were enclosures by agreement that did not result in depopulation or the destruction of villages. The third were enclosures in which manorial lords evicted their tenants and converted the fields to grass. It is impossible to assign precise proportions to the three kinds since it has not been possible to date the enclosure of all of the land in the south midlands. Depending on the chronology of enclosure of the undated deserted villages, the available information implies that one quarter to one third of the enclosing in the south midlands between 1450 and 1525 was not depopulating. The remaining two-thirds to three-quarters were either abandonment or eviction enclosures. The abandonment enclosures were the least important and probably occurred early in the first enclosure wave. Eviction enclosures were probably the most important type. The Tudor conviction that enclosure meant depopulation was fact, not fancy.

## Elizabethan and Seventeenth-Century Enclosures

In the seventeenth and eighteenth centuries, enclosure retarded the expansion of village populations. The check operated both in the new

Elizabethan and Stuart enclosures and in the villages that had been enclosed in the previous centuries. Unless the excess of births over deaths in these villages was dramatically lower than in the rest of rural England, enclosure was causing significant emigration from affected villages.

There are two possible explanations for the emigration from villages enclosed between 1575 and 1674. The first is that these enclosures led to the conversion of arable to pasture with the result that employment fell and the redundant labourers left. Even without detailed investigation, one must doubt this explanation since the conversion of arable to pasture in the eighteenth century did not lead to village depopulation.

The second explanation turns on the small number of owners in villages enclosed before the eighteenth century, and the incentives the Poor Law presented to them to restrict the populations of their villages. A succinct statement of this theory is found in a conversation Arthur Young had with some Northamptonshire gentry in 1785.

By Mr Ashby's good offices I was introduced to some considerable graziers and proprietors, very sensible and intelligent men, with whom I conversed on this interesting topic [enclosure and depopulation]. Mr Wade was of opinion, that inclosures by no means depopulate; that wherever such an effect is supposed, it is owing to a whole parish belonging to one or two men who take what steps they can to keep the poor out of it. But where there are more and smaller proprietors, the very contrary happens, and population is found to increase by inclosing; that this is the case in his own lordship of Clipston, which, he is confident, is more populous than it was when open.

Neither Mr Macro nor myself could understand, nor was it fully explained to us, how, in any case, the proprietors could drive the poor away. To act in that manner would be pursued in many parts of the kingdom if the law permitted, but the laws of settlement effectually prevent it. (Young 1786: 460–1)

The distinction drawn by Mr Wade was common in the nineteenth century: so-called 'open' parishes had many landowners, large populations, and heavy poor rates. In contrast, 'closed' parishes had very few landowners (often only one), small populations, and low rates. The owner or owners in closed parishes kept down the population by restricting the housing stock, while the owners in the open parishes expanded the stock to accommodate population growth.[12]

The different housing policies in open and closed parishes reflected the incentives created by the Poor Law and the associated laws of settlement. Those laws required every parish to support the poor *who had a settlement there*. The rules that determined settlements were complex and changed over the years, but, in practical terms the larger the population, the more

[12] See e.g. Holderness (1972) and Mills (1980).

people eventually qualified for support. The funds for assistance were raised by a parish property tax, the poor rate. The tax was paid, in the first instance, by the occupier of land, but, in the long run, it was borne by the owner of land, because of the competition for tenancies: 'The poor rates were in effect a tax on rent.'[13] An individual's tax assessment was proportional to the value of his property in the parish, so large landowners paid the biggest share of the tax. In so far as they were ratepayers, all landowners suffered if the population (and with it the number of poor eligible for support) increased. Table 3–6 shows that, indeed, poor expense per acre rose dramatically with population density around 1800.

Table 3–6. *Poor expense per acre and population density*

| Population density | Poor expense per acre (£) | Number of villages averaged |
|---|---|---|
| 0.0–0.1 | 0.0631 | 301 |
| 0.1–0.2 | 0.1194 | 708 |
| 0.2–0.3 | 0.1852 | 328 |
| 0.3–0.4 | 0.2419 | 69 |
| 0.4–0.5 | 0.3405 | 20 |
| TOTAL | 0.1317 | 1,426 |

*Note*: Population density is people per acre in 1801. Poor expenses are for 1803.

*Source*: Poor expenditure from *Abstract of Answers and Returns under Act 43 Geo. III, relative to the Expense and Maintenance of the Poor in England*, BPP 1803–4: xiii.

Despite the fact that a greater population density raised the poor rate per acre, small landowners could benefit from a policy of population expansion if they rented housing to the poor. A rough numerical example illustrates the principle. Suppose a village contained 1,000 acres, and a new poor family was added to its population. In the late eighteenth century that family required about £30 per year for its sustenance (Eden 1797). If the family was totally supported by poor relief, an additional rate of £0.03 per acre was needed to raise the funds. The family paid about £1.5 per year for the rent of its cottage. Suppose the cottage came with a quarter acre of land worth £0.25 per year as enclosed pasture. A labourer's shack cost about £10 to erect (Gooch 1813: 294). Allowing 10 per cent for interest and depreciation makes the annual capital cost £1. The net income of letting a house to the poor family was £0.25 per year, i.e. £1.5 − £1 − £0.25. A

[13] Clay (1985: 237). I examine the degree to which rents plus taxes came into alignment with the commercial value of land in Ch. 9.

person owning 8 acres (= 0.25 ÷ 0.03) *in total* in the village broke even in the sense that the net income from letting the house exactly equalled the rise in his poor tax payments. Anyone owning more land lost since the additional poor tax payments exceeded the rental income of the house; anyone owning less than 8 acres profited.

For this reason, the number of landowners influenced the rate of population growth. In villages owned by one or two large owners, the housing stock was limited to restrict the number of eligible poor. On the other hand, in villages with many small landowners, the housing stock expanded in response to demand. Table 3–7 shows the result: around 1800, poor expense per acre increased significantly as the number of owners in a village rose. With enclosed land letting for about £1 per acre, the tax burden was heavy in villages with many landowners and large populations. In such villages the Poor Law concentrated the burden on the large landowners: they paid most of the tax and most of the poor lived in houses owned by small proprietors. The expanded pauper population was a duct through which funds flowed from large proprietors to small.

*Table 3–7. Poor expense per acre and the number of property owners*

| Number of owners | Poor expense per acre (£) | Number of villages averaged |
|---|---|---|
| 1 | 0.0635 | 65 |
| 2 | 0.0693 | 57 |
| 3 | 0.0931 | 47 |
| 4 | 0.0894 | 40 |
| 5 | 0.0785 | 26 |
| 6–9 | 0.1001 | 101 |
| 10–19 | 0.1337 | 211 |
| 20–29 | 0.1402 | 122 |
| 30+ | 0.1850 | 306 |
| TOTAL | 0.1334 | 975 |

*Note*: The number of owners is from the land tax assessment *c.*1790.

*Source*: Poor expense per acre applies to 1803 and is taken from *Abstract of Answers and Returns under Act 43 Geo. III, relative to the Expense and Maintenance of the Poor in England*, BPP 1803–4: xiii.

Since the sixteenth century, large landowners felt the incentive to restrict population growth. As early as 1568 the court of Barham Manor in Linton, Cambridgeshire, forbade tenants from providing lodging to anyone but their family and servants. 'In 1581 it was ordered that none should take in married couples as under-tenants without giving surety to the parish' (*VCH*,

Cambs., vi. 93). Broad has shown that the Verneys followed similar policies in Middle Claydon, Buckinghamshire between 1600 and 1800, thereby limiting population growth. (Fox 1989: 111–12.) Clay (1985: 236–8) discusses the operation of these incentives in the period 1640–1750.

To find out how effective large landowners were in restricting population growth, I undertook a regression analysis of the population density of 690 south midlands villages. The results are shown in Table 3–8. The villages were classified into four categories according to their enclosure histories.

*Table 3–8. Regressions explaining 1801 population density (t-ratios in parentheses)*

| Variable | Equation | | | |
|---|---|---|---|---|
| | 1 | 2 | 3 | 4 |
| Constant | 0.12645 | 0.07514 | 0.09643 | 0.08767 |
| | (19.758) | (15.141) | (17.115) | (13.161) |
| P | | $6.46 \times 10^{-3}$ | $5.23 \times 10^{-3}$ | $4.78 \times 10^{-3}$ |
| | | (13.937) | (10.899) | (9.780) |
| $P^2$ | | $-7.89 \times 10^{-5}$ | $-6.00 \times 10^{-5}$ | $-5.49 \times 10^{-5}$ |
| | | (−7.959) | (−6.048) | (−5.545) |
| $P^3$ | | $3.07 \times 10^{-7}$ | $2.30 \times 10^{-7}$ | $2.16 \times 10^{-7}$ |
| | | (6.007) | (4.562) | (4.308) |
| DEEE | −0.06530 | | | −0.02130 |
| | (−6.076) | | | (−2.107) |
| DOOE | 0.06565 | | | 0.02804 |
| | (8.331) | | | (4.009) |
| DOOO | 0.05764 | | | 0.00162 |
| | (6.852) | | | (1.570) |
| DV | | | −0.05775 | −0.03748 |
| | | | (−7.176) | (−4.025) |
| $R^2$ | 0.24819 | 0.43176 | 0.47149 | 0.49420 |

*Variables*:
P, $P^2$, $P^3$ are the number of proprietors *c.*1790, and the number squared and cubed.
DEEE equals 1 if the village was enclosed by the mid-sixteenth century and 0 otherwise.
DOOE equals 1 if the village was enclosed betwen 1676 and 1801 and 0 otherwise.
DOOO equals 1 if the village was open in 1801 and 0 otherwise.
DV equals 1 if the village was a deserted village and 0 otherwise.

*Note*: The regressions exclude villages that went through a 'partly enclosed' stage.

Dummy variables in the table distinguish the classes: DEEE indicates villages enclosed before the middle of the sixteenth century, DOEE villages enclosed between the mid-sixteenth century and 1676, DOOE villages enclosed between 1676 and 1801, and DOOO villages still open then. In equation 1, there is a constant and the dummy variables DEEE, DOOE, and DOOO. The value of the constant (0.12645) measures the average population density of villages enclosed between the mid-sixteenth century and 1676, while the coefficients of the dummy variables indicate the differences in mean population density between the classes they represent and the constant. The large size of the t-ratios shows that the differences in population densities are statistically significant.

In equation 2, population density is regressed on a cubic function of the number of owners. The function shows that population density rises with the number of owners but at a diminishing rate. Equation 3 includes a dummy variable (DV) indicating that the parish was a deserted village according to the Beresford–Hurst list. That variable has a negative and significant coefficient indicating that the destruction of the village has a depressing effect on later population over and above the lowering effect due to the small number of owners. That result is not surprising.

Equation 4 includes all the variables. Notice that the values of the coefficients of the variables for the numbers of owners and the deserted village dummy variable are similar to the preceding regressions and still statistically significant. Notice also that the coefficients of DEEE, DOOE, and DOOO are all closer to zero. The coefficient of DOOO is insignificantly different from zero. These changes in the value and statistical significance of the enclosure period dummy variable coefficients show that the differences in the population densities of villages enclosed at different times are due to differences in the number of owners and the results of village destruction. The regression suggests that those are the mechanisms by which enclosure affected population.

Table 3–9 shows the relative importance of the number of owners and the destruction of villages to population in 1801. The table uses the equation in column 4, Table 3–8, to predict average population density using the mean values of number of owners, and so on, in the four categories of villages in the sample. The column headed 'effect due to number of owners' is computed from the constant in the regression and the terms involving the number of owners. The deserted village effect equals the coefficient of the

*Table 3–9. Decomposing 1801 population density*

| Enclosure period | Effect due to number of owners[a] | Effect due to deserted villages[b] | Residual effect[c] | Predicted population density[d] |
|---|---|---|---|---|
| Enclosed before *c.*1550 | 0.110351 | −0.02897 | −0.02130 | 0.060080 |
| Enclosed *c.*1550–1676 | 0.142067 | −0.00648 | 0.0 | 0.135585 |
| Enclosed 1676–1801 | 0.180622 | −0.00122 | 0.02804 | 0.207444 |
| Open in 1801 | 0.187846 | −0.00107 | 0.01162 | 0.198394 |

[a] 'Effect due to number of owners' was computed as the weighted average of the first four terms of col. 4, Table 3.8, evaluated at the average population density of villages enclosed in that period that were and were not deserted villages. The weights were the shares of villages enclosed in that period that were and were not deserted villages.
[b] 'Effect due to deserted villages' equals the coefficient of DV in col. 4, Table 3.8 (−0.03748) multiplied by the share of villages enclosed in that period that were deserted.
[c] 'Residual effect' equals the coefficient of the dummy variable for that period in col. 4, Table 3.8.
[d] 'Predicted population density' equals the sum of the three effects. Actual average population densities for the periods were 0.06115, 0.12645, 0.1921, and 0.18409 in descending order.
Actual averages differ from predicted averages since the regression equation is non-linear.

deserted village dummy variable in the regression multiplied by the proportion of villages in the category that were deserted. The residual effect equals the coefficient of the enclosure period dummy variable. The predicted population density is the sum of the three effects. It differs slightly from the average population density of the category since the regression is non-linear.

A comparison of the column 'effect due to number of owners' with the predicted average population density shows that variation in the number of owners explains most of the variation in village population. For enclosures before the middle of the sixteenth century, however, relying solely on the number of owners overpredicts population due to the high incidence of deserted villages in that period. The deserted village effect incorporates that feature of early enclosure and corrects about half of the overprediction. The residual effects are small compared to these factors and add little to an understanding of 1801 population levels.

The regression model proposed here explains population levels without any reference to the reduction in employment that followed from converting arable to pasture. Indeed, I can reject the hypothesis that such conversion was important in determining 1801 population levels by dividing the 690 villages into two groups—440 villages in the pasture district and 250 villages in the arable districts—and estimating the equation in Table 3–8, column 4, on each. The results for each region are similar to equation 4. This means that the village characteristics—the number of proprietors and the desertion of villages—were influencing population in the same way irrespective of the conversion of arable to pasture. That conversion affected what people did, but it did not affect where they lived.

## Conclusion

Enclosure did play a more important role in destroying the English peasantry than historians have been inclined to allow. In the fifteenth and sixteenth centuries, many enclosures conformed to the eviction model— lords evicted farmers, seized their land, amalgamated it into large farms, and converted it to pasture. That pattern was much less common in the seventeenth century since copyholders and leaseholders had acquired legal remedies to defend possession of their property. (That great revolution in landownership is the subject of the next chapter.) Nevertheless, there was still a connection between enclosure, inequality, and population. To effect an enclosure, manorial lords had to secure the concurrence of their customary tenants or buy them out. Often the latter course was followed. The result was the elimination of peasant proprietorship and the concentration of property in great estates. The small number of landowners who remained found it in their interest to restrict the housing stock of the village with the result that population growth was checked, and the natural increase in the population emigrated. Thus, enclosure before the eighteenth century did lead to rising inequality and emigration.

My results, however, confirm the twentieth-century studies of the land tax assessments which show that parliamentary enclosure had little impact on inequality. The great advantage (to the landed gentry) of parliamentary enclosures was that they could proceed without the consent of small copyholders and freeholders.[14] It was, therefore, no longer necessary to buy up all the small properties prior to enclosure. As a result, there were many proprietors in villages enclosed in the eighteenth and nineteenth centuries, and enclosure in that period did not reduce their number.

---

[14] The creation of copyhold tenure—not parliamentary enclosure—'represented a major advance in the recognition of the rights of the small man' (Chambers and Mingay 1966: 88). Copyhold, however, was only a weak and qualified recognition and was undermined during the 18th cent. Nevertheless, by allowing enclosure without the concurrence of small proprietors, the parliamentary procedure unambiguously curtailed the rights small owners had enjoyed in the 17th cent.

# 4

# The Rise of the Yeoman

The practice of enlarging and engrossing of farms, and especially that of
depriving the peasantry of all landed property, have contributed greatly
to increase the number of dependent poor.

D. Davies, *The Case of Labourers in Husbandry*,
1795: 51–7

ENGLISH peasants disappeared in two phases. First, early enclosure led to
their demise for the reasons discussed in the previous chapter. Second,
peasant farmers also disappeared from the open fields in the late seventeenth
and eighteenth centuries. Since five-eighths of the south midlands remained
unenclosed in 1675, what happened in the open fields was more destructive
than the early enclosures.

The demise of peasant agriculture in the open fields was not a simple
decline from a medieval golden age—the peasantry waxed before it waned.
In the seventeenth century, royal governments supported peasant proprietors in Denmark, western Germany, and France. These policies consolidated the peasants' ownership of land at the expense of the feudal nobilities
(de Vries 1976: 58–66). Comparable policies were followed in England
with comparable results. While the details of these policies have been
discussed by English historians, the ultimate disappearance of the small
farmer has so dominated discussion that the rise of the peasant has not
received its due. The omission is doubly unfortunate since the reasons for
the peasantry's decline are not separable from the manner in which the
peasantry consolidated its control over land in the sixteenth century. This
chapter is concerned with the latter story.

## Peasants and Family Labour

'Peasant' has no generally accepted meaning.[1] Two issues, however—wage
labour and landownership—have animated English rural history, and they

---

[1] Many writers insist that peasants are subsistence farmers whose decisions ignore the
possibility of selling their produce and, hence, take no cognizance of the price it might fetch.

serve to define 'peasant' in the English context. Thus, 'peasants' were family farmers, while 'capitalist' farms depended on wage labourers. Further, 'peasants' had a proprietary interest in the soil, while 'capitalist' farmers had none.

To analyse historical documents, the size of family and capitalist farms must be specified in terms of acreage. Several types of information suggest that a family could operate a farm of up to 50 or 60 acres without much hired labour. A farm of more than 100 acres was run preponderantly with hired (or, in the middle ages, coerced) labour. These divisions are, of course, subject to many qualifications, but, roughly speaking, peasant farms were less than 60 acres, while capitalist farms were more than 100 acres. Farms of 60 to 100 acres were transitional, employing roughly equal amounts of family and hired labour.

Arthur Young (1770: i. 118, 129, 148–51, 177–81, 211–15) computed the labour requirements of various model farms, and these calculations map out the transition from family to hired labour. Young's arithmetic shows that farms of 8 and 25 arable acres did not exhaust the farmer's time. Except for the harvest, 36 acres of arable could be cultivated by one man. He supplied 93 per cent of the adult male labour time. That proportion dropped to 52 per cent for the farm with 50 acres of arable and 45 per cent with 60 arable acres; 36 acres of arable, therefore, marks the upper limit that could be farmed without much wage labour, while 50 acres marks equality of hired labour with the farmer's own time. Allowing for meadow and pasture, a 50-acre farm was the largest that could be cultivated mainly by one man, while hired labour exceeded family labour above 70 acres.

This calculation ignores the labour of women and children, and they were not fully employed on farms of less than 60 acres arable. Women, for instance, ran the dairy. In his *Tours*, Young reported that a dairymaid could manage ten cows, which was more than the number kept on these small farms. So the farm wife had time for other work as well. Children could also supplement adult labour. While precision is impossible, it can be stated that the labour of women and children allowed larger acreages to be tilled. The upper limit of a farm worked mainly with family labour was probably closer to 60 acres than to 50, while the size at which hired labour surpassed family labour was probably near 75 acres.

Several sources provide checks on these inferences. First, the survey data collected by Young on his tours in the late 1760s indicate that 77 acres was

While this behaviour may be important in some parts of the world, it was not common in early modern England. Even small farmers marketed much output at an early date, and they certainly used market prices as the basis for valuing crops when they drew up probate inventories. See Allen (1988b: 120–1) for evidence on valuation.

the size at which hired labour exceeded family labour on arable farms.[2] Second, 50 acres was often accepted as the upper limit of a family farm in England. Thus, the *Report on the Decline in the Agricultural Population of Great Britain, 1881–1906*[3] concluded that 40–60 acres was 'sufficient to employ the whole labour of a man and his family and not enough to necessitate the employment of hired labour' (p. 17). Third, family farms in nineteenth-century America were of a similar size. For instance, Danhof (1969: 137) estimated that, before the introduction of harvesting machinery in the 1850s, one man could manage 30 acres 'in tilled crops and mown grass' on the level lands of the midwest. David (1966: 38) estimated that the average farm in the major wheat-growing counties of Illinois in 1849–50 harvested 25.2 acres of wheat, oats, and rye. Such farms did not employ much wage labour. Allowing for fallow, pasture, and meadow on the English pattern implies that a family farm was about 50 acres.

## *Landownership*

The definition of peasant agriculture in England is also a question of ownership. In this context, a 'full' owner of land enjoys the following six rights:[4]

1. the right to use the land
2. the right to lend or lease the land
3. the right to sell the land
4. the right to security in exercising the above rights
5. the right to exercise the preceding rights for the rest of his life[5]
6. the right to transmit the property to other people after his death.

Sometimes there was a 'full' owner in the sense of this list. Consider the house and grounds of a manor which was not entailed or encumbered with mortgages. The lord could use or let the property as he chose; he could use the royal courts to defend his possession; he could sell the land or keep it for the rest of his life; and (by the eighteenth century) he could bequeath it. The

---

[2] Allen 1988c: 127. For pasture farms, equality was reached at 150 acres. A family could operate a larger pastoral farm without the assistance of hired labour since the labour requirements of a pastoral farm were less than those of an arable farm.

[3] United Kingdom, Board of Agriculture and Fisheries, Cd. 3273, BPP 1906: xcvi.

[4] This list is derived from Honoré (1961). Other theorists have proposed other lists of the essential features of ownership (e.g. Becker 1977: 18–23). This list emphasizes the duration and transferability of title. 'Exclusivity'—the right to control the use of the land without reference to others—is also emphasized by many theorists. There were, indeed, important changes in this regard, most notably the enclosure of open fields and commons.

[5] Provided he remains solvent.

lord owned the property. Conversely, consider a farm on the estate that was occupied by a tenant at will: if the lord ordered him to quit, the farmer had to leave. He could not lease the land, let alone sell it or bequeath it. If the lord expelled him, he could not use the royal courts to recover the land, although he could sue for the value of the crops he had planted. The farmer had no proprietary interest in the soil since all ownership rights belonged with the lord.

In many cases, no one was a full owner since ownership was fragmented among several parties. Thus, if the manor were strictly settled, so that the land was entailed and the lord was only a life tenant, then he was only a partial owner. He could exercise the rights of use, lease, and sale only for the duration of his life. Consequently, as pointed out in Chapter 2, the life tenant of a settled estate could not enter into an enclosure agreement unless the settlement made specific provision for such a property exchange or unless a parliamentary act amended the settlement. Moreover, he could not, on his own, affect the transmission of property after his death: extension of the settlement for another generation required the connivance of his son. Despite these limitations, no one would deny that the life tenant of a settled estate 'owned' his lands. The important implication is that partial ownership is still ownership, provided its limitations are recognized.

Just as a manorial lord could be a partial owner, so could a farmer. In the sixteenth and seventeenth centuries, large farms were often let with a beneficial lease. These leases usually lasted for many years, often for the life of the farmer, his wife, and his son. The farmer made a large initial payment and then only small annual or biannual payments over the duration of the lease. Such a farmer was like the life tenant of a settled estate. The leasee could use the land and appeal to the courts to preserve his possession of the lease if the lessor tried to expel him. If the lease was for his life, he could sublet or sell it. He could occupy the property for the duration of his life and, through the lease itself, control the descent. This control did not extend indefinitely into the future, so ownership was not 'full'. These powers do, however, indicate considerable proprietary interest in the land—much more than a tenant at will—so there is reason to call the beneficial leasee a partial owner.

## The Medieval Background

England in the century before the Black Death is often described as a peasant society. While there were many family farms, few farmers, in fact, owned their land. From the point of view of ownership, medieval England was a low point for peasant agriculture.

OWNERSHIP IN THE MIDDLE AGES

The Crown is the ultimate owner of all land in England, but, in practice, the ownership of most property was parcelled out to the nobility during the middle ages. The manor was the basic unit of seigneurial ownership, and all land in the kingdom belonged to a manor. In the midlands, a manor was most often coterminous with a parish or village although there were numerous instances of parishes containing more than one manor and a small number of manors encompassing more than one village. Typically, the manor contained three kinds of land: the demesne, which was under the immediate control of the lord, the free lands farmed by free peasants, and the villein lands, which were farmed by the manor's villeins or serfs. A royal inquisition called the *Rotuli Hundredorum* of 1279 indicates the relative importance of these types of land.[6] Demesnes amounted to 32 per cent of the total land, villein land to 40 per cent, and free holdings to 28 per cent (Kosminski 1956: 91).

There is no question as to who owned the demesne—it was the lord of the manor. He could use it or lease it as he wished. His ownership was not unlimited, however. Before the statute of *Quia Emptores* in 1290, the manor could not be conveyed without the consent of his superior lord. After 1290, however, such consent was no longer required except for tenants holding directly from the Crown. In 1327 they were allowed the right to alienate freely but were subject to a fine that was not abolished until the Tenures Abolition Act in 1660 (Megarry and Wade 1975: 31–2). Further, after the thirteenth century, inheritance followed rules that could not be altered by wills. However, the Statute of Wills, 1540, authorized the bequeathing of most freehold property by will. This right was extended to all freehold land after 1660 (Megarry and Wade 1975: 471–2).

In the eleventh century, ownership of the land of free tenants was split between lord and tenant, but, by the fourteenth century, the free tenant had become a full owner of his property. He could use or lease his land as he chose. His ability to bequeath land was at least as strong as that of his lord. After the middle of the twelfth century, the free tenant could defend his rightful possession of his land even against his lord by using the 'real actions'—the writs of novel disseisin (instituted in 1166), the grand assize, and mort d'ancestor (1176). These writs initiated actions in the royal courts which took precedence over manorial courts in establishing title and enforcing possession for free tenants (Pollock and Maitland 1968: i. 145–9). The statute *Quia Emptores* in 1290 was a turning point in establishing

---

[6] Most of the survey is lost, but the surviving returns overlap considerably with the south midlands as defined in this book. (Kosminski 1956.)

freedom to alienate since the lord's consent was no longer required.[7] By the thirteenth century, a free tenant was in most respects a full owner of his property.

The villein's position was far inferior. Throughout the middle ages, land held in villeinage belonged to the lord as did much of the villein's labour power. Thus, in law, the villein used his land only at his lord's will, although, in practice, the villein was undisturbed if he abided by custom. Generally, villeins were prohibited from leasing their farms without specific permission. A villein tenancy could be alienated or bequeathed only with the concurrence of the lord, for the conveyance was effected by the seller's surrendering the tenement to the lord who then admitted the buyer, if he chose. Sometimes the lord chose not to, and, if the conveyance was executed, a fine was imposed. The villein tenant could not avail himself of any action in a royal court to preserve possession of his tenancy if his lord expelled him. The villein tenant was often subject to tallage (arbitrary fines), merchet (a fine for marrying someone other than another villein of the same lord), and heriot (the lord took the best animal from a farm when it passed to an heir). The villein tenant was also subject to the performance of uncertain and arbitrary labour services for the lord. Hence, villein tenure included command over labour as well as ownership of land.

While the intrusion of royal courts into the affairs of the manor had strengthened the position of free tenants *vis-à-vis* their lords, the intervention of these courts increased the obligations on villeins during the eleventh and twelfth centuries. Initially, the customs of many manors were ambiguous as to whether lords could increase the labour obligations of their tenants. Peasants resisted increases in royal courts by claiming to be free and, hence, subject to fixed exactions. However, the slightest hint that the lord had discretion in setting his demands was usually enough for the courts to rule that the peasants were villeins subject to wholly arbitrary fines and services. Decisions like these, which were common, sharpened the line between freedom and servility and reinforced the ascendancy of the feudal aristocracy (Hilton 1949).

The disabilities just described were features of villein tenure—they thus applied to a free person holding land in villeinage. But villeinage was also a personal status, which was inherited and involved additional disabilities, including the following. A lord could beat or imprison his villein, although he could not maim or kill him. The villein could be sold as a chattel (Pollock and Maitland 1968: i. 414–16). A villein could not sue his lord in a royal court. The lord owned his villein's personal property. In the thirteenth

[7] *Quia Emptores* also provided that estates in fee simple could be alienated only by substitution, rather than subinfeudation as previously.

century, villein tenure and villein status usually went together, so there was little reason to distinguish their effects. By the fifteenth century, however, many free people held land in villeinage, as we will see.

FARM SIZE AND FAMILY LABOUR

While only the free tenants were 'peasants' in the sense of owner-occupying farmers, most of the villeins were also 'peasants' when family labour is the issue. Table 4–1 shows the size distribution of villein farms as indicated by the hundred rolls. Most land was in holdings of half a virgate (*c*.14 acres) or one virgate (28 acres). A half-virgate farm was just large enough to support a villein family (Titow 1969: 78–93), although neither it nor the one-virgate farm was large enough to exhaust the labour supply of a typical family. Scarcely any villein land was in farms large enough to require wage labour. Estate surveys confirm these results and suggest that the size distribution of villein farms was fairly stable from the eleventh century to the fourteenth (Postan 1975: 154–5; B. Harvey 1977: 203–8, 434–7; Dyer 1980: 88–90). Villein farms were, therefore, peasant farms by the family labour criterion.

*Table 4–1.  The size distribution of villein holdings, 1279*

|  | Average size (acres) | Number | Total acres |
|---|---|---|---|
| 1 or more virgates | 49 | 173 | 8,477 |
| 1 virgate | 28 | 3,940 | 110,320 |
| ½ virgate | 14 | 5,724 | 80,136 |
| ¼ virgate | 7 | 1,378 | 9,646 |
| Petty allotments | 1.5 | 4,687 | 7,031 |
| TOTAL |  | 15,902 | 215,610 |

*Source*: The number of holding is from Kosminski (1956: 216). Total acres equals the number multiplied by the average size. I computed average size with the following assumptions: first, the hundred rolls usually measure villein tenancies in virgates. Kosminski (1956: 91) indicates that in Oxon. the virgate was usually 25 acres, while in other counties surveyed it was 30 acres. Table 4–1 sets the average virgate at 28 acres because it is convenient and because that value emerges from the more complex, weighted calculations underlying Table 4–2. Second, petty allotments (i.e. cottages) were taken to be 1.5 acres each to maintain conformity with Table 4–2. Third, the average size of holdings in the category of 1 or more virgates' was set at 49 acres (1.75 virgates) since Kosminski (1956: 215) found that such holdings were usually 1.5 to 2 virgates.

Considered from the point of view of family labour, the free holdings reveal a mixture of peasant and capitalist farms. Table 4–2 shows Kosminski's reconstruction of the distribution. Compared with villein property, much less of the land was in one-virgate or half-virgate holdings although together they comprised 45 per cent of the total. Almost as much land was in the category 'more than one virgate'. The average size farm in this largest category was substantial (76 acres) and approximated the size at which hired labour equalled family labour. Hence about 22 per cent of the free tenant land in 1279 was in farms that could reasonably be called 'capitalist'. (The wage labourers on these farms were recruited from the freeholders who had small holdings and the cottagers without any farm land at all.) Kosminski accepted that large farms were more efficient than small farms, and argued that the growing independence of free tenants from manorial restrictions allowed enterprising peasants to amalgamate property and realize the profits of large scale. The inequality in the free land distribution is an early example of capitalism's emerging through peasant differentiation.

Table 4–2. *The size distribution of free holdings, 1279*

|  | Average size (acres) | Number | Total acres |
|---|---|---|---|
| 1 or more virgates | 76 | 521 | 39,596 |
| 1 virgate | 28 | 904 | 25,312 |
| ½ virgate | 14 | 1,083 | 15,162 |
| ¼ virgate | 7 | 775 | 5,425 |
| 3–5 acres | 4 | 620 | 2,480 |
| Petty allotments | 1.5 | 2,251 | 3,377 |
| TOTAL |  | 6,154 | 91,352 |

*Source*: Average size is from Table 4–1. The number of holdings is from Kosminski (1956: 223). Total acres equals average size multiplied by number.

Demesnes were much larger than peasant farms. Table 4–3 shows the distribution of 290 demesnes in Huntingdonshire, Bedfordshire, and Cambridgeshire compiled directly from the *Rotuli Hundredorum*. The average size of all demesnes was 165 acres,[8] which was the average size of a farm on a large estate in the early nineteenth century. Such a farm needed the labour of many men to cultivate.

Demesnes were large enough to be run as capitalist farms, although often they were not. On small estates with large demesnes and few villeins, much wage labour was used. Such demesnes can reasonably be called early capitalist enterprises. However, on large estates amply provided with villeins, much of the labour to cultivate the demesnes consisted of the compulsory services of the villeins. Their labour was available for conscription since the villein holdings were not large enough, on average, to

[8] Dyer (1980: 62, 122–3) and B. Harvey (1977: 428–30) confirm that demesnes were often several hundred acres.

*Table 4–3. The size distribution of demesnes, 1279*

| Farm size (acres) | Number | Acreage | Percentage of | |
|---|---|---|---|---|
| | | | Number | Acreage |
| 5–10 | 4 | 31 | 1.3 | 0 |
| 10–15 | 9 | 101 | 3.1 | 0.2 |
| 15–30 | 18 | 382 | 6.2 | 0.8 |
| 30–60 | 30 | 1,346 | 10.3 | 2.8 |
| 60–100 | 59 | 5,000 | 20.3 | 10.4 |
| 100–200 | 82 | 11,429 | 28.3 | 23.8 |
| 200–300 | 45 | 10,506 | 15.5 | 21.9 |
| 300–400 | 25 | 8,503 | 8.6 | 17.7 |
| 400–500 | 10 | 4,568 | 3.4 | 9.5 |
| 500–1000 | 7 | 4,634 | 2.4 | 9.7 |
| 1,000+ | 1 | 1,440 | 0.3 | 3.0 |
| TOTAL | 290 | 47,940 | | |

*Source*: *Rotuli Hundredorum*, Record Commission, i and ii, 1812–1818.

exhaust their labour power. Estates operated in this way were not capitalist enterprises.

Was medieval England a peasant society? The answer depends on the definition chosen. If a peasant farm is defined as one employing mainly family labour, then there were many peasant farms between the eleventh and the fourteenth centuries. These included virtually all of the villein farms and all the free holdings of a virgate or less. Together these farms accounted for 56 per cent of the farm land.[9] The rest consisted of demesnes cultivated by servile labour, or capitalist farms. On the other hand, if peasants are also required to be owners, then there were few peasants in the medieval economy, since the free peasants were the only farmers who owned their land.

## The Black Death, the Collapse of the Manorial System, and Enclosure

The manorial system, as just described, lasted from the end of the eleventh century until the late fourteenth. It collapsed under the impact of the Black Death of 1348–9 and the subsequent century of population decline. From a peak in excess of 5 million people in the early fourteenth century, the English population probably dropped to less than 2 million in the middle of the fifteenth (Hatcher 1977). In the long run, this shock was too great to be contained within the manorial system.

[9] Villein holdings were 40 per cent of the land; free holdings were 28 per cent, but only 57 per cent of that was in holdings of a virgate or less. $0.56 = 0.40 + 0.57 \times 0.28$.

The collapse of manorialism was not immediate. Since villeinage was a system for exploiting labour as well as land, manorialism should have remained profitable. Indeed, manorial lords tried to carry on after the Black Death as they had before, and they were successful for half a century. In the immediate aftermath of the Plague, vacant holdings were filled by promoting landless cottagers. The Statute of Labourers (1351) aimed to keep wages at their 1346 level. The act was unsuccessful—wages did rise—but it is significant that there was no increase in real wages until the end of the fourteenth century. By then, it was impossible to preserve the pre-Plague arrangements. The continuing fall in population meant that many lords had vacant farms in hand and were desperate for any tenant who would pay rent. The result was a revolution in status and tenure.

An early and important casualty was villeinage as a personal status. A peculiar feature of that status was that a villein was in a subordinate position only to his own lord; *vis-à-vis* other lords he was no different from a free person (Pollock and Maitland 1968: i. 419). Therefore, by successfully fleeing his lord, a serf could leave his servility behind him. As tenancies became vacant, villeins moved to the estates of other lords (Davenport 1906; Dewindt 1972: 130; Hoskins 1950*b*: 131–2; Howell 1976; Hilton 1969: 32–5; Dyer 1980: 271; Faith 1984: 131–2, 136; Raftis 1964: 153 ff. 1974: 116–21). Villeinage might have survived had there been a 'Fugitive Serf Act' to return villeins to their lords. The reverse was closer to the truth—a lord could not track down a runaway serf who was gone for more than four days. Instead, the lord had to proceed against him through the royal courts (Pollock and Maitland 1968: i. 417–18). As it was, lords lost track of their serfs. Villein status had almost completely disappeared from the midlands by 1485 (MacCulloch 1988: 94–5). Land tenure also changed; rents were reduced and many obnoxious disabilities of villein tenure were eliminated as lords scrambled to fill vacant holdings. Farmers secured the abolition of labour dues either by moving or by collective resistance (Hilton 1949, 1973; Dyer 1980: 275–9). The end of servile labour meant that there was no longer any advantage to the lords' operating demesnes directly, so they were usually leased.

The late fourteenth and early fifteenth centuries were a period of great fluidity in tenurial arrangements. On many manors, the villein tenure of the pre-Plague period was abandoned in favour of tenancies at will or for short terms of years. One can conjecture that these arrangements appealed to lords and farmers for opposite reasons. The lords hoped to return to the pre-Plague arrangements and regarded tenancies at will simply as stop-gap measures that did not set new precedents for the long term. To the farmers, on the other hand, a tenancy at will may have been attractive since it was the furthest possible remove from the hereditary bondage of serfdom. Very short-term tenancies were the norm on many manors in the late fourteenth and early fifteenth centuries. Only in the fifteenth century did farmers and lords again enter into new long-term arrangements.

Enclosure was the most creative seigneurial response to the fifteenth-century scarcity of labour. Shifting land out of corn, which required much labour, into pasture, which required little, sidestepped the rise in labour costs and preserved the rental value of landed property. In less fertile areas that farmers voluntarily abandoned, the reorganization of agriculture proceeded without social conflict. When the farmers would not go willingly, however, eviction enclosures were the obvious recourse. They were common in the late fifteenth and early sixteenth centuries.

The agriculture that was established in the new enclosures was capitalist agriculture in terms of both labour systems and landownership. The small family farms that existed before enclosure were thrown together into huge sheep pastures, often let to capitalist tenants.[10] Sometimes leases were for a long term, but, frequently, the farms were tenancies at will or were let for a short term of years, in which case the farmer had no proprietary interest in the soil (Hilton 1969: 44–6). Enclosure following this pattern represented a precipitous leap into capitalist relations.

## From Villein to Yeoman

About 10 per cent of the villages in the south midlands disappeared in the fifteenth and early sixteenth centuries. In many cases, enclosure was the immediate cause. After the accession of Henry VII, the Crown sought to check these depopulations. The most effective policy was judicial: eviction enclosures could proceed only because of the insecurity of customary tenure. Granting the tenants enforceable title forestalled enclosure. This change converted villein tenure into peasant proprietorship and created the English yeoman. Thus, in most of the midlands, the sixteenth and seventeenth centuries saw the consolidation—not the collapse—of the English peasantry.

THE DEVELOPMENT OF PEASANT PROPERTY RIGHTS IN LAND

Between 1300 and 1600 most open field farmers acquired a substantial proprietary interest in the soil. By the sixteenth century, three major tenures had emerged—copyholds of inheritance, copyholds for lives, and beneficial leases (Bowden 1967: 684–7). Together with freehold, these were the

---

[10] Dyer (1980: 17–21) discusses several examples in Warwicks.

tenures by which most yeomen farmers held their land in the late sixteenth and seventeenth centuries.[11]

Copyhold is the most famous of the new tenures. In the legal histories, the copyholder is treated as the successor to the free person holding land in villeinage. In both cases, land was held from the lord according to custom—in the seventeenth-century formula, the copyholder held his land 'at the will of the lord according to the custom of the manor'—and conveyancing was done through surrender and admission in the manorial court. Indeed, the name 'copyhold' derives from the practice, which dates from the fourteenth century (P. Harvey 1984: 337), of giving the tenant a copy of the manorial court roll recording his admission to the holding.

There were two major types of copyholds in the south midlands—copyholds of inheritance and copyholds for lives. The former were most prevalent in the eastern portion of the south midlands—in Huntingdon-shire, Cambridgeshire, and Bedfordshire. Heritable copyholds were also found in Leicestershire, Northamptonshire, and Rutland, but they were less common in those counties due to the widespread practice of holding customary land on beneficial leases. With copyholds of inheritance, the tenant paid a small annual rent and a more substantial fine when the property was sold or when it passed to an heir. Sometimes the fine was fixed by custom and sometimes it was arbitrary (at the will of the lord). This distinction was of great moment *c.*1500—since a lord could block an inheritance by setting a sufficiently high fine—but by the seventeenth century the law courts had limited fines, so the distinction between fixed and arbitrary fines was of little practical significance.

Copyholds for lives were found in the Thames Valley and in Warwick-shire. These copyholds were not indefinitely heritable. They were usually grants for three lives—often the farmer, his wife, and his son. Usually when the son established a family he surrendered his copyhold and was readmitted with a new copy specifying himself and his wife and son as the three tenants. He paid an arbitrary fine for the extension of the agreement. This form of copyhold represented a smaller interest in the soil than heritable copyhold since there was no automatic right to renewal. Copyholds for lives originated in the early fifteenth century.[12]

---

[11] Evidence bearing on the geographical distribution of these tenures is presented in Appendix I.

[12] B. Harvey (1977: 244–93, esp. 278–80) has discovered that Westminster Abbey began granting life tenancies in villeinage at the end of the 14th cent. when tenancies at will were common. By the earlier 15th cent. these tenancies had become grants for 3 lives. Dyer (1980: 293–5) has observed that the tenancies at will common on the estates of the Bishop of Worcester in the late 14th cent. were followed by a return to heritable copyholds in the early 15th cent. and then by copyholds for lives in the late 15th cent. The advantage to the farmer of

Intervention by royal courts expanded the proprietary interest of the copyholder during the early modern period. Freedom to sell is an important example. Since conveyancing was effected by surrendering the land to the lord who then admitted the new tenant, the lord, in theory, could prevent the sale by refusing admission. From the fifteenth century, however, Chancery would force the lord to admit the purchaser, so the lord became (in Chief Justice Coke's phrase) simply 'custom's instrument'.[13] By the eighteenth century, the common law courts would also compel the conveyancing of copyholds (Simpson 1986: 170–1).

The most fundamental way in which the copyholder's proprietary interest expanded was in his security of tenure. In the thirteenth century, villein tenancies were regarded by the common law courts as tenancies at will, and the tenants could obtain no redress from those courts if they were evicted by their lord in violation of the customs of the manor.[14] By 1600 copyholders could recover possession of their land under such circumstances. This change converted villeins into peasant proprietors.

The question, much debated by historians, is when and how this change occurred. Debate has centred on the common law courts (King's Bench, Common Pleas, and Exchequer) and Chancery since these courts came to offer an aggrieved copyholder recovery of his land and dominated litigation for centuries. During the early 1500s, the Courts of Request and Star Chamber also heard some cases concerning enclosure and the rights of copyholders, but these courts did not lead the others and ceased to be of importance by the end of the sixteenth century.[15] Hence, the legal position of the copyholder turned on the remedies available in the law courts and Chancery.

Copyholders received protection from these courts in two stages. In the first stage, the courts enforced manorial custom against lords. The customs

the new form of tenure was that he could specify the descent of the property, whereas he could not with a traditional villein tenancy where descent was governed by custom. But the progression to copyholds for lives was not automatic. At Great Abington, Cambs., for instance, customary land was let on leases of 3 to 10 years from the 1360s to 1420s. Afterwards, leases for life were instituted as they were in the Westminster estates. However, the next step was not leases for lives. Instead, by the 16th cent. copyholds had become heritable with descent to the youngest son! (*VCH*, Cambs. vi. 10.) Why hereditary copyhold tended to emerge in the east and copyhold for lives in the west remains a mystery.

[13] Holdsworth 1942: iii. 248. Dyer (1980: 295–7) discusses the decline in the lord's ability to affect the choice of tenants in the 15th and early 16th centuries.

[14] If an evicted tenant in villeinage was personally free, presumably he could sue for damages if any crops, livestock, or other personal property were destroyed or injured. If the tenant was personally a villein, he could not even obtain that minimal compensation.

[15] See Leadam (1898) on Requests and Guy (1977) on Star Chamber. Guy (1977: 51–8) shows that Star Chamber heard many cases about land titles before 1551 when it restricted itself to cases involving civil disorder.

that received legal blessing in the sixteenth century had emerged (like copyholds for lives) in the fifteenth century—not the thirteenth. Chancery was the first court to hear petitions from copyholders. Beginning in the 1490s, the Chancellor ordered their reinstatement when they were evicted in violation of custom. The common law courts began to hear petitions from copyholders in the mid-sixteenth century, but it was not until the end of the century that these courts offered copyholders specific recovery of the land. Copyholders had no recourse to these courts earlier since the writs that provided for recovery (like novel disseisin) were restricted to free tenants. The law courts never changed that stipulation, but in the sixteenth century the writ of ejectment was developed as a preferred means of establishing title to freehold land. By the late sixteenth century, copyholders were allowed to pursue cases with that writ.[16]

Initially, the Chancellor and the common law courts enforced manorial custom as defining the agreement between lord and tenant, so security was no stronger than custom. The second stage in the protection of the copyholder was the overruling of customs that were 'unreasonable'. The common law courts took the lead in this regard—they had to, if they were to hear copyhold cases at all. Thus, for ejectment to proceed, the copyholder had to be able to make a lease, but granting a lease violated the customs of most manors. So the courts decided that that custom was 'unreasonable' and declared that the right to grant a lease was a custom in all manors in the kingdom.[17] Of greater importance was the decision to cap entry fines for 'heritable' copyholds. This decision greatly increased the proprietary interest of many farmers in the east midlands by converting their life tenancies virtually into freehold estates. Copyhold tenure, as it existed in the seventeenth century, depended as much on judicial decision as it did on the fifteenth-century agreements between lord and tenant.

[16] This is C. Gray's (1963) chronology. It has since been attacked by Kerridge (1969) who maintains that both the Chancellor and the common law courts intervened earlier. Kerridge's views have not been widely accepted. See the reviews by Barton (1971), Bean (1971), the anonymous review in the *Times Literary Supplement*, 1970: 1135–6, and J. Baker's (1978: ii. 180–92) authoritative discussion of the evolution of the common law. P. Harvey (1984: 328) and Simpson (1986: 144–72) both accept Gray's chronology. The discussion has rightly turned on the legal evidence, but there is other evidence that supports Gray. Thus, Harvey (1984: 330–1) noted the absence before the 16th cent. of the usual tendency towards the standardization of local terminology and practice that followed the intrusion of national courts into local matters. The history of depopulation is also relevant. In his attack on Tawney, Kerridge (1969: 92) observed that if copyholders did not have security of tenure, then 'one would expect to find numerous instances of customary copyholders being ousted and left without remedy'. Kerridge is right that there were not many such evictions in the 16th cent., but the situation was different in the 15th, when there were 'numerous instances', as shown in the previous chapter. For all of these reasons, Gray's conclusions command support.

[17] Cf. M. Campbell (1942: 129–30).

Most discussions of the property rights of small farmers in the early modern period centre on the security of tenure of copyholders. This orientation is not entirely appropriate since more land in the south midlands was probably held on beneficial leases than on copyholds. With a beneficial lease, the annual rent was small, and most of the payment for the lease was made at the outset as a large fine. In the sixteenth and early seventeenth centuries these leases were often drawn up for three specified lives, although by the middle of the seventeenth century the leases were often written for long terms of years. Demesnes were usually let on beneficial leases throughout the midlands. In the western part of the district, lords were replacing copyholds for lives with leases for lives in the eighteenth century (Clay 1981: 91). In Northamptonshire and Leicestershire, the occupiers of customary land often held it by beneficial lease instead of copyhold throughout the early modern period. Finally, an Elizabethan act limited the length of leases made by ecclesiastical bodies and by Oxford and Cambridge colleges to three lives or twenty-one years (renewable every seven years), and this became the usual way by which church and college lands were held.[18]

The beneficial leasee's proprietary interest was almost as substantial as the interest of the copyholder for lives. From the fifteenth century, the law courts recognized that the tenant had the right to 'quiet enjoyment and title' (Holdsworth 1937: vii. 251, 254). The beneficial leasee could control the descent of his property through the choice of lives or by bequeathing the lease if it was for years. The beneficial leasee necessarily exercised ownership for his life if the lease was for lives that included his own. If the lease was for years, the leasee also enjoyed lifelong tenure provided the term was long enough, and in the seventeenth century, terms were usually long. The principal way in which the leasee's rights were more limited than those of a copyholder for lives was in a diminished ability to sublet. While copyholders were accorded the right to lease their land by the end of the sixteenth century, it was settled in 1606 that leases could bar leasees from assigning or subletting the lease (Holdsworth 1937: vii. 281–2).

The beneficial leasee acquired remedies to protect his property at an early date. A lease for lives was not different from any other grant of freehold land for lives, so, from the middle ages, leaseholders for lives could recover possession of their property by the real actions like novel disseisin (Holdsworth 1942: iii. 120–4; 1937: vii. 240, 245). A lease for years or a lease for lives determinable upon ninety-nine years (a common formula) did not create a freehold estate and so an expelled leasee could only avail himself of a personal action. Before the sixteenth century, the leasee could

[18] See E. P. Thompson (1976) and Clay (1980) on church and college leases.

not hope to recover possession of the land but only damages, which included part of the fine. A decision in 1499, however, allowed tenants to recover possession of the land for the duration of the lease with the action of ejectment (Holdsworth 1942: iii. 180–4; 1937: viii. 4; J. Baker 1978: ii. 180–3). The beneficial leasee's security of tenure was established at about the same time the Chancellor intervened to protect the copyholder.

The judicial changes that protected the copyholder and beneficial leasee were responses to the agrarian changes of the fifteenth century. Depopulating enclosures caused disquiet in legal circles. In 1481, for instance, Chief Justice Brian said 'that his opinion hath always been and ever shall be, that if such tenant by custom paying his services be ejected by the lord, he shall have an action of trespass against him.' In the same year Sir Thomas Littleton, in his important legal treatise *Tenores Novelli*, averred that 'the lord cannot break the custom which is reasonable in these cases.'[19] None of these espousals resulted in practical remedies.

By the end of the fifteenth century, the Crown proceeded on many fronts to protect peasant farmers.[20] In 1484, the Chancellor discussed enclosure and depopulation in parliament. An act in 1488 prohibited engrossing on the Isle of Wight, and, in the following year, a sweeping act 'agaynst pullyng doun of tounes' outlawed depopulating enclosures throughout the kingdom. It is against this legislative background that the judicial activism of the 1490s should be seen. While the common law courts could still not find a formula to protect the interests of customary tenants, those courts at least allowed evicted leaseholders recovery of their land for the duration of the lease. The Chancellor, whose judicial role was not as circumscribed by precedent as was the role of the law courts, intervened in the 1490s to protect copyholders by enforcing the customs of their manors.

These initiatives did not immediately check the first wave of enclosures. In 1515 parliament passed another act preventing the conversion of arable to pasture and the amalgamation of small farms into large. In 1517 Wolsey appointed a commission into depopulation that took testimony throughout most of the country; the information was used for prosecutions in Chancery and King's Bench for the next twenty years. The nationwide investigation and the resulting prosecutions alerted lords and tenants to the new legal state of affairs and perhaps spread awareness of the enforceability of customary and leasehold tenure. Shortly after Wolsey's Commission, the rate of depopulating enclosure declined in the south midlands.

[19] Both as quoted by Kerridge (1969: 139).
[20] Thirsk (1967*b*) provides a thorough discussion of the non-judicial aspects of government policy, which I follow.

Even after the main battle was won, the Crown continued to strengthen the position of peasant farmers. Half a dozen acts were passed by parliament in the sixteenth century prohibiting enclosure, the conversion of arable to pasture, and the engrossing of farms. The principal victories of small proprietors, however, were now won in the common law courts. Allowing copyholders to use the action of ejectment to assert their title and the application of the test of reasonableness to manorial custom were the most important advances. In terms of legal rights, the early seventeenth century marked the high point of peasant proprietorship.

The development of property law in the sixteenth century meant that copyholders and beneficial leasees acquired substantial proprietary interests in the soil. This interest was recognized for centuries in both official and unofficial ways. For instance, the copyholder and the beneficial leasee—not the lord of the manor from whom the property was held—were listed as the 'proprietors' on the land tax assessment form in the eighteenth century. Furthermore, copyholders and beneficial leasees were treated as owners in enclosure proceedings. While parliamentary enclosure acts usually voided leases of less than twenty-one years, longer leases were preserved and copyhold estates were specifically protected. In the seventeenth century, the term 'yeoman' was applied indiscriminately to farmers whether they were freeholders, copyholders, or beneficial leasees (M. Campbell 1942: 22–5; Mingay 1968: 68). When the Board of Agriculture reports discussed the tenures by which land was owned, the range included freehold, copyhold, and leasehold, the latter meaning beneficial leases. The copyholder and the beneficial leasee not only had a substantial proprietary interest in their land, the interest was large enough to warrant calling them 'owners'.

The acquisition of legal rights by copyholders and beneficial leasees represented an enormous democratization of landownership in England. In the sixteenth century, the acreage farmed by owner-occupiers greatly increased. These farmers were the English yeomen, and the seventeenth century was their golden age.

FARM SIZE AND LABOUR SUPPLY

Non-demesne farms were also peasant farms when peasants are defined to be farmers relying mainly on family labour. Tawney tabulated a large number of sixteenth-century surveys and found that the average size of copyhold or customary holdings was 34 acres.[21] These holdings were

[21] Tawney 1912: 64–5, 212. I retabulated the data in accord with the size categories of Tables 4–4 and 4–5. The calculations in the text ignore holdings of less than 5 acres. Farms in Lancs. and Northumb. were excluded. For farms between 5 and 100 acres, the total acreage of each category was obtained by multiplying the mean size of the retabulated category by the number of farms in the category. I assumed the 11 farms between 100 and 120 acres had an average size of 110 acres and the 17 farms over 120 acres had an average size of 150 acres.

therefore larger than the villein farms of the thirteenth century, but they were still family farms.

To follow farm sizes into the nineteenth century, I reconstructed the farm size distribution from a sample of surveys of south midlands estates taken between 1595 and 1850. The results are shown in Tables 4–4 and 4–5. Demesnes, copyholds, and leaseholds are all tabulated together. Open farms

Table 4–4.  *The distribution of farm acreage from estate surveys*

| Farm size acres | Acres | | | Percentage of acres | | |
|---|---|---|---|---|---|---|
| | Early 17th century | Early 18th century | About 1800 | Early 17th century | Early 18th century | About 1800 |
| Open farms | | | | | | |
| 5–10 | 174 | 189 | 6 | 0.9 | 0.7 | 0 |
| 10–15 | 180 | 393 | 40 | 0.9 | 1.5 | 0.2 |
| 15–30 | 1,123 | 1,769 | 256 | 5.8 | 6.8 | 1.2 |
| 30–60 | 5,018 | 4,063 | 1,304 | 25.9 | 15.7 | 6.2 |
| 60–100 | 6,623 | 5,634 | 1,596 | 34.2 | 21.7 | 7.6 |
| 100–200 | 4,233 | 10,712 | 5,959 | 21.9 | 41.3 | 28.3 |
| 200–300 | 704 | 1,984 | 6,476 | 3.6 | 7.7 | 30.8 |
| 300–400 | 301 | 300 | 3,704 | 1.6 | 1.2 | 17.6 |
| 400–500 | 492 | 886 | 0 | 2.5 | 3.4 | 0 |
| 500–1,000 | 513 | 0 | 1,691 | 2.7 | 0 | 8.0 |
| 1,000+ | 0 | 0 | 0 | 0 | 0 | 0 |
| TOTAL | 19,361 | 25,930 | 21,032 | | | |
| Average size | 59 | 65 | 145 | | | |
| Enclosed farms | | | | | | |
| 5–10 | 18 | 8 | 11 | 0.4 | 0.1 | 0 |
| 10–15 | 10 | 72 | 47 | 0.2 | 0.9 | 0.1 |
| 15–30 | 23 | 207 | 845 | 0.5 | 2.5 | 1.8 |
| 30–60 | 254 | 764 | 2,976 | 5.3 | 9.1 | 6.2 |
| 60–100 | 135 | 1,270 | 3,086 | 2.8 | 15.1 | 6.4 |
| 100–200 | 465 | 3,687 | 12,248 | 9.6 | 43.9 | 25.5 |
| 200–300 | 651 | 964 | 12,689 | 13.5 | 11.5 | 26.5 |
| 300–400 | 350 | 337 | 9,590 | 7.2 | 4.0 | 20.0 |
| 400–500 | 0 | 427 | 4,017 | 0 | 5.1 | 8.4 |
| 500–1,000 | 1,214 | 645 | 2,438 | 25.1 | 7.7 | 5.5 |
| 1,000+ | 1,716 | 0 | 0 | 35.5 | 0 | 0 |
| TOTAL | 4,836 | 8,401 | 47,947 | | | |
| Average size | 210 | 100 | 147 | | | |

*Sources*: Some surveys underlying this table are from H. Gray (1915: 444) and Lennard (1916). Most, however, are from the following manuscript sources:
*Bedford Record Office*: R. Box 792, CRT 100/34, LA 1/14–27.
*Bodleian Library, Oxford*: MS Top Berks e 21, MS Top Berks d 26, MS DD Bertie c. 18/3, MS DD Bertie c. 18/4, MS DD Bertie d 1/27, MS Top Oxon b 121, MS Top Oxon c. 381, MS DD Harcourt b. 37, MS DD Harcourt b. 34, MS DD Harcourt e. 7.
*Huntington Record Office*: dd M b 10/1–7 Hinch /5/70, C4/2/2/13, C4/2/5/5a, C4/2/6/11, C4/2/6/13, C4/2/7/5.
*Northampton Record Office*: Brudenell ASR 95, Brudenell ASR 96, Brudenell ASR 138, Brudenell B.ii.49, H.xi.26, H.xi.31, O.xxii.6, D(CA) 211, D(CA) 213, D(CA) 215, D(CA) 306, D(CA) 444, G. 1654, G. 3916, G. 3898, C(A) 5739, F(M) Misc. Vol. 201, F(M) Misc. Vol. 555.

*Table 4–5. The distribution of the number of farms from estate surveys*

| Farm size acres | Number of farms | | | Percentage of farms | | |
|---|---|---|---|---|---|---|
| | Early 17th century | Early 18th century | About 1800 | Early 17th century | Early 18th century | About 1800 |
| **Open farms** | | | | | | |
| 5–10 | 24 | 26 | 1 | 7.3 | 6.5 | 0.7 |
| 10–15 | 15 | 34 | 3 | 4.6 | 8.5 | 2.0 |
| 15–30 | 53 | 80 | 10 | 16.2 | 20.1 | 6.9 |
| 30–60 | 114 | 96 | 29 | 34.8 | 24.1 | 20.0 |
| 60–100 | 84 | 73 | 22 | 25.6 | 18.3 | 15.2 |
| 100–200 | 32 | 77 | 38 | 9.8 | 19.3 | 26.2 |
| 200–300 | 3 | 9 | 28 | 0.9 | 2.3 | 19.3 |
| 300–400 | 1 | 1 | 11 | 0.3 | 0.3 | 7.6 |
| 400–500 | 1 | 2 | 0 | 0.3 | 0.5 | 0 |
| 500–1,000 | 1 | 0 | 3 | 0.3 | 0 | 2.0 |
| 1,000+ | 0 | 0 | 0 | 0 | 0 | 0 |
| TOTAL | 328 | 398 | 145 | | | |
| **Enclosed farms** | | | | | | |
| 5–10 | 2 | 1 | 2 | 8.7 | 1.2 | 0.6 |
| 10–15 | 1 | 6 | 4 | 4.3 | 7.1 | 1.2 |
| 15–30 | 1 | 11 | 39 | 4.3 | 13.1 | 11.9 |
| 30–60 | 6 | 18 | 69 | 26.1 | 21.4 | 21.1 |
| 60–100 | 2 | 16 | 39 | 8.7 | 19.0 | 11.9 |
| 100–200 | 4 | 25 | 81 | 17.4 | 29.8 | 24.8 |
| 200–300 | 3 | 4 | 52 | 13.0 | 4.8 | 15.9 |
| 300–400 | 1 | 1 | 28 | 4.3 | 1.2 | 8.6 |
| 400–500 | 0 | 1 | 9 | 0 | 1.2 | 2.8 |
| 500–1,000 | 2 | 1 | 4 | 8.7 | 1.2 | 1.2 |
| 1,000+ | 1 | 0 | 0 | 4.3 | 0 | 0 |
| TOTAL | 23 | 84 | 327 | | | |

*Sources*: As for Table 4.4.

averaged 59 acres in the 'early seventeenth century' (based on surveys taken between 1595 and 1650). Five-eighths of the farms were family farms of less than 60 acres, while only one eighth were capitalist enterprises of more than 100 acres. These large farms were probably demesnes. (In terms of acreage, however, the family farm sector looks less substantial—about one third of the land was in family holdings, one third was in transitional holdings, and one third was in capitalist holdings.) There was probably little increase in average farm size from the sixteenth century into the eighteenth.[22]

Thus far, I have discussed the rise of the yeoman from the perspective of the manor—who held its lands, under what tenures, and in what amounts.

[22] Exact comparison with Tawney's results is impossible since the 17th-cent. surveys often failed to label demesnes as such. Consequently, Tables 4–4 and 4–5 tabulate demesnes and customary holdings together. The inclusion of demesnes means that the average size of early 17th-cent. farms described in Tables 4–4 and 4–5 (59 acres) exceeded the average size of Tawney's customary holdings (34 acres).

This procedure is vulnerable to the charge that the manorial tenants were not the actual farmers—the leaseholds and copyholds may have been sublet to others. As a result, the farms may have been bigger or smaller than the estate surveys indicate and the occupiers may have been only tenants at will rather than peasant proprietors. There are no doubt examples of subletting; the issue is whether it was widespread enough to call into question the picture of rural life that emerges from the estate documents. The only satisfactory resolution is to build up a picture of agriculture from the records of farmers themselves. The historians who have studied them for the sixteenth and seventeenth centuries have confirmed the view of agrarian relations that I have advanced here.

In *The English Yeoman Under Elizabeth and the Early Stuarts* Campbell (1942: 64–155) has studied the property of yeomen 'from wills, deeds, leases, and other documents touching their land dealings'. She concludes 'that they held their lands in diverse tenures', namely as freehold, copyhold, or by beneficial lease. Consequently, she concludes that 'their lands were their own, or directly under their control' and contrasts their proprietary interest with that of tenant farmers.[23] In the sixteenth and seventeenth centuries, the records of farmers suggest that most farmers held their land on long-term agreements, so owner-occupation was widespread.[24]

Several historians have also studied farm sizes from probate inventories. Since these indubitably apply to operating farms, the possibility of bias from subtenancy is avoided.[25] In the event, probate inventories confirm the average farm size shown by the estate surveys. Thus, Hoskins (1950*b*: 146) analysed many Leicestershire inventories and concluded that 'Though the average farm was one of roughly 40 or 50 acres in sixteenth century Leicestershire, half of the farms of the county were below this size. On the other hand, about four per cent of the farms were 100 acres and upwards in extent.' He showed that the same general conclusion applied into the early eighteenth century. (Hoskins 1951: 12–15). Havinden (1961*a*) showed that Oxfordshire was little different in this regard. Thus, the probate inventories confirm that the farm size distribution was fairly stable over the sixteenth and seventeenth centuries and like that shown for 'early seventeenth century' and 'early eighteenth century' in Tables 4–4 and 4–5.[26]

---

[23] Quotations from pp. 64, 118.

[24] H. Gray (1910) and Habakkuk (1940: 16–17) agree.

[25] Farmers' inventories recorded the acreage of the crops but not fallow, meadow, or pasture. To determine farm size, the tilled acreage must be inflated to allow for omitted land. The need for this adjustment is the principal weakness of inventories as a source for farm sizes, but it is offset by the fact that inventories apply to integral farms.

[26] Harrison (1979) has argued that estate surveys give a highly distorted picture of the rural population since they ignore subtenancy. The conclusion is based on a unique field book of Cannock, Staffs., made in 1554, that distinguished manorial tenants from the actual occupiers.

The open field system in the midlands in the sixteenth and seventeenth centuries approximated a peasant system by both the ownership and family labour criteria. Beneficial leases and copyhold created a regime of peasant proprietorship of long (but not unlimited) duration. Many farms and much land were operated mainly with family labour. There were certainly many small owner-occupied freehold farms as well. These farmers were the English yeomen.

## Conclusion

English agrarian relations followed two contrasting paths of development in the fifteenth and sixteenth centuries. The most famous path was enclosure, and it resulted in the elimination of peasant farming. The second path is less often recognized but was more common; that path was the consolidation of farmers' property rights in open field villages. That consolidation created the yeomen.

The establishment of the yeomen's property rights was the achievement of Tudor policies that anticipated the *bauernschutz* policies followed by would-be absolutist monarchs on the continent in the seventeenth century. Since Tawney's *The Agrarian Problem in the Sixteenth Century* (1912), the Tudor policies have been dismissed as ineffective in checking the extension of capitalist relations. Tawney's assessment was on the mark in so far as it related to the use of legislation to prevent enclosure. Such legislation was not in the interests of the gentry and aristocracy, who weakened the bills in parliament or failed to co-operate in enforcing the laws. Tudor policy was much more effective in the courts where property rights could be created by fiat. Here a revolution took place. Just as the eleventh-century courts created peasant property rights by protecting the free tenants of the manor, so the sixteenth-century courts protected the customary tenants. The

There was also a 1570 survey with which it could be compared. Harrison showed that there were many more occupiers than manorial tenants in Cannock. However, the size distribution he provides shows that most of the surplus was found in the smallest category (zero to 9 acres). The surplus were landless labourers who rented their cottages. Moreover, if we consider the farms with more than 10 acres (which comprised most of the land), the average size was the same in all three sources—51 acres for the 1554 list of occupations, 56 acres for the 1554 list of manorial tenants, and 55 acres for the 1570 survey of manorial tenants. Relying on the surveys is, therefore, not misleading about the average size of a farm.

Harrison did demonstrate two important limitations of estate surveys. First, they are likely to underreport the number of landless labourers by omitting subtenants. Second, not all of the manorial tenants were farming the indicated lands. Sometimes the tenants were leasing in land; sometimes the farms were occupied by children and other relatives, or by non-relatives. However, the details he provides do not indicate that surveys give a seriously distorted view of the farm size distribution, nor does he show that a substantial amount of copyhold land was occupied by farmers other than the copyholders or their relatives.

creation of copyhold tenure and the protection of the beneficial leasee amounted to an even greater democratization of property ownership. The redefinition of landownership laid the basis for yeoman agriculture in the late sixteenth and seventeenth centuries.

The early seventeenth century marked the high point of yeoman property rights. These rights, however, were not absolute and were not sufficient to preserve the yeoman social structure. Large landowners continued to resist royal efforts to protect peasants: an indicator of their opposition is the charge against Archbishop Laud in 1644 that 'he did a little too much countenance the commission for depopulations' (Thirsk 1967*b*: 237). During the Civil War, the Levellers, a radical movement, proposed several policies that might have maintained England as a yeoman society. These policies included the prohibition and reversal of enclosure, an 'agrarian law' to set an upper limit to the property that any individual might own, and the conversion of copyhold and leasehold into freehold. I will discuss these ideas in more detail in Chapter 15, but the most important point about them is that they were never implemented. The rejection of the Leveller demands and the vesting of power in a parliament controlled by large landowners laid the basis for the final destruction of the English peasantry in the late seventeenth and eighteenth centuries.

# 5

# The Disappearance of the Yeoman in the Open Fields

> To alter Farms, and to turn several little ones into great ones, is a Work of Difficulty and Time; for it would raise too great an *Odium* to turn poor Families into the wide World, by uniting Farms all at once, in order to make an Advance of Rents: 'Tis much more reasonable and popular to be content to stay till such Farms *fall* into Hand by Death, before the Tenant is either rais'd or turn'd out.
>
> E. Laurence, *The Duty of a Steward to his Lord*, 1727: 3

THE yeoman farms that were common in open field villages in the seventeenth century were destroyed in the eighteenth. This destruction usually preceded enclosure. Certainly, by 1800 the great landowner, the substantial tenant, and the landless labourer were just as common in open villages as they were in enclosed. The purpose of this chapter is to analyse the transition from yeoman to capitalist agriculture as it occurred in the open fields.

The plan of the chapter is as follows: I begin by reviewing the evidence for the end of peasant proprietorship and the concentration of land in large farms. Then I discuss explanations for these changes. While there are examples of a 'kulak' transition to capitalism as rich peasants bought out their neighbours, the acquisition of land by great estates was primarily responsible for the exceptional inequality in landownership and the predominance of the large farm in nineteenth-century England. The acquisition of peasant property by large landowners commenced after the Restoration and was complete before Waterloo. Manorial lords bought up small freeholds and heritable copyholds, and ran out (i.e. did not renew) beneficial leases and copyholds for lives. These real estate transactions, in turn, were the result of the emergence of the long-term mortgage, which increased the propensity of manorial lords to buy land.

## The Decline of the Yeoman in the Eighteenth Century

Whether peasants are defined as owner-occupiers or as family farmers, there is no question that the class had essentially disappeared from English agriculture by the late eighteenth century. From his study of the land tax assessments, E. Davies (1927: 110) concluded that

By 1780 the occupying owners, including freeholders, copyholders and leasees for lives, had ceased to be an outstanding feature of English rural economy. In 1,395 parishes situated in Derbyshire, Leicestershire, Lindsey, Northamptonshire, Nottinghamshire and Warwickshire, they contributed only 10.4 per cent. of the land tax, so that already nearly 90 per cent. of the land was in the occupation of tenant farmers.

Estate surveys show that most tenancies were merely tenancies at will, and the Board of Agriculture county reports establish the point in its generality. Thus, 'Very few leases are given in Bucks' (Priest 1813: 86). 'Many considerable proprietors grant no leases [in Oxfordshire]' (Young 1813: 64). 'The county [of Northampton] may be said to be principally occupied (with a very few exceptions) by tenants at will' (Pitt 1809b: 45). 'The farms in this county [Bedfordshire] are in general held only from year to year' (Batchelor 1808: 40). 'In great part of the county [of Cambridge] none [leases] are granted' (Gooch 1813: 38). 'There are no leases granted in the greater part of the parishes [of Huntingdon]' (Parkinson 1811: 45). 'The greatest part of the land [in Rutland] . . . is let to tenants from year to year' (Parkinson 1808: 35). Somehow, most farmers lost their land between 1650 and 1800.[1]

Peasant agriculture also disappeared from open field villages in the eighteenth century when peasants are defined as farmers relying on family labour. There had been some movement to large farms in the open fields in the seventeenth century, but the average size only increased from 59 acres to 65 acres. By 1800, however, the average size of an open field farm had exploded to 145 acres. (See Tables 4–4 and 4–5.) These farms required a large hired labour force. Since average farm size was almost the same in enclosed villages at the same time (147 acres), the eighteenth-century shift to farms cultivated mainly with wage labour cannot be attributed to enclosure.

The land tax assessments confirm that open field farms were large c.1790. (See Tables 5–1 and 5–2.) An important advantage of the assessments is that they include all farms (not just those rented from great estates) and that they

---

[1] Indeed, Parkinson's parish-by-parish surveys of Hunts. and Rutland allow more precision in this matter. Of 157 villages surveyed, he reported no leases in 99 and 'a few' in 28. In 6 parishes leases of less than 5 years were given, in 18 leases of 5 to 15 years, and in only 6 villages leases of 21 years. (Parkinson 1808: 2–4, 34; and 1811: 42–4.)

allow one to add up all of a farmer's land, no matter who owned it.[2] At the end of the eighteenth century, the average open field farm was 114 acres, while the average farm in a village enclosed after 1525 was 122–124 acres. (Farms averaged somewhat larger—157 acres—in villages enclosed between 1450 and 1524.)[3] The acreages implied by the land tax assessments are less than those indicated by the estates surveys since the land tax assessments included farms outside the system of great estates. None the less, the land tax assessments confirm the widespread shift to 'capital' farms in the open fields before the nineteenth century.[4]

The conclusion that the eighteenth century witnessed the shift to 'capital' farms in the south midlands is consistent with studies of other parts of England. Eighteenth-century surveys of the Kingston estates in Nottinghamshire and the Bagot and Gifford estates in Staffordshire show a decline in farms of 20–100 acres and a corresponding rise in farms of more than 100 acres in the eighteenth century (Mingay 1962: 480–3). Wordie's (1974) study of the very large Leveson-Gower estates located in Shropshire, Staffordshire, and Yorkshire showed that the average farm in 1714–20 was 62 acres, which was similar to the average open field farm in the south midlands at the same time.[5] Farm size increased substantially in the next

---

[2] The land tax assessments suffer from several defects as a source for studying farm size. For instance, it is usually impossible to identify and remove non-farm property such as woodlands and parks. Further, one must adopt some procedure to convert tax liability to acreage. In this study, the tax liability of each occupier was first computed. Acreage was estimated by apportioning the total acreage of the village across all the occupations in proportion to their tax liabilities. Grigg (1963) and Mingay (1964) raised several objections to attempts to infer acreages from tax assessments. J. M. Martin (1966) showed that many of the problems that troubled them could be avoided by careful attention to the documents. Mingay and Grigg both objected to E. Davies' (1927) using the same coefficient to convert tax liability to acreage for all villages in a county. In this study, different coefficients are used for every village. The same coefficient, however, was used for all properties within a village. As Mingay noted, that procedure probably overstates the acreage of cottages. The implications for interpreting the results are discussed in the text. Mingay was also concerned that small owner-occupiers might have been farming on a large scale by renting land. That did occur but does not impair Tables 5–1 and 5–2 as long as the rented land was in the village where the small owner lived, since those tables combine all land farmed by each person in each village.

[3] These averages are for farms of more than 10 computed acres. In Tables 5–1 and 5–2 I treat holdings of less than 10 acres as cottages. When using estate surveys I make the division between cottages and farms at 5 acres. I use 10 acres instead of 5 with the land tax assessments because of the method of computing acreage, which tends to overestimate the acreage of cottages since for them the ratio of the value of the structure to the value of the land was exceptionally high. Gooch (1813: 294) reports that a poor cottage cost £10 to erect. Allowing 10 per cent per year for interest, depreciation, and repairs implies the annual capital cost was £1. This is the rent of 2 or 3 acres of open field land. Better quality cottages would imply more land.

[4] Tables 5–1 and 5–2 also indicate that there were few cottages in early enclosures. This result is consistent with the analysis of village population in Ch. 3.

[5] This average is for farms over 5 acres and is computed from Wordie (1974: 605), table 1(b), on the assumption that the average size of farms between 0 and 5 acres was 2.5 acres.

*Table 5–1. The distribution of farm acreage from land tax assessments, c. 1790*

| Farm size acres | Acreage | | | | Percentage of acreage | | | |
|---|---|---|---|---|---|---|---|---|
| | EEE | OEE | OOE | OOO | EEE | OEE | OOE | OOO |
| 0–5 | 122 | 418 | 2,721 | 5,881 | 0.2 | 0.5 | 1.0 | 1.2 |
| 5–10 | 215 | 636 | 3,368 | 8,739 | 0.4 | 0.7 | 1.2 | 1.7 |
| 10–15 | 263 | 1,115 | 3,026 | 7,260 | 0.4 | 1.2 | 1.1 | 1.4 |
| 15–30 | 1,102 | 2,918 | 9,524 | 17,097 | 1.9 | 3.2 | 3.5 | 3.4 |
| 30–60 | 3,151 | 4,794 | 15,826 | 32,135 | 5.3 | 5.3 | 5.8 | 6.3 |
| 60–100 | 4,480 | 9,122 | 26,952 | 51,837 | 7.6 | 10.0 | 9.9 | 10.2 |
| 100–200 | 12,965 | 21,561 | 68,022 | 129,732 | 21.8 | 23.7 | 25.0 | 25.6 |
| 200–300 | 9,651 | 17,783 | 46,409 | 97,065 | 16.3 | 19.6 | 17.0 | 19.2 |
| 300–400 | 8,498 | 11,061 | 34,927 | 51,146 | 14.3 | 12.2 | 12.8 | 10.1 |
| 400–500 | 3,985 | 8,973 | 18,368 | 29,292 | 6.7 | 9.9 | 6.7 | 5.8 |
| 500–1,000 | 7,707 | 9,485 | 25,892 | 58,157 | 13.0 | 10.4 | 9.5 | 11.5 |
| 1,000+ | 7,230 | 3,106 | 17,444 | 18,562 | 12.2 | 3.4 | 6.4 | 3.7 |
| TOTAL | 59,369 | 90,972 | 272,479 | 506,903 | | | | |
| Average farm size | 157 | 124 | 122 | 114 | | | | |

*Note:* The columns refer to the period of enclosure, as follows:
EEE—before the middle of the sixteenth century;
OEE—between the middle of the sixteenth century and 1676;
OOE—between 1676 and the date of the land tax assessment (c.1790);
OOO—open at the time of the land tax assessment.
Average farm size is computed for holdings of 10 acres or more.

*Source:* Tables 5–1 and 5–2 are derived from a sample of land tax returns for 636 villages in the counties of Beds., Bucks., Cambs., Hunts., Oxon., Northants., Leics., and Warwick. The return for 1790 (or the nearest suitable year) was used. I calculated the totalled tax liability of each occupier in each village and then computed the acreage occupied by each person as the product of the acreage of the village and the person's share of the total tax liability of the village.

century, although not to the same degree as in my sample of south midland farms.

## Agrarian Revolution in the Eighteenth Century?

The conclusion that the eighteenth century marked the end of the English peasantry is consistent with the work of some historians (e.g. Johnson 1909; H. Gray 1910; Habakkuk 1965), but it is at odds with F. Thompson's (1966, 1969) and Mingay's (1968) influential[6] reassessments of the aggregate evidence: they deny that the eighteenth century saw the end of peasant agriculture in England. Recently, MacFarlane (1978) has advanced an even more extreme view; he has argued that England never had a

[6] According to Beckett (1984: 5). Respecting the small-scale owner-occupier, Beckett's (1984: 16) survey concludes that 'a stability thesis would seem to be as plausible for the eighteenth century as one of decline'. Brenner's (1976: 48) argument presumes 'by the end of the seventeenth century, English landlords controlled an overwhelming proportion of the cultivable land—perhaps 70–75 per cent—and capitalist class relations were developing as nowhere else'. Mingay (1963) and F. Thompson (1966) were Brenner's sources.

*Table 5–2. The distribution of the number of farms from land tax assessments, c. 1790*

| Farm size acres | Number of farms | | | | Percentage of farms | | | |
|---|---|---|---|---|---|---|---|---|
| | EEE | OEE | OOE | OOO | EEE | OEE | OOE | OOO |
| 0–5 | 47 | 150 | 1,162 | 2,050 | — | — | — | — |
| 5–10 | 31 | 88 | 472 | 1,237 | — | — | — | — |
| 10–15 | 21 | 88 | 247 | 593 | 5.6 | 12.0 | 11.0 | 13.4 |
| 15–30 | 49 | 134 | 442 | 800 | 13.0 | 18.2 | 19.7 | 18.0 |
| 30–60 | 69 | 110 | 360 | 753 | 18.3 | 15.0 | 16.1 | 17.0 |
| 60–100 | 57 | 115 | 338 | 656 | 15.1 | 15.6 | 15.1 | 14.8 |
| 100–200 | 94 | 148 | 470 | 928 | 24.9 | 20.1 | 21.0 | 20.9 |
| 200–300 | 38 | 72 | 191 | 397 | 10.1 | 9.8 | 8.5 | 8.9 |
| 300–400 | 25 | 33 | 101 | 148 | 6.6 | 4.5 | 4.5 | 3.3 |
| 400–500 | 9 | 20 | 41 | 67 | 2.4 | 2.7 | 1.8 | 1.5 |
| 500–1,000 | 12 | 13 | 40 | 87 | 3.2 | 1.8 | 1.8 | 2.0 |
| 1,000+ | 3 | 3 | 11 | 12 | 0.8 | 0.4 | 0.5 | 0.3 |
| TOTAL | 455 | 974 | 3,875 | 7,728 | | | | |

*Notes:*
The percentages are computed only for farms greater than 10 acres on the grounds that computed acreage up to that level are in reality cottages with 5 acres or less of land. These should be excluded in this table to allow comparison with Table 4–4 in which holdings of less than 5 acres are also excluded.

The columns refer to the period of enclosure, as follows:
EEE—before the middle of the sixteenth century;
OEE—between the middle of the sixteenth century and 1676;
OOE—between 1676 and the date of the land tax assessment (*c.*1790);
OOO—open at the time of the land tax assessment.

*Source:* As for Table 5–1.

peasantry at all, so it could never have disappeared. It is important to see why the views of Mingay, Thompson, and MacFarlane are not persuasive.

Some of the disagreement among historians results from dispute about the facts, but much of it reflects differences in the definition of 'peasant'. Evidently, the number of peasants counted depends on the definition adopted. If we require peasants to be landowners but count only freeholders as owners, the number of peasants is much smaller than if we count copyholders as well. If we require that peasants be freeholders and use only family labour, we get a smaller number still. If we follow MacFarlane (1978: 16, 18–21) and require that the *family* own the land, we eliminate the peasantry altogether since in English law families could not own land. This is the *reductio ad absurdum* that shows the degree to which the size and history of the peasantry depends on the definition adopted.

MacFarlane's work is stimulating but ultimately unsatisfactory since it deals with the problem of the English peasant by defining it away. The work of Thompson and Mingay directly confronts the problem of the English peasant by adopting definitions suitable for the early modern period. Their arguments remain unconvincing both because they depend on unreliable aggregate statistics and because of the definitions adopted.

Mingay (1968) was concerned with the putative disappearance of small farmers—not necessarily owner-occupiers. He claimed that their number remained stable over the eighteenth century. His argument involves comparing the total of freeholders and farmers (reduced by one third to allow for non-occupying owners) in Gregory King's 1688 social table of England and Wales (King 1936) with later estimates of the numbers of farmers. Mingay found little decline. However, in making this comparison, Mingay ignored Cooper's (1967) demonstration that King had no certain idea how many farmers there were in England. In his notebooks King toyed with much larger numbers, and they imply a precipitate drop in the number ·of farmers in the eighteenth century. Holmes (1977: 57) has reviewed the matter and shown that

With the 'freeholders' and 'farmers' King was for a long time utterly at sea, his estimates for their combined total oscillating wildly, in a series of entries [in his notebooks] from the spring of 1695 to the summer of 1696, between 240,000 and 780,000. . . . It seems highly likely that King's final choice of 330,000 . . . was reached by a process of elimination, working downwards from a total number of solvent householders.

Aggregate statistics like King's are seductive, but in this case they are unreliable and preference must be given to estimates based on the traditional documentary sources.

And the traditional sources tell a different story. I have already remarked that the estate surveys tabulated in Tables 4–4 and 4–5 indicate that the average farm size in open field villages increased considerably in the eighteenth century. Consequently, the number of farmers declined sharply. The total acreage in my early seventeenth-century sample (19,361 acres) is virtually identical to the acreage in the c.1800 sample (21,032 acres). Yet the total number of farms declined from 328 to 145, the number of farms of less than 100 acres (i.e. family farms and farms where family and hired labour were of equal importance) declined from 290 to 65, and the acreage held by farms of less than 100 acres declined from 67.7 per cent of the total to 15.2 per cent. There is no doubt that both the number of peasant farms and the share of land farmed by the peasantry declined in the eighteenth century, when peasants are defined as farmers using mainly family labour.

Thompson also denied that the eighteenth century saw a significant decline in peasant agriculture. While he allowed that 'the peasantry certainly underwent relative decline', he concluded that the process was gradual and prolonged—it happened 'not all at once, not all in one century'

(F. Thompson 1966: 514). Moreover, he suggested that the peasants never owned the majority of English farm land, so 'the decline of the peasantry becomes an inconsiderable affair' (Thompson 1966: 513). While previous historians saw cataclysmic collapse in a short time, Thompson saw a minor reorganization extending over centuries.

Thompson's revisionism depends on a particularly narrow definition of 'peasant'. Not only does he require peasants to own their land as well as farm it with family labour, but he also restricts the category of owner to freeholders. Despite their substantial proprietary interest in the soil, he dismisses small-scale copyholders and beneficial leasees from the category of 'peasant'. It is only for this reason that he can dismiss their decline as insubstantial.

The empirical foundation of Thompson's revisionism involves two estimates of the national distribution of property ownership—Gregory King's social table of 1688, and an estimate of Mingay's (1963: 24) for 1790.[7] From King's table, Thompson concluded that owner-occupying freeholders owned about one third of the English farm land. While King's estimate of their number is unreliable, many historians have accepted this estimate of their aggregate landholding.[8] From Mingay, Thompson learned that 'it is unlikely that in 1790 occupying freeholders owned more than 15–20 per cent of the land'. Thompson is right that a decline from one third (already a small number) to one sixth was hardly a social revolution.

Thompson's argument is vulnerable on two scores. First, Mingay's estimate that occupying freeholders owned 15–20 per cent of England's farm land is certainly an exaggeration. Mingay's procedure is unreliable,[9] and the estimate itself is larger than Davies' (1927: 110) finding that owner-occupying farmers—a category that includes occupying copyholders and beneficial leasees as well as the freeholders whose land Mingay is claiming to measure—owned only 10 per cent of the total. Occupying freeholders owned a lot less land in 1790 than Thompson gave them credit for.

---

[7] Thompson also uses Bateman's (1883) rearrangement of the 1873 survey of landowners for a 19th-cent. benchmark. Thompson (1966) speculated on the distribution of property in 1500 but without any evidence. Thompson (1969) began with King.

[8] For instance, while Cooper (1967) savaged some of King's estimates, he ended up accepting that the freeholders owned one third of the farm land.

[9] There are two reservations. First, the share of land owned by small freeholders is estimated as the difference between the total acreage of English farm land and Mingay's estimates of the acreage belonging to large owners. This procedure is highly unreliable—since the share of land owned by small freeholders is a small number that is estimated as the difference between two big numbers, even small percentage errors in the big numbers translate into a large percentage error in the small number. Since there are surely substantial errors in the estimates of the large numbers, the estimate of their difference is fraught with error. Second, no defence is given for the estimates of land belonging to large owners. All we know is that the calculations are very approximate since they are done as round numbers.

Furthermore, peasants—on a broader definition—owned much more than one third of the land in 1688. Thompson is probably right that small-scale freeholders owned only one third of the country, while the freehold of the remaining two-thirds belonged to the gentry and aristocracy. Their land, however, was not all country houses and game parks; most of it was let to farmers (probably the 'farmers' in King's table).[10] We know from estate surveys how that land was organized. About one third of it was enclosed and divided into large, capitalist farms; the rest was divided among small copyholders and beneficial leasees. While some of these were capitalist farmers, most were not. In 1688, the peasantry of England occupied not a third of the country but closer to two-thirds.

The number of peasants fell in the seventeenth century as land was enclosed, but the rate of decline accelerated in the eighteenth century as peasants also disappeared from the open fields. By the end of the eighteenth century only 10 per cent of England belonged to owner-occupying farmers, so the decline from 1688 was precipitous. The eighteenth century witnessed not only change, but revolutionary change.

## Explanations for the Decline of the Yeoman

Parliamentary enclosure used to be the popular explanation for the decline of the owner-occupying family farm. Thousands of private acts passed in the face of peasant opposition by a parliament dominated by large landowners seemed tailor-made for the culprit (Hammond and Hammond 1932). Yet today that view is untenable. Parliamentary enclosures did not result in a concentration of landownership or in depopulation.[11] Further, as shown earlier in this chapter, farms were as large in the open fields c.1800 as in the enclosures. Administrative studies have established that the surveying and land allocation were conducted in a legalistic—some claim professional —manner (Tate 1967). While many small-scale owners opposed enclosure and many poor people probably suffered from the elimination of commons, the parliamentary enclosures effected no revolution in social relations. The

---

[10] King's social table does not use the terms copyholder or beneficial leasee, and there is a debate as to whether they were tallied as 'freeholders' or as 'farmers'. Mingay (1963: 7) and Habakkuk (1965: 654) believed copyholders were included with freeholders, but neither adduced evidence in support of the view. Thompson (1966: 507) contrasted King's freeholders with copyholders and long leaseholders, so apparently he thought King tallied them with the farmers. Cooper (1967: 428) was sceptical that King's freeholders included many copyholders since, if a lot were included, there would not have been enough genuine freeholders left over to make up the electorate. MacPherson (1962: 288) thought it most plausible that copyholders and leaseholders for years and lives were included with the farmers, although he acknowledged the problem was difficult.

[11] See the discussion in Ch. 3 and the land tax studies summarized there.

modern consensus is that parliamentary enclosure did not destroy the English peasantry because it had already been destroyed.

Yeoman agriculture was eliminated as large estates embarked on a long-term policy of land acquisition towards the end of the seventeenth century. The aim of this policy was to increase farm size for financial reasons. (Here I anticipate results that will be developed in detail in Chapter 11). The chance to economize on labour meant that large farms were more profitable than small farms. Indeed, the most efficient farms were 200 acres or more. Farms of this size were 'rent maximizing farms' since their lower costs allowed them to pay more rent per acre than smaller peasant farms. Farmers were anxious to lease several small farms to make a large farm if they could keep the gains. Landowners wanted to amalgamate farms since a 1,000-acre estate yielded more rent if it were divided into four 250-acre farms than if it were divided into twenty 50-acre farms. Farm amalgamation was driven by the desire to realize the profits of large scale.

But this argument is not the whole story since the technological basis for the superior efficiency of large farms was present as far back as the middle ages. Indeed, several books on estate management were written in the thirteenth century, and their discussions of the number of plough teams necessary to cultivate a demesne show an awareness of the economies in livestock and implements that contributed to the economies of scale (Oschinsky 1971: 155–60). Also, demesnes were being operated as large farms throughout the early modern period, so some people must have noticed the advantages of large size. If these advantages were present throughout the early modern period and yet large farms did not often emerge before the late seventeenth century, then there must have been an impediment to that amalgamation in earlier years. That impediment was removed in the late seventeenth century. In this chapter I will argue that the major impediment was the primitive state of the law of mortgage.

My explanation for the end of yeoman farming emphasizes the role of the great estate. There is another approach to this question, which emphasizes the activity of peasants. Since large farms were more profitable than small, peasants had an incentive to buy out their neighbours, so the system of large farms might have emerged by a process of differentiation within the peasant community. Certainly, at the end of the nineteenth century, many Marxists expected that Europe's peasantry would be destroyed in this way. However, the German censuses of 1882 and 1895 were a shock, for there was no evidence that large farms were displacing small. Perhaps the Agrarian Fundamentalists' faith in the superior efficiency of the large farm should be abandoned? Kautsky (1900) reasserted the superiority of the large farm and explained the persistence of small-scale farming by the peasants' willingness to accept a continuously declining income in the face of capitalist competition. Lenin (1899: 70–330; 1908) agreed that the statistics did not prove the viability of peasant farming. Further, he asserted that in Russia large farms were expanding at the expense of small farms. Lenin's work raises the possibility that peasant differentiation also contributed to the rise of capitalist agriculture in England.

Historians of English agriculture have been investigating the possibility that large farms emerged through voluntary exchange among the peasants. Tawney (1912: 57–97) discussed the 'peasant land market' in the early modern period and argued that the free market in land increased the inequality of holdings. In 1935, Kosminski (1956: xvii–xviii, 197–255) showed that the Hundred Rolls of 1279 recorded much greater inequality in the holdings of free peasants than of villeins. He attributed the differentiation among the free peasants to the fewer restrictions on their right to alienate property. Studies of the 'peasant land market' have since proliferated.[12] Hoskins's classic study of Wigston Magna (1957) and Spufford's *Contrasting Communities* (1974) have shown that the peasant land market could produce large holdings in the early modern period. Spufford (1974: 49) contrasted the peasant land market with the acquisition of land by manorial lords as an explanation for the emergence of large farms.

## The Failure of Leasing

Before showing how peasant agriculture was destroyed by the great estate, I will analyse the 'kulak' model of the origins of capitalist agriculture. There are two ways in which the peasant land market could have effected the transition: large farms could have been formed either by the peasants' *leasing* up small copyholds, beneficial leaseholds, and freeholds, or by their *buying* them up. I start with leasing.

There is no evidence that leasing occurred on a substantial enough scale to eliminate family farms generally. I rejected this possibility implicitly in Chapter 4 when I argued that the distribution of manorial tenancies was similar to the distribution of farms. That discussion relied on data from the sixteenth and seventeenth centuries. The question is most acute, however, for the eighteenth century when peasant agriculture was being rapidly destroyed.

To see whether large farms were being formed by leasing up small properties *c.*1790, I analysed the patterns of landholding in my sample of land tax assessments. To understand the procedure, one must distinguish

---

[12] P. Harvey (1984) summarizes and extends the medieval literature.

*properties* from *farms*. I call all the land owned by one person in a village a property. Properties included demesnes farmed by tenants at will, copyholds, leases for lives, and so on. In contrast, farms were units of land occupation and were either owner-occupied or were held either at will or for a short term of years. It is important that farmers could and did occupy properties owned by more than one person. My coding of the land tax assessments allows me to sum all the land held by each farmer (even if the parcels were owned by several people), so the ownership basis of large farms can be studied.

I undertook the following inquiry with the sample of land tax assessments. First, I selected a subset of villages that contained a large number of small properties, for it is only in such villages that one could conceivably observe large farms being created by leasing up small copyholds and freeholds. Specifically, I identified all of the villages in the data set that contained twenty or more properties of 10 to 50 acres. These properties were of the same size as copyholds in seventeenth-century surveys. There were thirty-eight such villages. Second, for that group of villages, I divided the farms into two groups—one in which the largest constituent property was between 10 and 50 acres, and the other in which the largest constituent property exceeded 50 acres.[13] The first group included farms that contained only small properties; the second group consisted of farms based around large properties. Third, by comparing the size of farms in the two groups, I could test whether large farms were created by leasing small properties or whether they presupposed large properties.

The data emphatically reject the possibility that large farms could in practice be formed from small properties. The average farm was only 27 acres for the farms containing only small properties. (See Table 5–3). Only two of those 541 farms exceeded 100 acres. In contrast, the average farm was 174 acres for the set of farms made from properties exceeding 50 acres. Most of the land in this set was arranged in farms big enough to yield the maximum rent. The conclusion is clear—large farms were not being created by leasing up many small properties. Instead, large farms were the creation of large proprietors. As we will see, they were usually the owners of great estates.

Given the financial incentives for farm amalgamation, this result is surprising since one would also expect peasants to expand the size of their farms. The explanation, however, is not difficult. Eighteenth-century commentators warned that farmers who cultivated their own land as well as

---

[13] I do not include holdings of less than 10 (computed) acres in these tabulations since they were in reality cottages with small gardens.

*Table 5–3. Were big farms made by leasing small properties? The distribution of farm size depending on property size*

| | Land tax assessments, c.1790 | | | |
| | Largest property in the farm (10–50 acres) | | Largest property in the farm (over 50 acres) | |
| | Number | Acres | Number | Acres |
|---|---|---|---|---|
| 5–10 | | | | |
| 10–15 | 264 | 3,235 | | |
| 15–30 | 418 | 8,826 | | |
| 30–60 | 232 | 9,394 | 41 | 2,251 |
| 60–100 | 48 | 3,550 | 129 | 10,321 |
| 100–200 | 2 | 243 | 180 | 25,238 |
| 200–300 | | | 66 | 16,366 |
| 300–400 | | | 27 | 9,218 |
| 400–500 | | | 9 | 4,010 |
| 500–1,000 | | | 14 | 10,680 |
| 1,000+ | | | 6 | 13,299 |
| TOTAL | 964 | 25,248 | 472 | 91,383 |

rented land were likely to enrich their land at the expense of the leased land. Thus, Laurence (1727: 34) cautions:

Neither should a Steward suffer any of his Lord's Lands to be let to the *Freehold Tenants* within or near his Lord's Manor, because 'tis a natural Contrivance, or piece of *Cunning*, in them to lay all or most of their Manure upon their own Land, to the great *Improvement* of the one, and *Beggary* of the other.

Abuses were difficult enough to police when a farmer owning 10 or 20 acres of pasture leased 200 acres of open field arable from a large estate. But a farmer hiring half a dozen mixed farms to add to his own would be very difficult to control. A small owner ran a risk in letting his farm to such a farmer. For that reason, leasing failed as a means to create large farms.

## The Manor and the Peasant Land Market

Buying up small properties was the other way in which large properties could be assembled. Potentially, there were two classes of buyers: enterprising peasants who bought up their neighbours' small copyholds and freeholds, or the gentry and aristocracy who were building up their estates. Yelling (1977: 112) sharpened the issue when he wrote: 'In many parishes there was a more or less aggressive expansion of seigneurial control; in others the main drive for consolidation came from among the tenants themselves.'

To see which process was more important, I developed another sample of land tax assessments. This sample is based on the returns for 690 villages in the south midlands. For these villages I computed the acreage owned by each proprietor and the acreage occupied by each farmer. There are 16,131 properties in the sample, counting all the land owned by each proprietor in each village as a property. Further, I used county histories to determine which landowners were manorial lords.[14] By comparing manorial and non-manorial land, I can distinguish the 'the expansion of seigneurial control' from 'consolidation [by] the tenants themselves'.

Manorial lords owned a great deal of land, especially large properties. Only 5.5 per cent of the properties were owned by lords (Table 5–4), but that land amounted to almost half of the acreage in the 690 villages studied (Table 5–5).

*Table 5–4. Manorialism and large holdings: the number of properties*

| Owner | Property size | |
|---|---|---|
| | Large | Small |
| Manorial lord | 696 | 196 |
| Not a lord | 667 | 14,572 |

*Notes:*
The table shows the number of properties owned by each group.
Large properties are 200 acres or more, small properties are less than 200 acres.

*Table 5–5. Manorialism and large holdings: the distribution of land*

| Owner | Property size | |
|---|---|---|
| | Large (%) | Small (%) |
| Manorial lord | 48.3 | 1.4 |
| Not a lord | 18.1 | 32.2 |

*Notes:*
The table shows the proportion of land owned by each group.
Large properties are 200 acres or more, small properties are less than 200 acres.

Furthermore, manorial lords owned most of the acreage in large properties. Since 'capital' farms at the end of the eighteenth century were 200 acres or more, and since they were built around holdings of approximately that size, I classified properties as large or small depending

[14] I relied on the manorial histories in the *Victoria County History* volumes for Beds., Bucks., Cambs., Hunts., Leics., Northants., Oxon., and Warwick. I also used the manorial histories in Baker (1822–41) and Nichols (1795–1815).

on whether they were greater or less than 200 acres. Clearly, most properties were small. (Table 5–4.) Two-thirds of the land, however, was organized in large properties, and manorial lords owned two-thirds of that. (Table 5–5.) Moreover, many of the non-manorial owners of large properties owned manors elsewhere, so the manorial system dominated landholding in the midlands. In the contest between engrossing lords and consolidating tenants, the lords were the winners.

But perhaps the manorial lords consisted of yeomen so successful that they bought manors? There are certainly examples from earlier periods of English history—the Spencers ran sheep before they bought Althorp in 1508 (Finch 1956: 38–9)—and there are a few suggestive possibilities even in the eighteenth century.[15] However, the vast amount of manorial property was owned by old landed families. This can be seen from Table 5–6, which breaks down the acreage of the large landowners (both manorial and non-manorial) by the title ascribed to them in the land tax assessment. Peers, baronets, and knights owned 40 per cent of the large properties, usually by virtue of being a manorial lord. These individuals were not rising yeomen. Nor were Oxford and Cambridge colleges, other institutions like hospitals, the few army officers, ecclesiastical bodies, or the individuals styled 'reverend', who together owned another 15 per cent of the large properties. Most of the land owned by reverends was non-manorial and consisted

Table 5–6. *The owners of large properties: lordship and status*

|  | Manorial lords (%) | Not lords (%) | Total (%) |
|---|---|---|---|
| Peer | 29.20 | 2.71 | 31.91 |
| Baronet | 5.60 | 0.43 | 6.03 |
| Knight | 1.03 | 0.57 | 1.60 |
| Esquire | 24.42 | 7.28 | 31.70 |
| Officer | 0.66 | 0.36 | 1.02 |
| Other layman | 2.02 | 6.36 | 8.38 |
| Women | 3.62 | 2.10 | 5.72 |
| College | 2.00 | 0.43 | 2.43 |
| Ecclesiastical body | 0.43 | 0.48 | 0.91 |
| Reverend | 3.27 | 6.21 | 9.48 |
| Other institutional | 0.50 | 0.32 | 0.82 |
| TOTAL | 72.75 | 27.25 | 100.00 |

*Note*: The percentages indicate the proportion of the acreage of large properties (those over 200 acres) owned by the indicated group.

[15] John Owsley, 'an apothecary of Hallaton', bought the manor of Blaston in 1750. (*VCH*, Leics., v. 25). 'In the beginning of the 18th century, [the manor of Totternhoe] reappears in the hands of John Mead, a wealthy London grocer.' (*VCH*, Beds., iii. 449). But neither of these individuals was a farmer.

mainly of enclosure allotments in lieu of tithe and glebe. Together, these groups owned 54.2 per cent of the properties over 200 acres.

If rising peasants owned large properties, they were tallied in Table 5–6 as 'esquire', 'other layman', or 'woman', the latter being mainly heirs of the former two.[16] These groups owned 45.8 per cent of the large properties. Since there was little control on someone's adopting the title 'esquire' it is possible that rising peasants could buy manors and be called esquire. However, the county histories show in case after case that these men were from old gentry families—there are few cases that could conceivably have been a differentiating peasant. This impression is buttressed by Stone and Stone's (1984) recent, exhaustive study of mobility into the class of English landowners. Using ownership of a country house as their index of membership in the gentry and aristocracy, Stone and Stone found very little entry of new owners in the eighteenth century. They found scarcely any rich industrialists—let alone yeomen—buying country houses.

The men without even the title of esquire are the obvious candidates for rising yeomen—although the non-titled also include small urban investors[17]—and, indeed, most of the land owned by these men was not owned through the possession of a manor. Furthermore, it is usually impossible to trace their ancestry in the county histories, which is consistent with their having humble origins. But no more than 10 per cent of the land in properties over 200 acres was owned by such men or their female heirs. Few large properties were created by the diligent accumulation of tiny holdings by rising peasants and even fewer were acquired by their purchasing manors.

Tables 5–7 and 5–8 shift our attention from landownership to its occupation, and they show that the accumulation of property by the manorial lords was critical for the creation of England's uniquely large farms. The tables show the farm size distribution for villages in the sample I have been analysing and break it down into two categories—those farms that included some land owned by a manorial lord and the other, non-manorial farms. The differences are striking. The average farm size in the manorial group was 200 acres, while the average non-manorial farm was 79 acres.[18] Three-quarters of the land in the manorial farms was in farms of over 200 acres, while only one third of the land in non-manorial farms was

[16] Women with titles are included with peers. My category 'women' includes women styled simply 'Mrs' or 'Miss'.
[17] Habakkuk (1987: 290) remarked on the growing importance of small absentee owners in the 18th cent. Table 5–6 shows their role was unimportant compared to the gentry and aristocracy.
[18] These calculations exclude holdings of less than 10 acres since they were probably cottages.

*Table 5–7. Manorial and non-manorial farms, c. 1790: number*

|  | Number of cottages and farms | | Percentage of number over 10 acres | |
|---|---|---|---|---|
|  | Manorial | Non-manorial | Manorial | Non-manorial |
| 0–5 | 202 | 2,593 | | |
| 5–10 | 153 | 1,450 | | |
| 10–15 | 104 | 742 | | |
| 15–30 | 154 | 1,127 | 26.0 | 58.5 |
| 30–60 | 158 | 940 | | |
| 60–100 | 219 | 718 | 13.7 | 15.0 |
| 100–200 | 398 | 844 | 24.8 | 17.6 |
| 200–300 | 269 | 262 | 16.8 | 5.5 |
| 300–400 | 123 | 103 | | |
| 400–500 | 62 | 33 | 18.8 | 3.5 |
| 500–1,000 | 93 | 30 | | |
| 1,000+ | 22 | 1 | | |
| TOTAL | 1,957 | 8,843 | | |

in these giant 'capital' holdings. Conversely, only 9 per cent of the land in manorial farms was in holdings of less than 100 acres, while one third of the non-manorial land was in such small farms. There is no doubt that the manorial system led to England's peculiar system of very large farms.

The allocation of most of the land owned by manorial lords to very large farms, however, did not mean that small farms disappeared. As Table 5–7 shows, almost two-fifths of the manorial farms were less than 100 acres, while 73 per cent of the non-manorial farms were so small. Even though most of England's farm land was in large farms, most farms were small. Historians who have been impressed by the survival of the small farm have been misled by this feature of the farm size distribution. Concentration on the distribution of the number of farms rather than on the distribution of their acreage obscures the revolutionary changes that occurred in England's rural life.

While the estate system led the way in creating exceptionally large farms, peasant differentiation did play an independent, if muted, role in the transition to capitalist agriculture. Even though the average size of a non-manorial farm *c.*1790 (79 acres according to Table 5–8) was less than the average size of a manorial farm (200 acres), the average non-manorial farm was more than double the average size of a customary holding (34 acres) in the sixteenth century.[19] Further, many of the small properties in *c.*1790

---

[19] There is much uncertainty in this comparison partly because the non-manorial farms in the land tax assessments include non-manorial freeholds, which are not included in the earlier surveys.

*Table 5–8. Manorial and non-manorial farms, c. 1790: acreage*

|  | Acreage of cottages and farms | | Percentage of acreage over 10 acres | |
|---|---|---|---|---|
|  | Manorial | Non-manorial | Manorial | Non-manorial |
| 0–5 | 575 | 7,236 | | |
| 5–10 | 1,083 | 10,306 | | |
| 10–15 | 1,262 | 9,128 | | |
| 15–30 | 3,389 | 23,958 | 3.6 | 19.4 |
| 30–60 | 6,865 | 40,970 | | |
| 60–100 | 17,215 | 56,417 | 5.4 | 14.8 |
| 100–200 | 58,488 | 116,994 | 18.2 | 30.7 |
| 200–300 | 66,654 | 63,784 | 20.8 | 16.7 |
| 300–400 | 42,088 | 35,430 | | |
| 400–500 | 27,591 | 14,359 | 52.1 | 18.4 |
| 500–1,000 | 62,446 | 19,425 | | |
| 1,000+ | 34,952 | 1,001 | | |
| TOTAL | 322,608 | 399,008 | | |
| Average size over 10 acres | 200 | 79 | | |

were farmed by tenants at will rather than owner-occupiers. This evolution is not surprising since yeomen faced the same incentive as manorial lords to amass land and enlarge farms—namely, the labour economies of large scale-production.

There are three reasons that the peasant land market did not account for the generality of large English farms in the nineteenth century. First, the value of a small copyhold to a small proprietor was less than its value to a large landowner. Thus, if 30 acres was on the market, it was worth more to someone who already owned several hundred acres—and who could therefore add the land to a farm that was already large—than to someone who owned only 30 acres and who, consequently, could only add the land to a small farm. For this reason, large proprietors were willing to pay more for smallholdings than were small proprietors. As a result, real estate transactions were mainly in one direction—from small owners to large— and the operation of the peasant land market accentuated the inequality of landownership.

Second, the functioning of the peasant land market was often inhibited by the insecurity of peasant tenures. When peasants held their land by beneficial lease and copyholds for lives, lords could impose arbitrary fines on purchasers and effectively charge them twice for their land. This interference killed the market for such properties.[20] The well-known

[20] B. Harvey (1977: 304) discusses this point as it applied to the Westminster estates in the 15th cent.

examples of the peasant land market producing large farms were all in villages where the peasants had secure titles. Thus, a large share of the land in Wigston Magna was in properties and farms of more than 100 acres in the mid-eighteenth century. The land in this village had been freehold since the early seventeenth century when the manorial rights were sold off to the tenants after they successfully defended their customary privileges in court (Hoskins 1957: 95–115, 216–27). In the Cambridgeshire villages of Chippenham and Orwell, where large peasant properties emerged in the seventeenth century, the copyholds were heritable (Spufford 1974: 75–6, 101).

Third, the large holdings accumulated by yeomen were only transitory phenomena since the engrossing yeomen were often bought out or run out by the engrossing lord. In Chippenham and Orwell, the large copyhold estates put together by seventeenth-century yeomen were bought up by the manorial lords at the end of the century. (Spufford 1974: 70, 101). It was the expanding estate—not the enterprising kulak—that produced England's peculiar nineteenth-century social structure.

## How Manorial Lords Enlarged Their Estates

When (and how) did the manorial lords acquire so much property? There are no records before 1780 as encompassing as the land tax assessments, so the story must be pieced together from diverse sources. One is Gregory King's 1688 social table, and it indicates that one third of English farm land was owned by freeholders, many of whom were small farmers. While many of King's figures are suspect, this one is accepted by many historians. The other two-thirds were held mainly by copyholders of inheritance, copyholders for lives, and beneficial leasees. Transactions in the different tenures were accomplished in different ways and generated different kinds of documents. This diffuseness of evidence has obscured the end of the peasantry. It can be traced, however, by considering the tenures in turn.

Freehold property was found throughout the midlands. Its accumulation by manorial lords can be followed in the deeds assiduously preserved by landed families. Habakkuk (1940) has exhaustively studied these for Northamptonshire and Bedfordshire. He concluded that the period 1680–1780 saw a sharp increase in purchases of small freeholds by large landowners.

Both the large old families and the newcomers [in Bedfordshire and Northamptonshire] purchased the holdings of a great number of freeholders. The Earls of Northampton had not made a single purchase of land for a hundred years, but between 1690 and 1710 the holder of the title bought three hundred acres from small owners on one of his estates alone. (Habakkuk 1940: 16)

Examples are common in other counties. For instance, the Dashwoods were lords of the manor of Kirtlington, Oxfordshire: 'Between 1684 and 1750 Sir Robert and Sir James Dashwood bought up most of the freehold farms, with the result that the substantial freeholder almost disappeared from the parish' (*VCH*, Oxon. vi. 226). Purchases like these eliminated the yeomen on freehold land at the same time as they effected the rise of great estates (Habakkuk 1965).

Copyholds of inheritance were common on the eastern edge of the south midlands and were scattered across Northamptonshire and Leicestershire as well. These copyholds conveyed a form of ownership that was almost as absolute as freehold. Manorial lords bought up copyholds at the same time as they bought up small freeholds.

From the late 17th century the area in Great Abingdon under independent yeoman owners diminished as the manorial estate was enlarged. John Bennet the younger began from 1683 to buy copyholds amounting to c.100 a., including land of the Smee, Amy, and Beteyn families. [His successor] Thomas Western (d. 1754) went on buying out the copyholds, and by 1800 the Hall estate included most of the parish. (*VCH*, Cambs., vi. 11–12)

In Whittlesford, Cambridgeshire, 'In the late 18th century Ebenezer Hollick and his nephew and namesake substantially enlarged the [manorial] estate, swallowing many smaller farms and leaving the farmsteads derelict' (*VCH*, Cambs., vi. 269). In the late seventeenth century, the lord of the manor of Orwell, Cambridgeshire, bought up hundreds of acres of copyhold. In 1696, the lord of Chippenham, Cambridgeshire, bought all the copyhold land in the parish (Spufford 1974: 70, 104). In Bedfordshire a large estate owned by the Earl of Ossory was assembled in the eighteenth century. Twenty-eight purchases were made between 1737 and 1806. Some vendors were prominent (i.e. Lord Fitzwilliam, Lord Spencer, and the Duke of Bedford) but most were humble. Most purchases were freeholds, but thirteen included copyhold cottages or small farms.[21] In this way great estates were assembled by buying out yeoman farmers.

Beneficial leases were the common tenure for much customary land in Northamptonshire and Leicestershire and for demesnes throughout the south midlands. Unlike freeholds, for which the purchase had to be agreeable to both the peasant and the lord, the reorganization of the leasehold land—like copyholds for lives—could proceed in the face of the

[21] Bedford RO, 'Note on Plan to Accompany R.O. and R.H. Catalogues', pp. 12–14.

tenants' objection. The lord's advantage arose since neither beneficial leasees nor copyholders for lives had unlimited interests in their property. When the last person named in the lease or copyhold died or when the term expired, the tenant's interest in the land ceased, and it reverted to the grantor. Before the eighteenth century (in villages that were still open at its commencement) lords usually did not take possession of the land because a new bargain was struck, a fine paid, and a new life added to the copyhold or a new lease drawn up. In either case, the yeoman social structure was perpetuated. In the late seventeenth, and especially the eighteenth, century, the policy of lords changed. They no longer renewed the old agreements but allowed them to 'run out'. When the copyholds and beneficial leases lapsed, the lords repossessed the land, amalgamated it into large, rent-maximizing farms, and let it to tenants at will for rack rents.

In the eighteenth century, landlords in the midlands stopped renewing beneficial leases. Mingay (1984: 115) reports that 'evidence from the Kingston estates in the region [Nottinghamshire] indicates that in the second and third quarters of the eighteenth century there was a tendency for leases for lives and leases for terms of years not to be renewed when they fell in'. They were replaced by leases for short terms of years or tenancies at will. Wordie's (1974: 599 n. 4) study of the Leveson-Gower estates in Shropshire and Staffordshire established that there was a

discontinuance of the seventeenth-century custom of taking a high entry fine and then charging a low rent under a lease for lives. After 1700, there was a swing towards shorter term leases and letting at will. Entry fines therefore diminished or disappeared and rents rose.

Habakkuk (1940: 16–17) noted the same change in the management of beneficial leases in Bedfordshire and Northamptonshire.

In these counties it is not possible to discover any attempt to change leaseholds for lives into leaseholds for terms of years in the years before 1690. Even in the nineties most of the great landowners were renewing leases for lives which fell in on the same basis. But after about 1710 the more usual practice is to replace them by a lease for some term of years at a rack-rent. [22]

The Grafton estates in Northamptonshire are an example of the reorientation of policy. In the sixteenth century, the leases were mainly beneficial for three lives. By the 1650s, the term had been changed, in most

---

[22] The abandonment of fines and the shift to short-term tenancies at full commercial rents (i.e. rack rents) began somewhat earlier in at least one village in Beds. The customary tenants of Blunham manor held their lands on beneficial leases and they were all renewed in 1615/16 for 21 years. A general reletting of the estate occurred in 1655, and those leases were all for 9 years at rack rents. Beds. RO, catalogue to de Grey papers, 1/15–1/50.

cases, to thirty-one years (Lennard 1916: 32–3, 120). A 'Particular of Leaseholds' in 1705 summarized the leases on the estate as of 1703. Most had about twenty years left (Northants RO, G. 3881). A survey of 1757 detailed the tenants on the estate. Most held their land at will; a few held for short terms of years at rack rents (Northants RO, G. 3966). The Particular in 1705 probably marks the date when leasing policy was reconsidered. The decision to replace beneficial leases with tenancies at will had been accomplished by 1757. It abolished yeoman agriculture by eliminating the farmer's proprietary interest in the land.

Beneficial leases were also the normal tenure for land owned by ecclesiastical bodies and Oxford and Cambridge colleges in the early modern period. These institutions also converted their beneficial leases to short-term tenancies, but the change did not come until the middle of the nineteenth century. St John's College, Cambridge, for instance, ran out most of its beneficial leases between 1863 and 1869 (Howard 1935: 179). The yeomen lasted longest on these estates.

Copyholds for lives were a common form of yeoman tenure in the west of the south midlands. Landlords also stopped renewing these agreements in the eighteenth and early nineteenth centuries. Oxfordshire provides several examples.

The first example is the Dillon estate in Spelsbury, Enstone, Tastone, and Fullwell. Copyholds were being renewed on the customary terms in the early eighteenth century, for instance:

At Enstone Court held Aprill 2: 1703 Lett a Cottage house & Garden in Little Enstone to Richard Wickens, for his owne Life and his wifes widdowes Estate & his Sone Arthur in Revershon by Coppy it being fallen into the Lords hands upon the Death of John Siburr. (Oxon. RO, DIL II/a/4c)

A document called 'A Particular of the Copyholders and Leaseholders Within the Manor of Spelsbury Taken March 11 1705' (Oxon. RO, DIL II/a/4e) may mark the change in policy, for it listed all the copyholders and indicated the size, annual value, and quit rent of each estate. More sinisterly, it recorded the number of remaining lives in the agreements and the ages of those future tenants. The document is covered with later notations of their deaths.

The passing of the copyholders can be followed in detail in the quit rentals. In 1705 there were forty-one copyholders in the Manor of Spelsbury. Most held a yardland (about 30 acres) or a half yardland although a few were cottagers and one held as much as three yardlands (DIL II/a/4e). In 1823 there were ten copyholders. At most, three of these were farmers, the rest only cottagers (DIL II/b/28). One large copyhold estate was

'Given up on the Spelsbury Inclosure at Michaelmas 1802 for an annuity granted by Lord Dillon to Mrs Walker', the owner (DIL II/b/29), but the rest all reverted to Lord Dillon at the death of the last named tenant. The quit rentals mark the end of the receipt of the quit rent with observations like 'John Cross Dead (his Death presented)', or 'In My Lords Hands from ye Expiration of the Deads year at Michms. 1767', or 'Widdows Deads year Expired Michms. 1764. Let by the year to Mr Walker'.

The policy of running out copyholds produced a social revolution on the Dillon estates. A rental of 1813 and a survey of 1817 show that most of the land was arranged in farms of several hundred acres. Some of these farms (e.g. 'Late Rooks') were still designated by the names of the former copyholders. Most of the people lived in cottages with only gardens attached. A society in 1705 that had been composed mainly of half yardlanders and yardlanders—yeomen—had been transformed in a century into a society of capitalist farmers and landless labourers.

We can also witness the passing of the yeoman in Wytham, Berkshire (now Oxfordshire). The manor was the property of the Earl of Abingdon. It was surveyed in 1728 and again in 1814, two years before it was enclosed. Table 5–9 summarizes the results.

The 1728 survey describes a yeoman village. Ninety per cent of the cottage and farm land was held as copyholds. Most were small, family farms; only a few were large enough to employ any hired labour. Two-thirds of the tenancies had at least 5 acres of farmland attached. The category 'In Hand' is a miscellany with two interesting components. First, the common is tallied here. Second, there are two smallholdings labelled 'late Mathews' and 'late Flixon'. I shall return to them later.

The 1814 survey describes an altered world. There were no copyholds, only tenants at will. The number of farms had fallen from twenty-two to eight, and most of the land was in a few large farms. The three largest were capitalist enterprises. The number of houses in the manor had not increased much, but now 76 per cent of the houses were without any attached land other than gardens. Their occupants were the new proletariat of the village. Notice also that the common had disappeared; that land was now designated 'pasture, etc.'. Since there were no longer any copyhold tenants of the manor, only the Earl of Abingdon had a legal right to depasture stock on the 'common', so it was *ipso facto* enclosed.

What happened between 1728 and 1814? The Earl of Abingdon ran out his copyholds just as Lord Dillon was doing. The two parcels 'in hand' in 1728 that are labelled 'late Mathews' and 'late Flixon' are suggestive. The designations referred to copyholders who had died and whose land had thereby reverted to the Earl. (The Wytham parish register shows Margaret

*Table 5–9. Wytham in 1728 and 1814*

Part A. Cottages and Farms

|  | 1728 | | | | | | 1814 | |
|  | Copyhold | | At will | | Total | | At will | |
|  | Number | Acres | Number | Acres | Number | Acres | Number | Acres |
|---|---|---|---|---|---|---|---|---|
| 0–5 | 3 | 3 | 7 | 11 | 10 | 12 | 26 | 7 |
| 5–10 | 3 | 26 | 2 | 15 | 5 | 41 | 3 | 24 |
| 10–15 | 1 | 11 | 1 | 14 | 2 | 25 | 1 | 12 |
| 15–30 | 10 | 198 | 1 | 16 | 11 | 214 | 0 | 0 |
| 30–60 | 2 | 88 | 0 | 0 | 2 | 88 | 1 | 46 |
| 60–100 | 2 | 135 | 0 | 0 | 2 | 135 | 1 | 96 |
| 100–200 | 0 | 0 | 0 | 0 | 0 | 0 | 1 | 128 |
| 200–300 | 0 | 0 | 0 | 0 | 0 | 0 | 1 | 291 |
| Total cottages | 3 | 3 | 7 | 11 | 10 | 12 | 26 | 7 |
| Total farms | 18 | 458 | 4 | 45 | 22 | 503 | 8 | 597 |
| TOTAL | 21 | 461 | 11 | 56 | 32 | 515 | 34 | 604 |

Part B. Land in hand (acres)

| In 1728 | | In 1814 | |
|---|---|---|---|
| Arable | 42 | Arable | 92 |
| Late Mathews | 19 | | |
| Late Flixon | 6 | | |
| Woods | 218 | Woods | 212 |
| Common | 350 | Pasture, etc. | 201 |
| Waste | 14 | | |
| TOTAL IN HAND | 649 | | 505 |

C. Total surveyed

| | *1728* | | *1814* |
|---|---|---|---|
| Total land | 1,164 | | 1,109 |

Mathews buried on 1 August 1727 and Rebecca Mathews buried 7 March 1726. William Flexon was buried 15 June 1714. These may have been the copyholders.) In the same way, one of Lord Dillon's farms was called 'Late Rooks' in the 1817 survey. Probably the Earl's policy of running out copyholds had been decided upon shortly before 1728 and Mathews' and Flixon's holdings were the first fruits.

The social transformation that occurred in Wytham was not unusual in the eighteenth century. Whether one studies freeholds, copyholds, or beneficial leases, one observes a polarization. How the acquisitions were accomplished depended on the form of peasant tenure, but in many villages small-scale peasant proprietorships disappeared as great estates assumed direct control of the land and let it at will to substantial farmers. The result was the three-tiered system of wealthy landowners, large-scale tenant farmers, and landless labourers that epitomized English rural society in the nineteenth century. The land acquisitions of the great estate may not have been 'force and fraud', but they were primitive accumulation none the less.

## Habakkuk's Theory of the Rise of the Great Estate

Habakkuk (1940, 1950, 1960, 1965) has advanced several complementary explanations for the purchase of small freeholds by the great estates. By the late seventeenth century, large landowners were more interested in buying for two reasons. First, after the Civil War and Restoration, landownership, rather than royal favour, was the source of political power. The social prestige of landownership also increased with the result that the wealthy were willing to pay more for land than was warranted by its value in agriculture. Hence, the yeomen could make money by selling out. Second, the development of the strict settlement prevented the dispersal of large estates and fostered their creation through the provision of generous marriage portions (dowries). These were financed by the bride's parents' mortgaging land. The new couple invested the portion in property. Consequently, 'the landowning class as a whole was mortgaging to buy more land, in effect, "raising itself by its own boot straps"' (Habakkuk 1950: 28).

The desire of large landowners to buy was matched by a new desire on the part of small owners to sell. There were two reasons. First, the depressed (and erratic) prices of the late seventeenth century were particularly difficult for small proprietors. Second, the land tax instituted in 1692 was an additional, heavy burden on them.

However, while these developments may have played some role, they were not decisive. There are two difficulties with the argument that the gentry was willing to pay more for land than its economic value. First, as Habakkuk (1952–3) pointed out in another connection, the premium was even greater in the first half of the seventeenth century. That is when the yeomen should have been selling! Second, Habakkuk's calculation of the return to buying land understates the true return by leaving out capital gains. When they are included, the price of land did not exceed its commercial value (Allen 1988*a*). The argument that the strict settlement led to land acquisitions has been denied on the grounds that marriage portions were not used to acquire property (Beckett 1984: 9). Moreover, it is not clear why small landowners should have been particularly disadvantaged by low prices and the land tax. Those developments affected large landowners as well. And, as

F. Thompson (1968: 48) pointed out, peasants were better placed to withdraw from the market. Other developments must have contributed to the concentration of property ownership in the late seventeenth and eighteenth centuries.

## The Mortgage Market

The evolution of modern mortgages was the major development that led to the elimination of yeoman farming in the eighteenth century. This financial innovation increased the propensity of lords to buy up freeholds and heritable copyholds, and it led directly to the running out of copyholds for lives and beneficial leases.

There was a connection between mortgages and agricultural tenure since copyholds and beneficial leases were, in effect, loans from peasants to their lords. At the commencement of the agreement, the peasant paid the lord a large sum and the lord repaid the peasant by permitting him to farm the land and retain the proceeds. Whether the lord borrowed money from his tenants or someone else depended on the costs and risks involved.

'Selling' land for fines generated little income. Usually the fines were less than the sums that could be raised by mortgaging the land at current interest rates. Stone (1965: 318–19) argued on the basis of evidence from several estates that this was the case in the sixteenth and early seventeenth centuries. Clay (1985: 202) came to the same conclusion in his survey of landownership in the period 1640–1750. 'A landowner who could afford to allow his outstanding leases to expire without taking any fines for renewing them would . . . benefit very largely from a change to rack rents.'[23] In the 1720s, the fines for renewing church leases were so low that the ecclesiastics were implicitly paying their tenants 9 per cent on the advances at a time when the mortgage interest rate was 4.5 per cent (Clay 1980: 150).

The question is why lords ever 'sold' their land to their farmers for fines. The answer is that in the sixteenth century there was no alternative way to use land as security to raise long-term capital. It was possible to mortgage freehold land, but the arrangements were so unsatisfactory that money was raised by mortgage only in dire circumstances. In such mortgages the freehold in the property was conveyed to the mortgagee, who advanced the money. Repayment was normally to be made in six months. If default occurred by even so much as one day, the title remained permanently with the mortgagee. Moreover, the mortgagor was still indebted to the mortgagee for the principal of the loan. The mortgage was thus not a device

---

[23] Clay (1981: 85). Further, Clay (1985: 202 n. 205) argues that land let under a regime of fines and long leases sold for a lower price than rack rented land.

for long-term finance and was at best a risky procedure for raising short-term funds.

The only way for a landowner safely to use his land to raise long-term funds was to let it on a beneficial lease either for a term of years or for lives or as a copyhold for lives. Under any of these arrangements, the landowner received a large initial fine and small annual rents for the duration of the agreement. The landowner always retained the freehold of the property, and the land reverted to him at the end of the term or on the death of the named individuals in agreements for lives. Since landowners at some time wanted to raise long-term finance (to build a house, for instance), beneficial leases were adopted. Moreover, once the practice of granting such leases began, it was hard to reverse since that involved a sharp reduction in income for a generation as fines were forgone.

The first steps in developing the modern law of mortgage were taken around 1600, but the development was far from complete even in the middle of the century. As with the protection of copyholders, it was again Chancery that led the way. In the late sixteenth century, the Chancellor accepted petitions from mortgagors who had defaulted. Mortgagees were ordered to reconvey the property once interest and principal were paid. Initially, the Chancellor intervened only when exceptional circumstances had prevented repayment, but, by 1625, Chancery intervened often to protect the mortgagor's (newly created) right to redeem (Turner 1931). 'Although in theory landowners had at last obtained a reliable instrument of long-term borrowing without endangering their estates, there is in fact no evidence for more than a modest increase in the use of the mortgage for borrowing before the Civil War.'[24] One reason is that, even in the middle of the century, the right to redeem was still limited only to those contracts which specifically mentioned it (Melton 1986: 128–9). Such mortgaging as occurred was often done without invoking the right to redeem. Loans using land as security were often effected by the lender's immediately occupying the land and receiving its income until the debt was discharged. Modern mortgages in which the owner retained possession of the land while paying off the mortgage were still not common.

The Civil War and the Restoration marked a turning point in the development of the mortgage. When parliament sold confiscated Royalist estates in the 1650s, the most important buyers were the dispossessed owners who bought back their land through trustees. The purchases were financed with mortgages, and some, no doubt, relied on the right to redeem.

---

[24] Stone (1965: 527). However, Finch (1956: 168–9) was convinced that landowners began mortgaging heavily before the Civil War and does give some examples.

The matter was contentious—in 1654 parliament passed an ordinance limiting the right to redeem to one year. The ordinance was not re-enacted after the Restoration; instead, the Crown promoted the interests of its suppporters. The mortgaged lands of many of the Royalists were in the hands of their creditors. To aid their recovery, Chancery was allowed to elaborate its doctrine of the equity of redemption. The law, however, remained confused at least into the 1670s. (Thirsk 1954a; Simpson 1961: 227; Clay 1985: 145, 149–51; Melton 1986: 127–8). It was not until the end of the century that mortgages became automatically and indefinitely extendable so long as the mortgagor regularly paid interest. Only then did the mortgage become a routine device for using land to raise long-term finance.

The volume of finance raised by mortgage grew rapidly in the late seventeenth century. Petty (1927: i. 247) estimated *c.*1660 that capital raised with land as security equalled only 2.5 per cent of the value of English real estate. In the mid–1690s Davenant estimated that loans secured on land amounted to £20 million (Habakkuk 1980: 207), which equalled about 8 per cent of the value of English and Welsh property.[25] These numbers, of course, are very imprecise but may indicate the order of magnitude of debt. The tiny volume in 1660 and its increase by the 1690s are not surprising in view of the legal history. But 8 per cent was still a small number. The landed classes had hardly begun to take advantage of the mortgage.

They did so in the eighteenth century. There are no contemporary estimates of the volume of mortgages, but the opinion of contemporaries and historians has been that the gentry and aristocracy heavily encumbered their estates. 'The general consensus seems to be that total aristocratic indebtedness increased as the [eighteenth] century advanced, especially after the 1770's as consumption became ever more conspicuous and family charges accumulated.'[26] What did they do with this money? It scarcely matters. The purpose of mortgaging was to allow consumption to be maintained as copyholds and beneficial leases were run out and as freeholds were purchased. Borrowed money replaced entry fines until the flow of annual rack rentals equalled the income that entry fines had formerly provided. Ultimately, the rack rents exceeded the entry fines due to the economies of large-scale farming, and at that point debt could be retired even as consumption was increased. Conspicuous consumption financed by debt was consistent with prudent financial management in the eighteenth century.

[25] King (1936: 35) had estimated the rental of the country at £12 million per year. Clay (1985: 173) reports that from 1690 to 1703 land sold at 20–22 times its annual value. $0.08 = 20/12 \times 21$.

[26] Cannadine (1977: 628). See also Mingay (1963: 36). Offer (1981: 139) found that agricultural land was mortgaged to 27.4 per cent of its value in 1904–14.

# PART II
## *Enclosure and Productivity Growth*

# 6

# The Adoption of Modern Methods

Nothing is so necessary to a good system of Agriculture as to determine
a ROTATION OF CROPPING as will enable the farmer to raise throughout a
series of years the greatest quantity of produce, whether of corn or
meat, with the greatest return, and at the least expense of money and
time. To obtain these important objects, the general character of the soil
and its situation,—(for climate will have its effects)—its capability for
good drainage, propinquity to a large town, and length of tenure, are
amongst the most needful considerations.

Richard Noverre Bacon, *The Report on the Agriculture of
Norfolk*, 1844: 201 (capitals in original)

THE major argument in favour of the landlords' agricultural revolution is
that it raised farm productivity. In so far as enclosures are concerned, it is
often claimed that they led to the adoption of modern farming methods. In
the sixteenth and seventeenth centuries, it was generally accepted that
enclosure led to the conversion of arable to pasture, a change that was
profitable but that open field farmers rejected. In the eighteenth century, this
observation was broadened into a general denunciation of open field
farming. In a famous passage, Arthur Young contrasted 'the Goths and
Vandals of open fields' with 'the civilization of enclosures' (Young 1813:
35–6). Since then, Agrarian Fundamentalists have condemned open field
farmers for keeping too much land in tillage, for ignoring new crops, for
failing to alternate their land between pasture and arable, for having too few
livestock, for not properly draining—in short, for ignoring all aspects of
capital-intensive agriculture. In Lord Ernle's words: 'That open-field
farmers were impervious to new methods is certain'. (Ernle 1912: 199).

Crop choice, however, has also been the issue on which revisionist
historians (like Havinden, Kerridge, Jones, and Yelling) have defended the
open fields. But while there are undoubted examples of open field villages
adopting new crops, it is far from clear that this modernization was general.
Yelling (1977: 202) concluded that 'the step from acknowledging that

common fields were not totally inflexible to determining the extent of their flexibility in comparison with enclosures is a very great one, and requires an altogether more sophisticated type of evidence, which is largely not forthcoming at present'. Given the limited evidence underlying the revisionist case, this agnosticism is appropriate.

More evidence is needed to judge between the Agrarian Fundamentalists and their revisionist critics. I will pursue the matter for the south midlands. In each of its three natural districts, there was an improvement or package of improvements that raised productivity in the early modern period. I will assess the performance of open and enclosed farmers by comparing how rapidly they adopted that package.

## The Three Natural Districts

The south midlands encompasses three main divisions in which broadly similar farming methods generated the most profit. The three principal divisions were the heavy arable district (401,886 acres), the light arable district (706,681 acres), and the pasture district (1,742,299 acres). Map 2–1 shows their location.

The heavy arable district was mainly located in Cambridgeshire, Huntingdonshire, and northern Bedfordshire, although there was a scattering of parishes elsewhere with the same characteristics. In the main portion of the district, the subsoil is Oxford clay or gault, and the surface soil is boulder clay deposited by glaciers (Chatwin 1961: plate I, pp. 11–24, 59–73). This succession of clays produces a soil for which drainage has always been a serious problem. In his *General View of . . . Bedford*, Batchelor (1808: 10–11) described several varieties of clay, each with unattractive properties. In some districts 'we frequently meet with cold, thin stapled clays, which are so tenacious as to "hold water like a dish"'. Another variety 'is as sticky as glue'. On other sorts of Bedfordshire clays 'both winter and spring corn, which always come up luxuriantly turn yellow and sickly in May, particularly if the season be wet'.

Economic historians are inclined to say that clays were put down to pasture in the eighteenth century (e.g. Turner 1980: 150–1), but that was not true of the heavy arable district, for it supported terrible grass. In 1939, Stapledon and Davies surveyed the grassland potential of England and Wales and ranked pastures in the heavy arable district as among the least productive in the country (W. Davies 1941: 112–21; 1960: 93–110). Two hundred years before, the heavy arable district produced similarly miserable grass. In his *General View of . . . Bedford*, Batchelor (1808: 10–11) observed of some species of clay that 'this soil is also productive of the worst grasses; so coarse, that most part remains untouched by the cattle all winter'. Vancouver (1794: 199) described the same sort of grass in Cambridge: 'very coarse, sharp, sour grass and herbage, vegetating very late in the spring from wet, cold, and compact clays'. Gooch (1813: 184–6) remarked that 'These [pastures] are dispersed chiefly over the upland part of the county on heavy, wet soils; they are miserably poor, and abounding with everything but what they ought . . . These pastures afford a miserable support to various kinds of lean stock'. Since the grassland potential was so limited, the object of enclosure in this district was never to convert arable to pasture. Instead, the land was kept under the plough, but the soil was improved by installing subsurface drains.

Light arable villages were scattered across the south midlands on several kinds of geological formations. Bands of these villages stretched from Cambridgeshire to Oxfordshire on the outcrops of structural beds like the Upper and Lower Greensand, the Portland and Purbeck beds, and the Northampton sand. Young (1813: 12–13) noticed the outcrops in Oxford. 'There is a tract of reddish sand, three miles wide and four or five miles long, in the parishes of Beckley (where it lies on a rock), Eldsfield, Heddington, Shotover, Standon, and Forest-hill, upon which turnips succeed greatly'. The light arable villages along the Nene, the Ise, the Ouse, and the Avon were situated on glacial drift or gravel terraces.

From Atherstone to Stratford-on-Avon, in the direction of Wellesbourn, Tidington, Alveston, Hampton Lucy, and Sherborne, is all fine dry red clay loam, and sandy loam, mostly in tillage. In the strath of the Avon, the soil is equal to that of any county in England. This district is excellently adapted for the turnip husbandry. (Murray 1813: 19)

While the pasture potential of these soils was usually better than the heavy arable district, the land remained in tillage after enclosure. Usually a Norfolk rotation (turnips–barley–clover–wheat) was introduced and the increased production of forage allowed improvements in the breed and management of sheep.

In the pasture district, grazing was the farming activity that maximized the value of the land. Much of the tillage in the district produced winter forage for the livestock. The district is divisible into a lowland and an upland division.

The lowland pasture district includes the clay plain that covers much of Buckinghamshire north of the Chilterns with slight extensions into Northamptonshire and Oxfordshire, the western end of the Vale of White Horse, southwestern Cambridgeshire, a variety of riverside meadow parishes, and some light soil parishes near the fens and on the Lower

Greensand formation. The predominant farm activity was dairying. Fattening of beef cattle was rarely undertaken (Priest 1813: 5, 245–8; Pitt 1809*b*: 199). In 1939 Stapledon and Davies found that the lowland pasture district supported much better grass than the heavy arable district, but not as good grass as the better parts of the upland pasture district where fattening was carried on. The Board of Agriculture reporters noted the same thing. Thus, Priest (1813: 234) judged the Buckinghamshire pastures to be 'numerous, extensive, and though not in general so rich as in some counties, yet very valuable'. They were sufficiently rich that they fully supported dairy cattle—'in Bucks no farmers give any other food to milch cows but hay, except in a very few instances'—but not so strong as to permit profitable fattening: 'although many acres in this Vale [of Aylesbury] are rich, yet there are but very few and very small spots of pasture, where a bullock of 9 stone (8lbs. to the stone) can be grazed upon an acre' (Priest 1813: 6, 8–9, 246).

The upland pasture district included the higher country in Leicestershire, western Rutland, eastern Warwickshire, northern Oxfordshire, and central and western Northamptonshire. Eighteenth-century commentators noted that soil changed frequently: 'The nature of the soil is very liable to vary much in short distances, respecting its strong or friable qualities' (Pitt 1809*a*: 4). Two sorts of soil were usually distinguished: 'The upland surface very generally consists of a grey or brown loam, and sometimes of a snuff-coloured or reddish loam' (Pitt 1809*b*: 6–7).

The type of soil reflects the structural and drift geology of the district. The western edge of the upland pasture district is located on Kemper marl, but most of the district is underlain by Jurassic rocks that dip from northwest to southeast. The rocks are layers of limestone, marlstone, clay, and sand (Hains and Horton 1969: 66–72). The red loam is produced by the decomposition of the ferruginous stratum called Northamptonshire sand. The clay soils are either outcrops of the bedrock layers of clay or else boulder clay deposited by glaciers. The elevation and dip of the underlying rocks, however, prevent drainage from being the problem it was with the clays of the heavy arable district. As Pitt (1809*b*: 12) noted, 'the loose and open under stratum' was 'acting as natural hollow drains to carry off stagnant, and superfluous moisture, and thus putting land naturally in that state which, in many other counties, can only be imperfectly done by the expense of hollow draining'.

Throughout the upland pasture district, grass quality has always been high. The 1939 Stapledon–Davies survey showed that this district includes some of the best pastures in England. In the fifteenth and sixteenth centuries, sheep were the principal animal kept, but by the eighteenth century a more diverse livestock husbandry was practised. Fattening was an important activity, and its practice shows that the pastures in this region were richer than those in the lowland part of the pasture district. Pitt (1809*a*: 154) estimated 'that from 128 to 160 lb. per acre, of beef or mutton, is as much as can be bred and fatted on good pasture land' in Leicester and Northampton.

## The Light Arable District

To determine whether enclosure had any impact on farming methods, I compare the responsiveness of open and enclosed farmers to the new technologies available in the early modern period. I begin with the light arable district where the new methods included some of the most famous innovations of the eighteenth century. In the light arable district the package of improvements included the Norfolk rotation (turnips–barley–clover–wheat), the abandonment of sheepfolding, and improvements in the breed of sheep. These innovations were introduced independently, and were combined into a unified system only in the middle of the eighteenth century.

Turnips and clover were introduced into East Anglia from Holland in the late sixteenth century (Chambers and Mingay 1966: 56, 60). Probate inventories show that the proportion of farmers in Norfolk and Suffolk growing turnips increased from less than 10 per cent in 1680 to over 50 per cent in 1710. At the same time the share of farmers growing clover increased from 10 to 17 per cent. (Overton 1977: 325–6). The imbalance between turnips and clover shows that the Norfolk four-course rotation was not yet the norm. It became so by Young's visit in the late 1760s (Young 1771*b*: ii. 1–73, 83–163).

In the midlands, the order of adoption of the new crops was reversed. Cultivated clover, lucerne, trefoil, ryegrass, and sainfoin were introduced into Oxfordshire in the seventeenth century, but the cultivation of turnips was delayed until the mid-eighteenth. 'The earliest reference to turnips which I have come across in examining thousands of probate inventories for Oxfordshire is in 1727' (Havinden 1961*b*: 77). The adoption of turnips and of the Norfolk four-course rotation occurred during the period of parliamentary enclosures, and so, indeed, may have been linked to it as Agrarian Fundamentalists claim.

Turnips and clover were cultivated in order to provide winter forage for livestock, especially sheep. More winter feed allowed more mutton and wool, but to realize that potential it was necessary to change the management and breed. In the open fields, sheep were usually run as folding flocks. They were grazed on the common during the day and driven to the fallow field at night. They manured that field and, more importantly, ate it

clean of weeds. (Hoeing turnips accomplished the same objective.)
However, all commentators agreed that folding reduced weight gain among
the sheep. 'The sheep of the fallow fields never obtain a sufficiency of food
to improve their condition in any considerable degree, and are employed, as
it were, for no other purpose than to carry the produce of the commons, &c.
upon the arable land in the shape of manure' (Batchelor 1808: 560).
Parkinson (1811: 61, 66, 71, 77) estimated the average sale value of wethers
(castrated male sheep) at 25 shillings for folding flocks but 45 shillings if
sheepfolding was abandoned. For ewes, the corresponding receipts were 30
shillings and 50 shillings. Sheepfolding was alleged to reduce wool
production as well. Lawrence (1809: 270) concluded: 'Folding sheep . . . is
not generally deemed a part of improved husbandry.'

Were enclosed farmers more likely than open farmers to adopt the
Norfolk rotation and to upgrade their sheep? The most helpful sources are
Richard Parkinson's reports to the Board of Agriculture on Rutland and
Huntingdon (Parkinson 1808, 1811). In 1806 he conducted a veritable
agricultural census and reported the results on a parish-by-parish basis; the
details are summarized in Table 6–1. It establishes two important
conclusions. First, enclosed villages practised sheepfolding much less than
open field villages. Of the villages enclosed by act, 74 per cent had
abandoned the practice as had 80 per cent of the anciently enclosed villages.
Only 14 per cent of the open villages had ceased to fold their sheep. Second,
the cultivation of clover and turnips was also linked to enclosure. The most
archaic farms were the open field farms that folded their sheep. The share of
arable that was fallow (23.5 per cent) was higher for these parishes than for
others, and the share of land under turnips and clover (16 per cent) was
lower. Even these villages, however, had come a long way from triennial
fallowing. None the less, the enclosed farms were the most progressive.
They had replaced the fallow with clover and turnips. The innovations in
cropping were most extensive in the villages that had also abandoned
sheepfolding.

It is unlikely that the differential behaviour of open and enclosed farms in
Parkinson's surveys was simply a matter of differences in rates of response
to a new opportunity since similar differences are observable a generation
later. The surveys of farming prepared pursuant to the Tithe Commutation
Act of 1836 allow the comparison of land use patterns among open and
enclosed villages around 1840.[1] Table 6–2 summarizes the cropping
patterns of the thirty-three villages with uniformly light soil that had usable
returns.[2]

[1]  See E. J. Evans (1976) and Kain (1986) for a discussion of this source.
[2]  Since the light arable district was mainly enclosed by act, most of the tithe reports for that
district do not record cropping patterns. All that do have been transcribed.

*Table 6–1. Land use patterns in the light arable district, c. 1806*

| | Open (%) | Enclosed by act (%) | Anciently enclosed (%) |
|---|---|---|---|
| | | Sheepfolding villages | |
| *Major divisions* | | | |
| Arable | 81.31 | 73.82 | 71.72 |
| Meadow | 7.22 | 14.95 | 4.14 |
| Pasture | 10.24 | 11.22 | 10.34 |
| Common | 1.23 | 0.00 | 13.79 |
| *Allocation of the arable* | | | |
| Fallow | 23.50 | 14.00 | 33.00 |
| Wheat | 22.50 | 18.00 | 34.00 |
| Barley | 14.17 | 19.20 | 0.00 |
| Oats | 5.00 | 2.00 | 0.00 |
| Beans/peas | 18.83 | 22.00 | 33.00 |
| Clover | 6.33 | 13.40 | 0.00 |
| Turnips | 9.67 | 11.40 | 0.00 |
| Rape/coleseed | 0.00 | 0.00 | 0.00 |
| *Number of villages* | 6 | 5 | 1 |
| | | Non-sheepfolding villages | |
| *Major divisions* | | | |
| Arable | 77.54 | 75.71 | 83.71 |
| Meadow | 19.46 | 9.62 | 4.77 |
| Pasture | 2.99 | 13.96 | 11.52 |
| Common | 0.00 | 0.71 | 0.00 |
| *Allocation of the arable* | | | |
| Fallow | 0.00 | 8.21 | 8.25 |
| Wheat | 25.00 | 16.14 | 18.50 |
| Barley | 25.00 | 18.29 | 19.00 |
| Oats | 0.00 | 8.29 | 6.00 |
| Beans/peas | 0.00 | 10.43 | 11.25 |
| Clover | 25.00 | 21.71 | 24.75 |
| Turnips | 25.00 | 14.57 | 12.25 |
| Rape/coleseed | 0.00 | 2.36 | 0.00 |
| *Number of villages* | 1 | 14 | 4 |

*Note*: With the exception of the number of villages, all entries are percentages.

*Sources*: Parkinson (1808: 2–4, 11–16, 45–9, 83–93; and 1811: 2–5, 9–15, 103–7, 166–9, 251–7). The arable acreage recorded for each parish was divided among the crops as indicated by the rotation. When different rotations were specified for different soils in a parish, I presumed the arable was divided among soil types in the same proportion as the whole parish. When alternative crops were indicated for a year in a rotation, I presumed each was grown equally. 'Light arable' villages had more than half of their land arable and less than half clay.

The enclosed villages present a consistently modern appearance. The classic Norfolk rotation or a close variant was general. Fallow was rare, and with only one exception, all the villages cultivated turnips. This crop averaged 20 per cent of the arable and fallow only 3 per cent. In contrast, the open villages were very backward. Six of the fifteen open villages grew no turnips at all, and their cropping pattern was archaic. Even among the open villages that did grow turnips, the share of land that was fallow was

*Table 6–2. Allocation of the arable, c. 1840: villages with uniformly light soil*

|  | Enclosed villages (%) | Open villages | |
|  |  | Growing turnips (%) | Not growing turnips (%) |
| --- | --- | --- | --- |
| Fallow | 3.39 | 11.22 | 24.83 |
| Wheat | 22.94 | 21.56 | 29.33 |
| Barley | 21.06 | 22.67 | 16.50 |
| Oats | 5.67 | 9.11 | 4.17 |
| Beans, etc. | 4.11 | 7.22 | 21.83 |
| Clover | 22.89 | 14.33 | 4.17 |
| Turnips | 20.33 | 14.22 | 0.00 |
| Number of villages | 18 | 9 | 6 |

*Source*: Public Record Office IR/18 with the following file numbers:
*Enclosed villages*: 13175, 13107, 13157, 13250, 13253, 13207, 13116, 13173, 7752, 13305, 13504, (two reports), 13604, 13515, 6660, 7907, 7913, 7878.
*Open villages—turnips cultivated*: 7615, 13609, 13603, 13588, 13652, 6504, 6695, 7893, 7901.
*Open villages—no turnips cultivated*: 7654, 7698, 13415, 13000, 13615, 7897.
See Allen (1989) for a list of the village names.

higher, and the share under turnips lower, than in most of the enclosed villages. John Pickering, the Assistant Commissioner for Morcott, Rutland, noted that 'the soil is chiefly useful turnip land', but that 'the system of husbandry is rather peculiar'. Turnips were not cultivated as much as they might have been. 'The rotation is kept up for the sake of adhering to the old open field course of two crops and a fallow, the seeds being substituted for the alternated [turnip] Fallow season' (PRO IR 18/7893). Norfolk rotations did not quickly or easily penetrate the farming routines of open field villages.

Growing turnips and clover allowed farmers in the light arable district to increase their production of livestock products. The gain was realized by improving the stock rather than by increasing the number of animals. Table 6–3 shows that enclosure resulted in modest *reductions* in the numbers of cows, calves, hogs, horses, and foals. Neither open nor enclosed farmers kept any considerable number of beef cattle. Sheep numbers were maintained after enclosure.

The increased livestock output was achieved by abandoning sheepfolding and upgrading the breed of sheep. The Board of Agriculture reporters distinguished three types of sheep in the south midlands: '1. the common-field sheep; 2. the antient-pasture sheep; and, 3. the improved-pasture sheep, by crossing with the new Leicester breed' (Pitt 1809*b*: 203).

The common-field sheep were universally denigrated. According to Pitt (1809*b*: 203), they 'were meant for the fold, and if they would endure that . . . little attention was paid to other circumstances'. T. Stone (1794: 31)

*Table 6–3. Livestock densities, light arable district, c.1806: number of animals per 100 acres of farm land*

|              | Open  | Enclosed by act | Anciently enclosed |
|--------------|-------|------|------|
| Cows         | 4.74  | 3.24 | 1.36 |
| Calves       | 2.33  | 1.31 | 0.87 |
| Hogs         | 11.28 | 7.09 | 4.34 |
| Store beef   | 0.95  | 0.23 | 0.00 |
| Fatting beef | 0.00  | 0.24 | 0.36 |
| Sucklers     | 0.00  | 0.03 | 0.00 |
| Horses       | 4.49  | 2.86 | 2.34 |
| Foals        | 0.72  | 0.38 | 0.32 |
| Sheep        | 54.35 | 50.71 | 50.34 |
| Lambs        | 25.94 | 21.32 | 22.83 |

Note: Farm land means the total of arable, meadow, pasture, and common.

claimed they were 'a very unprofitable quality'. 'The provision intended for their maintenance, is generally unwholesome and scanty.' These flocks suffered high mortality from the rot and were being continuously replenished by purchases from itinerant jobbers. 'The healthiness of the animal when purchased, is the first and almost the only object of consideration with the farmers' (T. Stone 1794: 31). Hence, the designation of these folding sheep as 'mixed' breed. Parkinson (1811: 245–6) called them 'dunks'.

In the important grazing districts in the south midlands, most of the sheep kept in enclosures in the early modern period were known as Leicesters. They derived from, and were very similar to, the Lincoln (Trow-Smith 1959: 59). Defoe (1724–6: 408–9) was impressed by these sheep: 'The sheep bred in this county [Leicester] and Lincolnshire, which joins to it, are, without comparison, the largest, and bear not only the greatest weight of flesh on their bones but also the greatest fleeces of wool on their backs of any sheep of England.'

The New Leicester was the third sort of sheep found in the south midlands. It was developed in the middle of the eighteenth century by the renowned stockbreeder Robert Bakewell through crossing the Lincoln with the Ryeland (Trow-Smith 1959: 61–2). His aim was to tilt the genetic balance from wool to mutton. The main advantage of the New Leicester was that it attained its full weight in two years rather than the three taken by the Old Leicester. 'The New Leicester . . . possess an aptitude to become fat at an early age' (Batchelor 1808: 557). Wool quality and volume, however, declined. 'The reason assigned for not liking the intire breed is, that it does not produce so much wool as the old Leicester sort' (Crutchley 1794: 14).

The mutton of the New Leicester also tended to be excessively fatty, the sheep folded poorly and had low lambing rates, and the ewes produced little milk for the lambs (Batchelor 1808: 557–8).

Bakewell did not sell New Leicesters but hired rams for stud. In the 1780s and 1790s enormous prices were paid by graziers anxious to breed their own rams for hire. Bakewell once let his ram Two Pounder for 800 guineas for a season.

Such high prices are given only by tup [stud] breeders. This profitable branch of husbandry has so much increased, since the days of Bakewell, that it is averred, there are now not less than ten thousand farmers, in the Midland counties, each of whome, either lets or hires a tup for the season, at ten pounds. (Lawrence 1809: 331)

That was the sort of price an ordinary farmer paid to service his ewes. Through the seasonal hiring of rams, the genes of the New Leicester were propagated away from Bakewell's farm at Dishley.

Some farmers kept New Leicester flocks, but others sought to realize the breed's rapid weight gain while avoiding its disadvantages through crossing it with Leicesters, Lincolns, and other kinds of sheep. 'The New Leicester breed has been introduced into many parishes [in Bedfordshire], and various crosses of this breed with the Wiltshires, and other kinds, have found their way even into the open fields' (Batchelor 1808: 537).

Some people are very partial to this breed [the New Leicester]; and all to having some cross of it in their stock . . . the superiority of it in form of carcase, and inclination to fatten, is universally allowed; and as a proof of the estimation in which it is held, it is certain that all the ram merchants who profess dealing in the old Leicestershire breed, have had a cross of this breed. (Crutchley 1794: 14)

'Specimens of the pure original breed of Midland long woolled polled sheep yet remain in Leicester, Northampton, and Warwickshire . . . For the most part, however, the sheep of those counties have a mixture, more or less, of Dishley blood' (Lawrence 1809: 332). As a result, 'the Lincolns are now so generally improved by new Leicester tups, that they are, probably, in a great measure free from those defects of the old breed' (Lawrence 1809: 325). In his summary table of breed characteristics, Lawrence (1809: 324) showed Lincolns being slaughtered at two years—the same age as the New Leicester and younger than the slaughtering age fifty years earlier—and yielding more mutton (25 lb. per quarter versus 22) and more wool per fleece (11 lb. versus 8). Crossing the New Leicester back on the old had produced a better sheep than the New Leicester itself.

Enclosed farmers in the light arable district were much more likely than open farmers to have improved breeds. Parkinson's census shows that, out of eight open field villages, one had New Leicesters, two had Lincolns and Old Leicesters, and five had mixed flocks. In contrast, out of nineteen villages enclosed by act, five had New Leicesters, nine had Lincolns and Old Leicesters, and five had mixed flocks. The improved breeds were kept in 74 per cent of the enclosed villages but only 38 per cent of the open villages.

The evidence of Parkinson's census and the Tithe Commutation surveys support an adverse assessment of open field farming. While Ernle went too far in saying that open field farmers were 'impervious' to new methods, they did not adopt them to nearly the degree that enclosed farmers did. Moreover, open field farmers rarely adopted the whole 'package' of improvements, so they missed the advantages of the modern system.

## The Heavy Arable District

Agriculture in the heavy arable district was severely hampered by poor drainage. Consequently, hollow draining—not new crops—was the essential technique in the 'package' of improvements in this district.

While open field farmers in the light arable district could be condemned for inflexibility in cropping, the charge is pointless in the heavy arable district where enclosure led to only minor changes. Table 6–4 summarizes the results of Parkinson's census. For both open and enclosed villages, about 80 per cent of the land was arable. There was scarcely any common, even in open villages. This commitment to corn production is understandable in view of the miserable quality of grass the district supports. Cropping looks old-fashioned. The usual cropping pattern was fallow–wheat–spring crops, predominantly beans. A bit less than a third of the arable was fallow. Its purpose was the destruction of weeds by repeated ploughing—'the farmers that fallow their strong land every third season with proper attention, have frequently very few weeds to contend with' (Batchelor 1808: 328–337).

The only sign of more advanced husbandry was a modest cultivation of clover, to be found mostly in the enclosed villages. The crop was not particularly successful, however.

On cold clays which prevail in the north of the county [of Bedford], and several other places; the farmers say it [clover] frequently fails, though never sown before, and the woodland clays have frequently a thin plant in consequence of its being thrown out of the ground by the winter frosts. (Batchelor 1808: 427–8)

The poor performance of clover was probably related to the generally poor quality of grassland in the district.

Just as enclosure did not lead to much increase in grass, so it did not lead to more livestock. (See Table 6–5.) There were virtually no fatting beasts,

*Table 6–4. Land use patterns in the heavy arable district, c. 1806*

| | Open (%) | Enclosed by act (%) | Anciently enclosed (%) | Partly open (%) |
|---|---|---|---|---|
| *Major divisions* | | | | |
| Arable | 83.67 | 79.98 | 76.01 | 71.27 |
| Meadow | 7.16 | 8.21 | 8.87 | 14.47 |
| Pasture | 7.00 | 11.81 | 7.62 | 13.93 |
| Common | 2.17 | 0.00 | 7.50 | 0.33 |
| *Allocation of the arable* | | | | |
| Fallow | 31.00 | 25.48 | 23.50 | 30.40 |
| Wheat | 25.52 | 24.03 | 22.50 | 23.60 |
| Barley | 9.40 | 14.10 | 6.75 | 7.00 |
| Oats | 9.08 | 2.14 | 10.50 | 14.20 |
| Beans/peas | 23.04 | 19.17 | 15.50 | 20.80 |
| Clover | 1.04 | 12.21 | 15.00 | 4.00 |
| Turnips | 0.92 | 2.86 | 0.00 | 0.00 |
| Rape/Coleseed | 0.00 | 0.00 | 6.25 | 0.00 |
| *Number of villages* | 25 | 29 | 4 | 5 |

*Note*: With the exception of the number of villages, all entries are percentages.

*Source*: As for Table 6–1. 'Heavy arable parishes' were those with at least half of the land arable and at least half of the land clay.

*Table 6–5. Livestock densities, heavy arable district, c.1806: number of animals per 100 acres of farm land.*

| | Open | Enclosed by act | Anciently enclosed | Partly open |
|---|---|---|---|---|
| Cows | 3.67 | 2.35 | 2.44 | 2.65 |
| Calves | 1.28 | 1.14 | 1.86 | 1.52 |
| Hogs | 8.32 | 6.74 | 3.71 | 8.89 |
| Store beef | 0.00 | 0.19 | 0.00 | 0.00 |
| Fatting beef | 0.00 | 0.36 | 0.00 | 0.00 |
| Sucklers | 0.24 | 0.00 | 0.00 | 0.00 |
| Horses | 3.78 | 2.46 | 2.14 | 3.17 |
| Foals | 0.33 | 0.35 | 0.52 | 0.46 |
| Sheep | 58.53 | 49.74 | 35.88 | 83.19 |
| Lambs | 25.42 | 21.63 | 15.59 | 36.15 |

*Note*: Farm land means the total of arable, meadow, pasture, and common.

store cattle, or sucklers (for veal) in either open or enclosed villages. A typical farm of 200 acres had the half dozen horses necessary for cultivation and a dairy herd of half a dozen cows and a couple of calves. Pigs were reared on the skim milk from the dairy. There was also a flock of a hundred sheep. If anything, enclosure lowered livestock densities.

Enclosure only slightly improved the management of sheep. In open fields the flocks were usually folded—according to Parkinson's census twenty-one of the twenty-five open field villages in the heavy arable district did that. The practice was curtailed in enclosed villages, but still twenty of the thirty-two enclosed by act practised sheepfolding. Furthermore, the improved breeds made little progress. According to Parkinson's census, there were no New Leicester flocks in the district. In only 39 per cent of the villages were Lincolns or Old Leicesters kept, and in the remaining 61 per cent of villages the flocks were 'mixed'. Enclosed villages were more likely than open villages to have Lincolns and Leicesters. All breeds of sheep were equally likely to have been folded.

Enclosure in this district would have been pointless if it had only led to a bit more clover and a bit less folding. The big benefit of enclosure was that it allowed hollow draining. The open fields were usually drained by the furrows between the strips. This system was unsatisfactory; much good soil was eroded away and the furrows were often full of water 'so that not more than half the land is worth occupying, viz. from the ridges half way down to the furrow. The remainder is so chilled with water, and so robbed of manure, from its being washed into the furrows, that the grain produced thereon is of little or no value' (T. Stone 1785: 42–3).

Since the furrows formed boundaries, the responsibility of scouring them fell equally on the owners of the adjoining strips. Co-ordination was difficult, and the furrows were not always cleaned. With the furrows useless, there was little incentive to scour the main drains into which they emptied. Hence they too were clogged. Blocked main drains further reduced the incentive to scour the furrows. These reinforcing external diseconomies meant that the open fields were badly drained. When Young (1771*b*: i. 27–8) crossed the open fields on the heavy clay near Buckingham, he found them still wet in midsummer. 'Let me observe that the furrows were under water: if so in June, what must they be in winter?'

Better drainage was not achieved by improved management of the ridge and furrow system but by the adoption of a new technology—hollow draining. Hollow drains were originally dug in the furrows. The procedure was to plough them as deeply as possible and then dig down a further foot with a spade. Stones, blackberry bushes, sedges, or straw were placed in the trench; a layer of straw was added; and the ditch was filled with dirt. Sometimes furrow drains emptied directly into open ditches at the side of the field. In that case, however, each drain was quite long and there was considerable danger of a blockage somewhere that would flood a great length of furrow. This problem was alleviated by digging deeper and more substantial hollow drains across the field at right angles to the furrows to act as intermediate drains which conveyed the water to the open ditches on the side of the field (Pusey 1843–4). Systematic hollow draining was probably developed in East Anglia in the eighteenth century (Darby 1964; 1976: 37).

Observers around 1800 were impressed by the improvement effected by hollow draining. John Foster (1807: 19–20) of Brickhill recounted:

The benefit is almost incalculable in every respect. Land which before used to bake and crust at the surface, in case of heavy rains succeeding its being worked fine, by which many crops have been lost, is gradually cured of this defect, by becoming more porous. The furrows, in which before, the crop was weakened or destroyed to the breadth of four or five feet, now even exceed the ridges in produce. Where before, the general wetness rendered the ground unfit for any tool to work, or the drought made it too hard to be impressed, now the land can be worked in almost any kind of weather.

Vancouver (1795: 139) believed 'There is no improvement to which the heavy land husbandry of this county [Essex] owes so much as to the fortunate introduction, and continuance of the practice of hollow-draining'.

The installation of hollow draining usually followed enclosure. Vancouver's *General View of . . . Cambridge* (1794) permits the comparison of draining systems in open and enclosed villages. The tally for forty-four villages containing heavy clay is shown in Table 6–6. Almost three-quarters of the open field villages had bad drainage whereas only one fifth of the enclosed villages were in such straits. Fully half of the enclosed villages were hollow drained compared to only 20 per cent of the open villages. In the decade after Vancouver wrote, many heavy arable villages were enclosed, and observers noted that they were then usually drained. Lord Hardwick enclosed Whaddon, Cambridgeshire, about 1800 without an act. He constructed a system of open main drains 'for the tenants to direct their hollow ones into . . . They have not neglected the use of them, for the number of drain-mouths, which I remarked as I examined the lordship, show how much they have exerted themselves in this most capital step to all following good management' (Young 1804*b*: 486). In Bedfordshire, Foster (1807: 48) observed that 'The benefits of inclosure are in some parishes already conspicuous, and are everywhere in preparation . . . Great things

*Table 6–6. Propensity to drain*

| Drainage | Enclosure status | |
|---|---|---|
| | Open | Enclosed |
| None | 25 | 2 |
| A little | 5 | 3 |
| Good | 4 | 5 |

Note: The table shows the number of villages in each case.

have been done and are now doing, in drying the land by deep ditches, by under-furrow, and plow draining'. Young (1804*b*: 486, 492, 502), Gooch (1813: 242–5), Foster (1807: 19), and Batchelor (1808: 475) discussed seventeen villages which were hollow drained and all but one were wholly or mainly enclosed.

Enclosure was usually a prerequisite for hollow draining since the externalities that inhibited the regular scouring of furrows in the open fields also checked the installation of hollow drains. The full dangers of relying on the open field community to solve these problems were shown in Eversden:

The obstinacy of some of the farmers in this parish has defeated the very laudable and spirited exertions, of a very industrious and intelligent young man, by stopping the passage of the water in the leading drains, into which his hollow drains in the open field discharge their water. His drains in consequence have blown up, and a considerable expense has been incurred to produce only a mortifying disappointment. (Vancouver 1794: 99)

Parliamentary enclosure, in particular, was a good device for solving these co-ordination problems. Enclosure awards in the heavy arable district often empowered a group of landowners to maintain the main drains and finance the cost by levying a rate on all the property owners in the parish. The enclosure award for Houghton and Wyton, Huntingdonshire, empowered 'the five Proprietors possessing the largest Quantity of Land . . . or any three of them' to construct and maintain 'Publick Brooks, Ditches, Drains, Watercourses, Banks, Cloughs, Watergates, and Bridges'. If any proprietor failed to pay his share of the costs within fourteen days of being notified, the proprietors who incurred the expenses were empowered to seize and sell his chattels and retain proceeds equal to his share of the costs plus the costs of levying the distress 'in the same manner as Landlords are by Law authorized and Impowered to recover rents in arrear' (Huntingdon RO, Houghton and Wyton enclosure award, 335).

## The Pasture District

In the pasture district, the conversion of arable to grass was the modernization 'package' that increased efficiency. The nature of the changes is shown in Table 6–7, which contrasts land use in 1806 in the arable districts with various subdivisions of the pasture district. In the arable districts, only 21 per cent of the farm land was pasture, meadow, or common. Throughout the pasture district (virtually all of which was enclosed), the percentage of grass was far higher: about two-thirds in the lowland pasture district and close to 90 per cent in the upland pasture district.

*Table 6–7. Land use, Huntingdon and Rutland, 1806 (acres)*

|  | Heavy arable | Light arable | Pasture district | | | |
|---|---|---|---|---|---|---|
|  |  |  | Lowland clay | Lowland light | Upland clay | Upland light |
| Arable | 1,331 | 1,295 | 495 | 462 | 140 | 237 |
| Pasture | 204 | 179 | 818 | 684 | 951 | 970 |
| Meadow | 135 | 145 | 280 | 225 | 220 | 256 |
| Common | 6 | 23 | 28 | 0 | 0 | 0 |
| Total farm land | 1,676 | 1,642 | 1,621 | 1,371 | 1,311 | 1,463 |
| *Allocation of arable* |  |  |  |  |  |  |
| Fallow | 370 | 121 | 109 | 86 | 32 | 6 |
| Wheat | 336 | 220 | 102 | 115 | 33 | 43 |
| Barley | 169 | 248 | 84 | 100 | 20 | 44 |
| Oats | 68 | 60 | 20 | 0 | 16 | 19 |
| Beans | 257 | 190 | 75 | 69 | 22 | 14 |
| Clover | 108 | 258 | 80 | 38 | 5 | 63 |
| Turnips | 23 | 198 | 13 | 54 | 12 | 48 |
| Rape/coleseed | 0 | 0 | 12 | 0 | 0 | 0 |

In the pasture district, crop choice depended on soil type. In the clay divisions, cropping was similar to the heavy arable district and probably little different from what it had been a century earlier. A fifth of the arable was still fallow and the rest was planted with corn and beans. In contrast, there had been extensive modernization in the light soil parts of the pasture district. Clover and turnips were widely cultivated. On the light soils in the uplands, the fallow had been virtually eliminated. Something like a Norfolk rotation was the norm.

The greater share of grass in the pasture district meant that more animals were carried. Table 6–8 shows the stocking densities implied by the Parkinson census. Fewer cows, calves, hogs, horses, and foals were kept on these farms than in open field farms in the arable districts. (Cf. Tables 6–3 and 6–5.) Flocks were larger. The biggest difference, however, was in beef cattle. Hardly any rearing or fattening of beef was carried on in the arable districts, but such activity was characteristic of the enclosed villages whose agriculture is summarized in Table 6–7. This activity was most extensive in Rutland in the old enclosures.[3]

Not only were there more sheep and beef cattle per acre in the pasture district, but their management was superior. Generally, sheepfolding was abandoned. '*Folding,* of sheep is not practised in Leicestershire, or at least

---

[3] It should be noted that Table 6–8 gives only a partial view of livestock husbandry in the pasture district since it is limited to villages in Rutland and Hunts. Thus, it contains few light pasture villages and no villages from the really productive lowland dairy regions. A census covering such places would show many more cows.

Table 6–8. *Livestock densities in the pasture district (animals per 100 acres of farm land)*

|  | Huntingdon | | Rutland | |
|---|---|---|---|---|
|  | Parliamentary | Non-parliamentary | Parliamentary | Non-parliamentary |
| Cows | 1.71 | 2.55 | 3.47 | 2.67 |
| Calves | 1.05 | 1.79 | 2.31 | 0.94 |
| Hogs | 3.93 | 5.39 | 4.87 | 2.58 |
| Store beef | 0.00 | 1.72 | 0.00 | 1.40 |
| Fatting beef | 3.43 | 0.40 | 2.16 | 9.77 |
| Sucklers | 0.00 | 0.03 | 0.00 | 1.10 |
| Horses | 2.02 | 2.04 | 2.48 | 1.34 |
| Foals | 0.13 | 0.28 | 0.33 | 0.36 |
| Sheep | 71.96 | 56.54 | 78.49 | 97.37 |
| Lambs | 31.27 | 24.57 | 34.63 | 34.88 |

Note: Farm land means the total of arable, meadow, pasture, and common.

but little, except in the few remaining common fields' (Pitt 1809a: 269). Moreover, the breeds had been modernized. Of twenty-nine enclosed pastoral villages in Rutland, six had New Leicester sheep, fifteen had Lincolns and Old Leicesters, and only eight had 'mixed' flocks. In Huntingdon, only six of the twenty-two enclosed pastoral villages had 'mixed' flocks. The rest had Lincolns and Old Leicesters except for one village with New Leicesters.

While there were many variations in execution, the value of land in the pasture district was maximized by keeping all or most of it under grass. Did open field farmers respond to this opportunity by converting tillage to pasture? We know that some strips were put down to permanent pasture in the open fields (Yelling 1977: 148–53). Stock could be tethered on them when the field was under crops. When it was fallow, the village herd could be depastured on it, and the grass strips or leys provided nutrition far better than the weeds available on the fallows of the arable districts. Thus, in the pasture district, open field farmers practised a more livestock-intensive husbandry than did farmers in the arable districts.

The question is whether the open field villages had gone most of the way towards a pastoral economy or whether they had made only token concessions. A table of the acreages of grass and arable under the two agrarian systems would settle the matter. Estate surveys provide a start. Table 6–9 is derived from descriptions of 157 enclosed farms and 52 open farms in the pasture district. The enclosed farms were 73 per cent grass and 27 per cent arable.[4] In contrast, open field farms averaged 84 per cent

[4] This division presumes that 78 per cent of the common field land on these farms was arable.

*Table 6–9. Land use patterns from estate surveys, pasture district (percentages)*

|                 | Enclosed (%) | Open (%) |
|-----------------|:------------:|:--------:|
| Enclosed arable | 27           | 1        |
| Enclosed pasture| 62           | 8        |
| Enclosed meadow | 9            | 2        |
| Common field    | 2            | 84       |
| Open meadow     | 0            | 5        |

*Note*: The table is based on estate surveys of 52 predominantly open and 157 predominantly enclosed farms in the pasture district. Buildings and farm yards, always a tiny fraction of the total land, are generally tallied as enclosed pasture.

*Sources: Bedford Record Office*: R Box 792.
*Bodleian Library, Oxford*: MS Top Berks e 21, MS DD Harcourt b. 37.
*Huntingdon Record Office*: C4/2/5/5a.
*Northampton Record Office*: D(CA) 213, D(CA) 215, D(CA) 444, G. 1654, G. 3898, G. 3916, Brudenell ASR 95, Brudenell ASR 96, C(A) 5739.

common field and 16 per cent grass. Table 6–9 suggests that land use in open field villages was radically different from that of enclosed farms.

There are two difficulties in the way of this inference. First, Table 6–9 excludes the acreage of commons and, to that extent, undercounts the pasture of open field farms. Descriptions of four open villages and maps of sixteen others were examined in order to determine the share of the land that was common pasture.[5] Many villages had no commons, and, overall, common pasture amounted to only 6 per cent of the land of these villages. This was considerably more than open field villages in the arable districts but was still not a large proportion.

Second, and much more important, Table 6–9 does not indicate the balance of arable and grass within the open fields. The glebe terriers that list the possessions of the parish church provide a means of exploring this issue. For many open villages in Northamptonshire, Rutland, and Leicestershire in the late seventeenth and early eighteenth centuries, the terriers distinguish

[5] All information derives from the 18th or early 19th cent. Villages with more than 25 per cent of the land enclosed were not included since the possession of much enclosed pasture might have led to unusually small amounts of common.
Maps in the Northants Record Office included those of Denford (map 5156), Dusten (map 583), Desborough (map 4642), Broughton (map 3576), Rushden (map 5440), Wicken (map 3145), Wollaston (map 4447), Overstone (map 564), Yardley Hastings (map 4155–6). Maps in the Warwick Record Office included those of Cherrington (Z83(u)), Great Wolford (Z183L), Little Wolford (Z183L), Thurlaston (Z8(u)), Upper Eatington (CR224/117/1,2,3). Printed sources included Rothwell, Northants (Pitt 1809b: 64–71), Towersey, Marsh Gibbon, Cheddington, Bucks. (Priest 1813: 367–72), Woodford, Northants (Humphries and Humphries 1985: 34–5), Naseby, Northants (Mastin 1792: 24).

between arable and grass strips in the fields. For thirty-nine parishes, the overall distribution was 78 per cent arable and 22 per cent grass.[6]

By combining these percentages with the distribution of open field land shown in Table 6–9, I compute the land use pattern of open field villages. Six per cent was commons. On the presumption that the grass in the fields was pasture rather than meadow, 63 per cent of the farmland was arable, 7 per cent meadow, and 24 per cent pasture. Open field farmers in the pasture district pursued a much more grass-intensive agriculture than their counterparts in the arable districts but still devoted much less of their land to pasture than did enclosed farmers.

## Convertible Husbandry

Agrarian Fundamentalists conjure up before us great herds of prize animals fertilizing corn lands and raising yields. Convertible husbandry—the systematic alternation of land between pasture and tillage—is often suggested as the management scheme. The inflexibility of land use in the open fields ruled out this alternation. That is a major indictment of open fields in most accounts of the agricultural revolution (e.g. Chambers and Mingay 1966: 52).

Despite the enthusiasm of historians, convertible husbandry was not practised in the arable districts of the south midlands. There was little grass in those districts even after enclosure, so the scope for alternating land between tillage and pasture was tiny. And there is no indication that farmers in the arable districts did shift land between arable and grass. For instance, Batchelor (1808: 449) reported that 'Laying Land to Grass', 'has not been practised to any extent in Bedfordshire', which was mainly an arable county. Likewise, Young (1813: 8) remarked of the open fields at Milton, Oxfordshire, in the light arable district:

Milton Field is one of the finest soils I have met with in the country: a dry, sound, friable loam on gravel—convertible land, as they call it in Oxfordshire; but I know

---

[6] Glebe terriers for the following places were used to determine the proportions of arable and grass in the common fields. Terriers used were all from the late 17th or 18th centuries and applied to villages in the pasture district enclosed in the 18th and 19th centuries with no more than 25 per cent of the land enclosed prior to the final enclosure. The Rutland and Northants terriers were in the Northampton Record Office and the Leics. terriers in the Leicester Record Office. *Rutland*: Whissendine, Normanton, Braunston, Wing. *Northants*: Maidford, Stoke Albany, Ashley, Green's Norton, Barby, Kilsby, Whilton, Heyford, Harleston, Staverton, Blatherwick, Wilbarston, Yelvertoft, Scaldwell, Tansor, Thrapstone, Heldon, Wappenham, Tiffields, Aldwinkle All Saints, Aldwinkle St Peter, Wollaston, Hannington. *Leics.*: Harston, Braunston, Ashfordby, Carlton, Harby, Bottesford, Ab Kettleby, Scalford, Somersby, Alathern, Redmile.

not why, unless that it is generally good for everything: it would not be converted to grass, were it enclosed.

The pasture district was the only part of the south midlands where convertible husbandry was practised, and even there only in some enclosures.

There is dispute as to when convertible husbandry was widespread. There is no evidence for its practice on a substantial scale before the sixteenth century—the first wave of enclosures was to convert open fields to permanent pasture mainly for sheep (Wordie 1983: 492). Dr Kerridge, in many respects a maverick among agricultural historians, none the less regards convertible husbandry as the fundamental innovation in English agriculture. 'The backbone of the agricultural revolution was the conversion of permanent tillage and permanent grassland . . . to permanently cultivated arable alternating between temporary tillage and temporary grass leys' (Kerridge 1967: 181). He has traced its spread to the first half of the seventeenth century:

> In the early sixteenth century, up-and-down husbandry was confined to the north-west and to a few farms elsewhere. It expanded and spread rapidly after 1560 and fastest between about 1590 and 1660, by which time it had conquered production and ousted the system of permanency from half the farmland. In this expansion, up-and-down husbandry progressed at the expense of permanent tillage, mostly in common fields, and of permanent grass, partly in common fields, partly in enclosures. Much permanent grass was converted to up-and-down land between 1590 and 1660. (Kerridge 1967: 194)

According to Kerridge, convertible husbandry remained the usual farming system in the midlands into the nineteenth century.

Recently, Broad (1980: 77–89) has reassessed the significance of convertible husbandry at the end of the eighteenth century. He observed that the region I have called the pasture district 'had evolved a system of specialized livestock farming on improved permanent pasture in preference to alternate husbandry' (p. 79). This judgement is founded in part on the descriptions of the Board of Agriculture reporters. They make three observations about the incidence of permanent pasture and convertible husbandry. First, permanent pasture was the norm on the clay soils of the upland pasture district. Thus Pitt remarked that 'much of the greater proportion of the land of this county [Northamptonshire] is of a strong heavy staple, applied to the culture of beans and wheat before enclosure; and, when enclosed, generally laid down to permanent pasture' (Pitt 1809*b*: 9). This generalization is borne out by Parkinson's census. Second, Priest, Vancouver, and Mavor all remark that the dairy regions of Cambridgeshire,

Buckinghamshire, and the Vale of White Horse were largely permanent pasture. Third, the light soils in the pasture district were the only regions likely to be farmed by convertible husbandry. Pitt, for instance, noted that there was much red loam between Rothwell and Northampton. 'This soil is pretty extensive, including a portion of the common fields as well as enclosures; in which the latter state it is generally kept in tillage in the up and down system' (Pitt 1809*b*: 13). Thus, in the early nineteenth century, convertible husbandry was practised principally on lighter soils in Northamptonshire and Leicestershire.

While Broad downgraded the importance of convertible husbandry *c.*1800, he was willing to concede its importance in earlier centuries. 'Up until about 1650 Dr Kerridge is probably right in stressing the role of up-and-down husbandry' (Broad 1980: 77). Here he gives up too much, for there are several reasons to doubt that convertible husbandry was ever practised on the scale that Kerridge believed. First, the sources are susceptible to interpretations other than Kerridge's. Thus, in spite of his own admonitions that a presentment of enclosure does not prove that an enclosure occurred (Kerridge 1955), he accepts at face value the pleas of accused depopulators that they were 'really' practising convertible husbandry and not laying land down to permanent pasture (Kerridge 1967: 190–1). Much of his case rests on interpreting documents to mean other than what they say—in particular, surveys (pp. 181–3) and leases:

For enclosed farms, however, it was usual to forbid ploughing up 'pastures' under a penalty of £5 a year per acre. These penalties were not intended to prevent ploughing up: on the contrary, landlords were solicitous to persuade their tenants to do just this. The purpose of covenants was to ensure the landlord an increased rent and prevent over-ploughing.

The interpretation is certainly wrong: £5 per acre exceeded the gross value of agricultural output.[7] The fine was prohibitive and vastly more than the increased rental value of the land.[8] If Kerridge is right that a penalty of £5 per acre per year was usually imposed for ploughing up pasture on enclosed farms, then one can be sure that few farmers were doing it.

Second, the system of farming that Kerridge described was such an unsatisfactory solution to the technical problems involved in alternating

---

[7] See Tables 7–5 to 7–11 for the gross value per acre of agricultural output *c.*1806 and Tables 11–1 and 11–2 for comparable figures for *c.*1770. These tables show that the gross value of agricultural output was less than £5 per acre. The agricultural price level in *c.*1770 and in *c.*1806 was higher than in the period Kerridge was discussing, so the value of agricultural output was even lower then.

[8] See Table 9–1 for open and enclosed rents per acre covering the period Kerridge was discussing.

arable and grass that it is hard to believe that the alternation had the beneficial effects he claimed for it. There was never any insurmountable problem in ploughing up pasture, but there was a difficult problem in laying land down to grass. How did one get grass growing on the former arable? The modern solution, and the solution followed since the late seventeenth century (Mingay 1984: v. part I, p. 96), is to sow grass seed. That was not an alternative before the Restoration (Kerridge 1967: 199) since there was as yet no market for grass seed in England. 'Natural regeneration', that is letting the wind carry seed from adjoining grassland, was unsatisfactory since it 'took between seven and twelve years or more years to produce good turf' (Kerridge 1967: 195–6) and during that time the farmer was earning little or no income from the land. Kerridge's proposal, which was Blith's recommendation in *The English Improver Improved* (1652), was to plough the land very little, and particularly not in summer, so that the grass plants were never killed.

> The secret was to preserve the grass 'turf' in the soil throughout. It was for this that the turf was simply inverted for the first crop of oats, which thus grew in the very roots of the grass. It was for this that cultivations were restricted to the minimum and summer fallow stirrings eschewed. The new ley of natural grass was raised on the ruins of the old, for radical particles of the ploughed-up grasses were still there when the land came to be laid down again. (Kerridge 1967: 200)

While this scheme no doubt accelerated the reappearance of pasture, it would have been a disaster for the corn. The natural grasses growing in the field may have been nutritious for livestock, but from the point of view of the corn they were competitors for light, moisture, and nutrients like any weeds. Blith's prescription for alternate husbandry was a recipe for poor farming that could only have hurt corn yields.

It is likely that in the first half of the seventeenth century some farmers were experimenting with convertible systems, and writers like Blith were groping for a way to regenerate pasture quickly. The solution was not found until the second half of the seventeenth century, however, when grass seed was widely marketed. Convertible husbandry probably diffused during the eighteenth century on the lighter lands in enclosed villages in the pasture district. Otherwise, it was not an important feature of midland farming.

## Conclusion

The propensity of open field farmers to adopt new methods has been disputed by Agrarian Fundamentalists and their revisionist critics. In all natural districts there are examples of open field villages shifting to the advanced techniques—planting clover and turnips in the light arable district, digging drains in the heavy arable district, laying arable down to grass in the pasture district. Hence, Lord Ernle went too far in describing open field farmers as 'impervious to new methods'. Nevertheless, open field farmers rarely adopted the new methods to the same extent as enclosed farmers. This backwardness was true of all the advances individually, and was even more pronounced when the modern methods comprised a system. Thus, while some open field farmers in the light arable district grew turnips or clover or gave up sheepfolding or bought New Leicesters, hardly any did them all. The open field farmers missed most of the gains to improvement.

# 7

# Yields and Output

I particularly put the question in the county [Northampton]: Is not more grain raised in open fields than upon the same land after enclosure?

W. Pitt, *General View of the Agriculture of the County of Northampton*, 1809: 62

ENCLOSURE did accelerate the adoption of new crops, livestock, and draining. This is one of the principal props supporting Agrarian Fundamentalism. But this evidence is insufficient to establish that enclosure was an important cause of productivity growth in English agriculture. The main manifestations of that growth were a doubling of yields and labour productivity; the productivity of livestock production also increased. Changes in methods were important in so far as they contributed to these advances. They are also the yardstick for judging enclosure.

This chapter begins with corn yields and addresses the following questions: How much did yields rise between the middle ages and the nineteenth century? How much of that increase was due to enclosure? Did the techniques of capital-intensive agriculture—the new crops, fertilizing, draining, convertible husbandry—raise yields? I show that enclosure and capital-intensive agriculture were far less important in explaining productivity growth than the histories of turnips and land use suggest.

Perhaps enclosure made some other contribution to boosting output besides raising yields. I investigate that possibility by computing the changes in farm output that followed enclosure. I combine my estimates of the impact of enclosure on yields with the estimates I made in the last chapter of the effect of enclosure on the acreage of corn to compute changes in the total production of wheat, barley, oats, and beans. I also examine the implications of enclosure for the output of livestock products, taking into account changes in animal numbers as well as improvements in breeds. In a few cases, Tory optimism is vindicated, but more frequently enclosure gave no boost to output and may, indeed, have resulted in a decline.

## The Growth of Yields in England

The broad outline of yield growth is clear. The accounts of demesnes in southern England show that wheat yielded about 10 bushels per acre between the thirteenth and the fifteenth centuries. There was, of course, variation from manor to manor and over time, but only in northeastern Norfolk were yields systematically higher at about 20 bushels per acre (B. Campbell 1987). Norfolk's exceptional performance ended with the labour shortages following the Black Death and yields declined to more typical medieval levels. Barley and bean yields were similar to wheat yields; oats produced about 15 bushels per acre.

There is not much doubt about yields in the late eighteenth and early nineteenth centuries. The principal sources are Arthur Young's *Tours* of England in the 1760s, the 1801 crop returns (together with some other official inquiries in the 1790s), and the Board of Agriculture county reports published between 1793 and 1816. All of these sources report yields in hundreds of villages or give informed estimates of the average yields in the counties concerned.[1] Typically the yield of wheat in England *c.*1800 was 20–22 bushels per acre—twice the medieval level. Yields of the other grains had also doubled.

The growth in yields did not make England unusually productive. English yields were about equal to yields in Ireland (22 bushels per acre) and northeastern France (22 bushels) (Allen and O Grada 1988: 107, 111). Dutch yields were similar (21 bushels).[2] Yields in Belgium and northwestern Germany were probably of the same order.[3] Compared to the rest of northwestern Europe and using grain yields as the measure of productivity, English agriculture was not especially efficient.

Yields in northwestern Europe as a whole, however, were radically higher than yields in the rest of Europe—indeed the rest of the world—at the beginning of the nineteenth century. Wheat yields outside of northwest Europe were probably about 10 bushels per acre at this time—that is, the same as medieval yields in England.[4] Indeed, until the 1950s, wheat yields in

[1] These sources have recently been compared and evaluated by Turner (1982*a*) and Allen and O Grada (1988).

[2] Van Zanden (1985). Weighted average of the yields of the sample provinces in 1812–13 where the weights are the areas of wheat cultivated in each.

[3] For Germany see Chorley (1981: 83).

[4] The yield–seed ratios in Slicher van Bath (1963) suggest this conclusion. Zytkowicz (1971) presents new yield–seed ratios of the same order for Poland, Bohemia, Hungary, and Slovakia. Davico (1972: 92) reports a yield of 11 bushels per acre for the Italian piedmont. Young (1794: ii. 157–217) reports similar figures for northern Italy. For many European countries, the earliest official returns in the 19th cent. show wheat yielding about 10 bushels per acre. Examples are Hungary, Italy, Portugal, Romania, Spain, Russia, and Yugoslavia. The earliest

many less developed countries were of this order. The Indian subcontinent between the 1890s (when official returns begin) and Independence (Blyn 1966: 258–61) and China in the mid-1950s are important examples.[5] Yields ranging from 8 to 12 bushels per acre were recorded in other Asian countries, in much of Latin America,[6] and, indeed, in Australia, Canada, and the United States during the nineteenth century.[7] In very arid regions yields were somewhat lower (only 6–9 bushels) since the lack of soil moisture indicated very light sowing rates.

Ten bushels per acre—but less in dry regions—is a typical yield for traditional farmers. Hanson, Borlaug, and Anderson (1982: 4) have constructed a 'yield ladder' for wheat in Third World countries. The bottom rung is 'traditional wheat varieties grown in normal weather under dryland conditions, without fertilizer, and with little weed control'. That technology produces 400–600 kilograms per hectare, i.e. 6–9 bushels per acre. It has been conjectured that farmers in Roman Britain reaped yields of this order (Bowen 1961: 13). Certainly, in the middle ages, few English farmers, or indeed farmers anywhere else in the world, had progressed much beyond that standard.

There are three notable exceptions to this generalization. The first were the farmers in northeastern Norfolk already referred to. The second were farmers in Flanders and northeastern France. As with Norfolk, the evidence is mixed: there are examples of French and Flemish farmers reaping only 10 bushels per acre of wheat in the thirteenth and fourteenth centuries,[8] but there are also examples of such farmers realizing much higher yields (Morineau 1968, 1970; Van der Wee 1963: ii. 298 n. 42). As with the the exceptional Norfolk demesnes, these high yields were often not maintained throughout the early modern period. The third exception were farmers in

official returns for Austria, Germany, and Scandinavia, however, report considerably higher yields, which probably indicate that productivity growth had occurred before the collection of agricultural statistics was begun. Where Slicher van Bath reports yield–seed ratios for these countries in the early modern period, the implied yields are low.

[5] Chen and Galenson (1969: 108). Surveys carried out between 1929 and 1937 suggest higher yields, but Perkins (1969: 268) regards those figures as unreliable.

[6] For most countries in Asia, Latin America, and Africa official agricultural statistics begin early in the 20th cent. For most Asian and Latin American countries wheat yields at that time were 8–12 bushels per acre. Japanese yields, however, were higher. See, for instance, Institut Internationale d'Agriculture, *Annuaire internationale de statistique agricole*, 1910: 50–3, and 1924/5: 102–5.

[7] For Australia see Dunsdorfs (1956: 534), for America see Gallman (1972: 191–210) and Parker and Klein (1966: 542), and for Canada see *Census of the Canadas*, 1851–2, ii. 60–3, 160–3.

[8] The yield–seed ratios abstracted by Slicher van Bath (1963) point to this conclusion for France. See Mertens and Verhulst (1966) for Flanders.

the Nile basin. Usually they were reaping 20 bushels per acre early in the nineteenth century[9] and probably also in antiquity (N. Lewis 1983: 122). These high yields are indicative of the exceptional fertility of the Nile, already legendary in ancient times.

We can now assess the achievement of English farmers in a broader perspective. Medieval English farmers reaped yields of the same order as farmers in most poor countries—about 10 bushels of wheat per acre. Sometime between 1400 and 1800 this yield was doubled in England. That increase was impressive and important, but it did not distinguish English agriculture from farming elsewhere in northwestern Europe—or Egypt— where similar yields were ordinarily realized.

## Enclosure and Yield Growth

Yields doubled in early modern England. To see how much of that growth was due to enclosure, one must compare the yields reaped by open and enclosed farmers. The principal sources of evidence are the same as those for the level of yields late in the eighteenth century—Young's *Tours*, the official inquiries, the Board of Agriculture reports. While these are well-known sources, they have only recently been analysed systematically.

The earliest evidence was collected by Arthur Young. In 1768–70, early in his career, he travelled throughout England. His observations of farming practice and rural economy were embodied in nine volumes totalling 4,500 pages (Young 1769, 1771a, 1771b). In these books Young presented detailed and uniform descriptions of agricultural practices in the villages he visited. In general, his information appears reliable (Allen and O Grada 1988).

Among the information Young collected were crop yields. He used them in the first (and only) attempt to test statistically the effect of farm size on yield. (I will review his effort in Chapter 10). The data can also be used to compare the yields of open and enclosed villages, although Young never made that comparison. The summary tables in the *Northern* and *Eastern Tours* are incomplete, so new summaries have been abstracted from all of the village reports in the volumes. The information is available for several hundred villages scattered from Northumberland to Hampshire and Somerset to East Anglia. To compare open and enclosed farming, it was necessary to determine which villages were open and which enclosed, and that has been done for most of the sample. Table 7–1 compares average

[9] Crouchley (1939: 142). I used the conversion rates in Whitaker's *Almanac*, 1936: 560, to change ardebs of wheat per feddan into bushels per acre. Richards (1982: 145) reports yields of 25–30 bushels per acre for 1920–39.

*Table 7–1. Corn yields and enclosure, early results*

| | Open (bushels per acre) | Enclosed (bushels per acre) | Enclosed relative to open (%)[a] | Enclosure gain relative to medieval–19th century advance (%)[b] |
|---|---|---|---|---|
| *Part A: The Arthur Young sample, c.1770* | | | | |
| Wheat | 22.6 | 24.5 | 8.4 | 13.8 |
| Barley | 31.0 | 33.2 | 7.1 | 13.4 |
| Oats | 34.5 | 38.7 | 12.2 | 15.6 |
| Beans | 27.6 | 29.5 | 6.9 | 9.7 |
| Peas | 21.1 | 22.7 | 7.6 | 12.6 |
| Average | 28.3 | 30.9 | 9.2 | 14.2 |
| *Part B: The 1801 crop return sample* | | | | |
| Wheat | 18.2 | 23.0 | 26.4 | 39.0 |
| Barley | 25.2 | 30.6 | 21.4 | 39.1 |
| Oats | 27.8 | 34.9 | 25.5 | 30.6 |
| Average | 22.9 | 28.5 | 24.5 | 35.7 |

[a] 'Enclosed relative to open (%)' equals the enclosed yield divided by the open yield minus one and multiplied by 100.
[b] 'Enclosure gain relative to medieval–19th century advance' equals the difference between the open and enclosed yield divided by the difference between the enclosed yield and medieval yields. These were taken to be 10.7 bushels per acre for wheat, 16.8 for barley, 11.7 for oats, and 10.0 for peas and beans. For the 'average' medieval yield, I computed a weighted average using the same weights as for the 'average' yield at the end of the 18th century.

*Sources:* The open and enclosed yields are taken from Turner (1982*a*: 500) and Allen and O Grada (1988: 98).
   The average yields are weighted averages where the weights are the relative proportions of land devoted to the crops whose yields are averaged. I used land shares for enclosed farms. These proportions are computed from Allen (1982: 949) for the Young sample and Turner (1986: 692) for the 1801 sample.

yields for the principal field crops. In every case the table shows a higher average yield for enclosed than for open villages. However, the differences are never statistically significant and the percentage differences in the means are quite small.

What is most important about Young's yields is that the difference between open and enclosed yields was very small compared to the growth in yields between the middle ages and Young's tours. For a medieval base line, I used the average yields of wheat, barley, and oats reaped on the Winchester estates and the average yield of beans on similar estates.[10] These yields are representative of demesnes yields in southern England.[11] On

[10] The yields for wheat (10.7 bushels per acre), barley (16.8), and oats (11.7) are derived from Titow's (1972: 121–35) summary of the yields for the Winchester demesnes. First, I computed the grand average for each crop of the 'overall average 1209–1349' for the 39 manors, and then divided the grand average by 0.9 to add the tithe back on. The yield for beans (10) is from the records of demesne cultivation tabulated by Rogers (1866–1892: i. (1866), pp. 38–45) and by the *VCH*, Cambs., ii. 60–1. The yield–seed ratios in these sources are similar to the broader sample in Slicher van Bath (1963: 214–29).

[11] These yields are less than demesnes yields in northeastern Norfolk. There is no direct evidence about non-demesne yields. Some historians believe they were lower than demesne

average, open field farms accomplished 86 per cent of the advance realized by enclosed farms. Enclosure played a very small role in the yield increase.[12]

Harvest failures and the wars with France prompted the British government to undertake several inquiries into food production. The last of these was the 1801 crop returns which provides another cross-section of yields. Turner (1982*a*) has abstracted them and compared open and enclosed villages. (See Table 7–1.) He found that the yield of wheat in enclosed villages was 26.4 per cent higher than in open villages. The differential was 21.4 per cent for barley and 25.5 per cent for oats. These increments are considerably larger than those implied by Young's data. Nevertheless, Turner's measurement of the open–enclosed differential is still only 36 per cent of the growth in yields between the middle ages and the nineteenth century.

The investigations of Young's *Tours* and the 1801 crop returns both suffer from the same problem—they fail to take account of the effect of soil and climate on yields. For instance, Turner recognizes the importance of these variables, and so carries out yield comparisons within groups of counties that he regards as homogeneous. These groupings, however, encompass considerable environmental diversity which affects the results. To control for that variation, I have assembled a large sample of yields for south midland villages around 1800. The main sources are the Board of Agriculture reports, the 1801 crop returns, and the 1795 inquiry for Northamptonshire.[13] By comparing open and enclosed yields within natural districts, the effect of environmental variation can be controlled.

Table 7–2 reports the results. In the heavy arable district, barley yields in enclosed villages exceeded yields in open villages by 19 per cent, for oats the differential was 39 per cent, and for beans 10 per cent. The differential for wheat yields was a small amount—only 3 per cent. This disproportion in the yield increase between wheat and other crops was noted by contemporaries. Thus, Foster (1807: 11) commented:

It must be owned that in the open field state, the wheat crops [grown on heavy, wet clays] were frequently fine, by no means inferior to what we at present see . . . Except upon the unconquerably rich soils, the spring crop hardly paid seed, labour, and expenses . . .

On average, open field farmers in the heavy arable district achieved over

yields on the grounds that peasants probably had less livestock. Other historians think peasants may have reaped higher yields since they could have applied surplus family labour to the cultivation of their plots.

[12] See Allen and O Grada (1988) for a fuller discussion.
[13] I am grateful to Dr Turner for allowing me to use his photocopies of this source.

*Table 7–2. Corn yields and enclosure, new results c. 1800*

| | Open (bushels per acre) | Enclosed (bushels per acre) | Enclosed relative to open (%)[a] | Enclosure gain relative to medieval–19th century advance (%)[b] |
|---|---|---|---|---|
| *Part A: The heavy arable district* | | | | |
| Wheat | 19.7 | 20.2 | 2.5 | 5.3 |
| Barley | 26.5 | 31.8 | 20.0 | 35.3 |
| Oats | 23.5 | 33.0 | 40.4 | 44.6 |
| Beans | 18.8 | 22.2 | 18.1 | 27.9 |
| Average | 21.2 | 24.1 | 14.7 | 23.8 |
| *Part B: The light arable district* | | | | |
| Wheat | 20.0 | 19.7 | −1.5 | **** |
| Barley | 27.0 | 29.3 | 8.5 | 18.4 |
| Oats | 26.5 | 32.5 | 22.6 | 28.8 |
| Beans | 19.9 | 18.1 | −9.0 | **** |
| Average | 23.4 | 24.7 | 5.6 | 10.9 |
| *Part C: The pasture district* | | | | |
| Wheat | 20.9 | 21.9 | 4.8 | 8.9 |
| Barley | 28.0 | 32.2 | 15.0 | 27.3 |
| Oats | 36.9 | 38.1 | 3.3 | 4.5 |
| Beans | 22.4 | 23.4 | 4.5 | 7.5 |
| Average | 24.7 | 26.7 | 8.1 | 14.2 |

[a] 'Enclosed relative to open (%)' equals the enclosed yield divided by the open yield minus one and multiplied by 100.
[b] 'Enclosure gain relative to medieval–19th century advance' equals the difference between the open and enclosed yield divided by the difference between the enclosed yield and medieval yields. These were taken to be 10.7 bushels per acre for wheat, 16.8 for barley, 11.7 for oats, and 10.0 for peas and beans. For the 'average' medieval yield, I computed a weighted average using the same weights as for the 'average' yield at the end of the 18th century. Asterisks signify that enclosure made a negative contribution to the advance in yields.

*Source:* The average yields are weighted averages where the weights are the relative proportions of land devoted to the crops whose yields are averaged. I used land shares for farms enclosed by act for heavy arable district, for non-sheepfolding farms enclosed by act for the light arable district, and a simple average across all subdivisions of the pasture district for that district. These proportions are computed from Tables 6–4, 6–1, and 6–8.

three-quarters of the progress over medieval levels that was posted by enclosed farmers. For wheat, open field farmers accomplished 95 per cent of the advance. Even for spring corn in the heavy arable district, open field villages had accomplished two-thirds of the difference between medieval yields and early nineteenth-century enclosed yields.

In the other districts the difference between open and enclosed yields was much smaller, and open field farmers had accomplished almost all of the advance over medieval levels that had been achieved by enclosed farmers. In the light arable district, the average enclosed yield was only 5.6 per cent higher than the average open yield, and open field farmers achieved the highest yields for wheat and beans. In the pasture district, enclosed yields exceeded open yields by an average of 8.1 per cent. Open field farmers in the light arable district had realized 89 per cent of the advance over medieval levels claimed by enclosed farmers. In the pasture district, the corresponding figure was 86 per cent. Enclosure was not necessary for progress since open field villages had increased their yields almost as much as had enclosed.

These findings differ from those implied by the 1801 crop returns and Young's data. In the case of Turner's study, the difference arises since he controlled for county rather than soil type in making his comparisons. Around 1800 most of the enclosed land in the south midlands was in the pasture district, and most of the open land was in the arable districts. As Table 7–2 shows, enclosed yields (and indeed open yields) in the pasture district exceeded open yields in the arable districts by a considerable margin. Not only did the pasture district support lush grass, but it was more fertile for corn than the arable districts. Turner's comparisons confound differential fertility with the effects of enclosure, thereby overstating the importance of the latter.

The very modest yield differential in the data collected by Arthur Young has a similar explanation. When he undertook his tours, Young was enamoured with clover and turnips, so he tended to report on agriculture in villages where they were grown or might have been grown. Also he visited many regions that specialized in livestock production. Thus, his sample is dominated by villages like those in the light arable and pasture districts, where the differential between open and enclosed yields was small.

## Enclosure, Farm Methods, and Yields

Enclosure, *per se*, did not raise yields. If there was a connection, it was because farmers in enclosed villages could pursue more productive methods; in particular, they could practise capital-intensive agriculture. Depending on the environment, they could plant turnips and clover, or increase livestock densities, or install hollow drains. One must doubt the importance of some of these practices—notably the Norfolk rotation and higher livestock densities—since open field farmers in the light arable and pasture districts realized yields almost as high as enclosed farmers. None the less, the question is important enough to warrant specific investigation.

I have explored the importance of some aspects of capital-intensive agriculture with a regression analysis of Parkinson's data. He reports the details of farming methods as well as yields for each village, so the importance of methods can be assessed. I tested the importance of legumes, livestock density, and various fertilizers on crop yields. I do not report the results in detail since they were monotonously repetitive: neither more legumes, nor more cattle, nor exotic fertilizers raised the yields of wheat, barley, oat, or beans. By this reckoning, 'improved agriculture' was not very powerful.

## Draining and Yields

One technique of improved agriculture was effective in raising yields—hollow draining. Indeed, the adoption of draining by enclosed farmers explains the difference between their yields and the yields of open field farmers.

The relationship between drainage and yield cannot be explored with Parkinson's census since he did not inquire about draining. However, Vancouver's *General View of . . . Cambridge* (1794) reports the details of soils, enclosure status, crop yields, and drainage systems village by village. I used this information in the last chapter to show that enclosed villages were more likely to use hollow draining than were open villages. Here I use the data to measure the effect of that practice on yields.

Table 7–3 reports the results of regressions of crop yields on dummy variables denoting enclosure status and drainage design. Open and enclosed villages are distinguished, with DE in Table 7–3 signifying the latter class. Three drainage classes were distinguished: 'bad' or 'no' drainage, 'moderate' or 'a little' drainage, and finally an extensive system of hollow and

*Table 7–3. Drainage regressions (t-ratios in parentheses)*

|  | Constant | DE | DM | DH | $R^2$ |
|---|---|---|---|---|---|
| 1. Wheat | 18.759 (39.042) | 0.89051 (0.91698) | −1.3431 (−1.3584) | 3.0456 (2.9568) | 0.34 |
| 2. Wheat | 18.562 (45.168) | | | 3.6875 (4.0127) | 0.30 |
| 3. Barley | 22.922 (31.331) | 1.9416 (1.3131) | −2.3996 (−1.5940) | 2.8577 (1.8221) | 0.25 |
| 4. Barley | 22.625 (35.605) | | | 4.1250 (2.9031) | 0.18 |
| 5. Oats | 20.617 (25.690) | 1.1391 (0.74314) | 0.54410 (−0.36219) | 4.9570 (2.7223) | 0.33 |
| 6. Oats | 20.667 (32.082) | | | 5.6667 (3.7509) | 0.31 |
| 7. Beans | 15.987 (20.230) | −1.3552 (−0.81601) | −1.2292 (−0.76705) | 4.1310 (2.3519) | 0.20 |
| 8. Beans | 15.452 (23.158) | | | 4.2151 (2.5440) | 0.16 |

*Notes:*
The independent variables are dummy variables and equal 1 in the following circumstances: DE (enclosed villages), DM (a little drainage), DH (complete hollow drainage).
   There are 40 observations for the wheat and barley regressions, 33 for the oat regressions, and 37 for the bean regressions.

surface drains. The dummy variables DM and DH represent the latter two
categories in the regressions.

What is striking about the regressions in Table 7–3 is that it is only the
coefficients of DH that are significantly different from zero. (The coefficient
of DM in equation 3 might also be deemed significant, but note that it is
negative). Whether a parish was open or enclosed, in itself, had no impact
on yields. Likewise, doing only a little draining did not raise yields. A village
must have installed a complete system of drains to achieve any benefit. If it
did, yields rose sharply. The even-numbered equations in Table 7–3 show
increases in yields of 20 per cent for wheat, 18 per cent for barley, 27 per
cent for oats, and 27 per cent for beans when land was thoroughly drained.
Hollow draining was one example of improved agriculture raising yields.

This finding revises our understanding of the agricultural revolution on
the heavy clays. The present consensus among historians is that the drainage
problem was not alleviated until tile drains came into general use in the
middle of the nineteenth century. Thus, of the south midlands around 1800,
Darby (1976: 119) wrote, 'Draining was rudimentary or completely lacking
in most cases'. Yelling (1977: 145) concurred: 'it is generally agreed that the
real advance in this field came in the nineteenth century when pipes were
introduced'. Chambers and Mingay (1966: 65) came to the same
conclusion, and Collins and Jones (1967: 65 n. 3) agreed that there was little
progress in the first half of the nineteenth century. In the same spirit,
Sturgess (1966: 105) concluded: 'On the clays, the problem of increasing
production remained as intractable after enclosure as before.' This was not
the view of contemporaries. Marshall (1818: iv. 292), for instance, evinced
a common spirit when he wrote in 1812: 'The theory and practice of
draining has been raised to a degree of science, which I will presume to
suggest, few, if any of the rural arts have attained'. My regressions indicate
that Marshall was right, and that enclosure in the heavy arable district led to
greater output through better drainage.

## Convertible Husbandry and Yields

Convertible husbandry was another technique of improved agriculture that
many historians have argued raised yields. However, there are two reasons
for doubting that convertible husbandry had much to do with the growth of
yields. First, many farmers—including all those in the arable districts and
open farmers in the pasture district—did not follow the practice, but they
realized high yields anyway. Second, the evidence offered in favour of the
importance of convertible husbandry is not very persuasive. Dr Kerridge,
one of its major champions, has claimed it increased the productivity of

both pasture and tillage. Shifting land between crops and grass was supposed to have raised grass quality (since ploughing destroyed weeds and new grass was more nutritious than old) and crop yields (since the land was manured as pasture). Neither contention is plausible.

The claim that the system improved the productivity of pasture—'Up-and-down husbandry gave better grass blade for blade' (Kerridge 1967: 205)—is very doubtful in view of the testimony of the Board of Agriculture reporters. Thus, Pitt (1809a: 157) wrote:

*Breaking up grass lands.*—This is not often done in Leicestershire, at least not old grasslands; . . . The opinion of the principal graziers and breeders of the county is, that rich old feeding land broken up and converted to tillage could scarcely be brought to recover its former fertility and nutrititive qualities in 40 years . . . Rich old pasture land should never be broken up; if such wants improvement, it is best done by draining, watering, and top-dressing.

This statement directly contradicts Kerridge's (1967: 203) claim that 'even the best of these fatting pastures could be improved by ploughing'.

Reporters for other counties agreed with Pitt. Thus, Priest (1813: 250–1) related some examples of the ploughing up of pasture in Buckingham and then advised landlords not to permit their tenants to plough up old grass:

It is with extreme caution such indulgences as these should be granted to any tenants, and particularly to such as are ignorant of the management of arable land. If an estimate were made of the profit and loss attending the above circumstances, it would be found that the tenant has gained nothing; his expenses have exceeded the produce. But from the advantage of a little straw for winter to make manure and a few oats and beans for his horses, he is encouraged to ask for permission to break up more sward; and thus he would go on so long as sward remained and his landlord were indulgent, until he would ruin himself, and put the land into such a state, that it could not be recovered before it had undergone an expensive course of three or four years.

Mavor (1813: 238–9), discussing the pastures at the western end of the Vale of White Horse 'where the dairying farms chiefly lie', concurred:

The more upland pastures are sometimes managed on a principle radically wrong. The desire of obtaining a few crops of corn induces the farmer when he is not restrained by his lease, to break up his grass land and when it is exhausted, he wishes to lay it down again permanently for pasture. For this purpose he sows rye grass, and other seeds; but a short time convinces him that the plant is not only unprofitable in itself, but will not stand more than a very few years, and that weeds of every kind fill the ground. He must, therefore, make it what is called convertible land, however anxious he may be to bring it to a good turf. I speak of a practice not common with sensible and honest cultivators, but which deserves to be marked with censure, till it is wholly exploded.

Even if convertible husbandry failed to improve a farmer's grass, perhaps it increased the yields of his crops? Kerridge (1967: 208) claimed that 'it is reasonable to suppose that crops in "pastures" [i.e. convertible land] yielded twice as heavily as those in common fields'. Here he relies on Blith's optimistic recommendations and some seventeenth-century remarks on yield–seed ratios. Kerridge (1967: 208) accepts Plot's[14] claim that 'in common fields a tenfold increase of harvest over seed was considered good' and contrasts that with three other claims suggesting that up-and-down husbandry achieved yield–seed ratios of nineteen or twenty. That gives the doubling.

But this evidence is exceptional. First, it is not surprising that Plot thought ten was good since that was above the English average for wheat, barley, and oats a hundred years after he wrote (Slicher van Bath 1963: 16, 53–6, 133–7, 172–5). Indeed, a yield ratio of ten to one implies that the yield per acre of wheat was 25 bushels since the sowing rate was typically about 2.5 bushels per acre. And 25 bushels was above the national average *c.*1800. Indeed, the real significance of Plot's observation is that it indicates that by the late seventeenth century open field farmers had already realized all of the advance in yields that was achieved between the middle ages and the early nineteenth century. (This is a theme I will pursue in Chapter 10). Second, ratios of twenty to one were not routinely achieved anywhere in England or on the European continent even in the late eighteenth century. For instance, a yield–seed ratio of twenty implies that wheat was yielding 50 bushels per acre. Reference to Table 7–2 shows that English yields *c.*1800 were very much below these levels. It is thus incredible that farmers practising convertible husbandry in the seventeenth century were routinely realizing these yields.

Convertible husbandry may have made a modest contribution to the growth in crop yields. But the fact that so many villages—open and enclosed—reaped high yields without alternating grass and arable shows that convertible husbandry was not 'the backbone of the agricultural revolution'.

## Total Farm Output

Opponents of enclosure often claimed that it led to the conversion of land to pasture resulting in less corn production. Supporters of enclosure countered that the more intensive agriculture of enclosed farmers raised yields enough to compensate for any decline in corn acreage. My finding that enclosure

[14] Robert Plot was secretary to the Royal Society, Professor of Chemistry at Oxford, and first custodian of the Ashmolean Museum. See *Dictionary of National Biography*, xv. 1310–11.

and 'improved agriculture' had little effect on crop yields in most environments suggests that the opponents of enclosure were right. To pin the matter down, however, one must compute how enclosure changed farm output. This exercise requires estimating the production of animal products as well as corn. Enclosure did often improve the management of livestock, and I will quantify that effect as well as the impact on corn yields.

I compute farm output as the value of production (in constant prices) of four crops—wheat, barley, oats, and beans (or peas); eight animal products —butter,[15] pork, veal, beef, mutton, wool, calves, and foals; and firewood trimmed from hedges. I presume that other products like turnips and hay were consumed on the farm by the livestock. I distinguish gross from net output by estimating the on-farm consumption of the wheat, barley, oats, and beans used as seed and forage. In Appendix II, I document the Batchelor–Parkinson farm accounting model that generates the estimates of output as well as the employment and profit estimates discussed in the next two chapters.

Tables 7–4 and 7–5 contrast production per 100 acres of farm land for open and enclosed farms in the heavy arable district. Enclosure had scant effect on the production of wheat, but it led to higher net outputs of beans (24 per cent) and especially barley (72 per cent). Purchases of oats were also reduced.[16] Overall, the net value of corn was 28 per cent greater in the villages enclosed by act than in the open villages.[17] Enclosed villages also produced more animal products, but by a much smaller margin. So the overall effect of enclosure was to increase production (i.e. farm revenue) 12 per cent.

In the light arable district, output per acre was *less* in enclosed, non-sheepfolding villages than in open, sheepfolding ones. (See Tables 7–6 and 7–7.) This contradicts Young's claims about enclosure in his paradigm case. Wheat and bean production declined, although barley and oat output were higher in enclosed villages. Overall, the net output of corn was less in enclosed villages (£217.40 per 100 acres) than in open villages. (£226.57). This decline was the result of two features of enclosed farming: first, the large share of the arable planted with clover and turnips meant that the share planted with corn was less. Second, the share of arable was also

---

[15] Some milk was made into cheese, but I treat it all as converted into butter.

[16] Farmers were probably using beans as fodder rather than not buying as much oats as the tables indicate. Hence, the value of the net output of oats and beans combined is probably more reliable than the sales and purchases shown separately for the two crops.

[17] The net value of corn per 100 acres of farm land was £221.20 in open villages and £282.25 in villages enclosed by act. These net outputs are computed as the value of gross output less the value of corn used for seed and horses in the subtotal rows of Tables 7–4 and 7–5.

*Table 7–4. Heavy arable district: production per 100 acres in open field villages*

| | Values (£) | | | Net output (bushels) |
|---|---|---|---|---|
| | Gross output | Used for | | |
| | | Seed | Horses | |
| Wheat | 176.17 | 24.35 | | 346.63 |
| Barley | 42.10 | 7.20 | | 197.63 |
| Oats | 22.14 | 5.25 | 40.80 | −159.34 |
| Beans | 80.73 | 22.35 | | 231.34 |
| Subtotal | 321.14 | 59.15 | 40.80 | |
| | | | | Weight (lb.) |
| Butter | 17.62 | | | 302.04 |
| Pork | 12.48 | | | 374.35 |
| Veal | 4.25 | | | 113.38 |
| Beef | 15.42 | | | 528.57 |
| Mutton | 38.61 | | | 1,158.16 |
| Wool | 12.57 | | | 2,476.50 |
| Calves | −2.14 | | | |
| Foals | 3.31 | | | |
| Faggots | 0.00 | | | |
| TOTAL | 423.26 | | | |

*Table 7–5. Heavy arable district: production per 100 acres in enclosed villages*

| | Values (£) | | | Net output (bushels) |
|---|---|---|---|---|
| | Gross output | Used for | | |
| | | Seed | Horses | |
| Wheat | 180.14 | 23.08 | | 358.58 |
| Barley | 83.14 | 10.84 | | 339.47 |
| Oats | 8.25 | 1.28 | 26.52 | −130.33 |
| Beans | 90.80 | 18.36 | | 287.43 |
| Subtotal | 362.33 | 53.56 | 26.52 | |
| | | | | Weight (lb.) |
| Butter | 21.72 | | | 372.28 |
| Pork | 10.11 | | | 303.36 |
| Veal | 1.61 | | | 42.84 |
| Beef | 19.35 | | | 663.36 |
| Mutton | 35.69 | | | 1,070.75 |
| Wool | 13.95 | | | 2,748.77 |
| Calves | 0.77 | | | |
| Foals | 3.54 | | | |
| Faggots | 5.37 | | | |
| TOTAL | 474.44 | | | |

slightly less. Together, these reductions meant the acreage planted with corn was 18 per cent less in the enclosed than in the open villages.

Table 7–6. _Light arable district: production per 100 acres in open field, sheepfolding villages_

| | Values (£) | | | Net output (bushels) |
|---|---|---|---|---|
| | Gross output | Used for | | |
| | | Seed | Horses | |
| Wheat | 171.71 | 24.90 | | 335.18 |
| Barley | 75.04 | 10.26 | | 304.13 |
| Oats | 9.28 | 2.54 | 45.27 | −256.87 |
| Beans | 74.01 | 20.50 | | 212.34 |
| Subtotal | 330.04 | 58.20 | 45.27 | |
| | | | | Weight (lb.) |
| Butter | 21.77 | | | 373.20 |
| Pork | 16.00 | | | 480.00 |
| Veal | 3.39 | | | 90.48 |
| Beef | 23.49 | | | 805.37 |
| Mutton | 29.27 | | | 878.10 |
| Wool | 10.02 | | | 1,974.11 |
| Calves | −1.71 | | | |
| Foals | 6.41 | | | |
| Faggots | 0.00 | | | |
| TOTAL | 438.68 | | | |

Table 7–7. _Light arable district: production per 100 acres in enclosed (by act), non-sheepfolding villages_

| | Values (£) | | | Net output (bushels) |
|---|---|---|---|---|
| | Gross output | Used for | | |
| | | Seed | Horses | |
| Wheat | 116.83 | 15.44 | | 231.48 |
| Barley | 90.02 | 11.95 | | 366.53 |
| Oats | 34.90 | 5.22 | 32.94 | −21.73 |
| Beans | 50.43 | 9.23 | | 163.49 |
| Subtotal | 292.18 | 41.84 | 32.94 | |
| | | | | Weight (lb.) |
| Butter | 15.22 | | | 260.91 |
| Pork | 10.99 | | | 329.70 |
| Veal | 2.49 | | | 64.00 |
| Beef | 20.90 | | | 716.57 |
| Mutton | 51.87 | | | 1,556.10 |
| Wool | 14.78 | | | 2,911.91 |
| Calves | −0.09 | | | |
| Foals | 4.35 | | | |
| Faggots | 5.37 | | | |
| TOTAL | 418.06 | | | |

The production of butter, pork, veal, and beef were all somewhat smaller in enclosed villages. Only in the production of mutton and wool did enclosed villages exceed open ones. For these products, the output gains were substantial. This was the pay-off from abandoning sheepfolding and improving the breed. Notice, however, that wool and mutton did not loom large enough in farm output to reverse the declines elsewhere. In the light arable district enclosure reduced farm revenue.

Data limitations complicate the comparison of open and enclosed villages in the pasture district. Parkinson is virtually silent about open field farming since almost every place was enclosed. Consequently, I have computed output for a hypothetical open field village on the following assumptions: I have used the distribution of land (63 per cent arable, 7 per cent meadow, 24 per cent pasture, 6 per cent common) that I argued earlier was typical of the pasture district. I divided the arable among the crops in proportion to a weighted average of the crop distributions in open field villages in the heavy arable district (Table 6–4) and open field, sheepfolding villages in the light arable district (Table 6–1).[18] Crop yields were assumed to equal the average open field yields in the pasture district shown in Table 7–2.[19] I assumed farmers kept 4 cows per 100 acres, 1.6 calves, 10 hogs, 4 horses, 0.5 foals, 55 sheep, and 25 lambs. These values are typical of open field farms in the arable districts, and not implausible given the values of enclosed farms shown in Table 6–8. I assumed that the greater extent of grass in open field villages in the pasture district was used for fattening beef, and allowed each farm one beast per 100 acres. Finally, I assumed the sheep were folded so their productivity was low.

Contrary to the claims of Tory fundamentalists, enclosure reduced corn output in the pasture district. (Tables 7–8 to 7–12.) Net production of corn for a 100-acre farm was £163.80 (= £252.04 − £45.04 − £43.20) in the open field village compared to £29.19 in non-parliamentary enclosures in Rutland, £46.90 in parliamentary enclosures, £100.33 in non-parliamentary enclosures in Huntingdon, and £98.26 in parliamentary enclosures. The Rutland non-parliamentary enclosures were almost pure grazing farms. Even in the other villages, however, where more land was arable, the higher yields reaped by enclosed farmers did not make up for the decline in cropped acreage, so total corn production fell.

---

[18] To determine the weights, I first divided the pasture villages into two groups depending on whether the soil was more or less than 50 per cent clay. The weights were the fractions of the villages in each group.

[19] Seed rates were assumed to be the same rates as recorded by Parkinson. If he did not record a seed rate, I assumed that 2.5 bushels per acre of wheat were sown and 4 bushels of barley, oats, and beans. See Batchelor (1808: 107).

*Table 7–8. Pasture district: production per 100 acres in a hypothetical, open field village, c. 1806*

| | Values (£) | | | Net output (bushels) |
|---|---|---|---|---|
| | Gross output | Used for | | |
| | | Seed | Horses | |
| Wheat | 125.70 | 19.80 | | 246.35 |
| Barley | 38.90 | 6.33 | | 152.91 |
| Oats | 17.46 | 3.16 | 43.20 | −192.67 |
| Beans | 69.98 | 15.75 | | 215.20 |
| Subtotal | 252.04 | 45.04 | 43.20 | |
| | | | | Weight (lb.) |
| Butter | 19.20 | | | 329.14 |
| Pork | 15.00 | | | 450.00 |
| Veal | 2.94 | | | 78.40 |
| Beef | 40.80 | | | 1,398.86 |
| Mutton | 34.38 | | | 1,031.40 |
| Wool | 17.63 | | | 3,472.25 |
| Calves | −0.96 | | | |
| Foals | 5.00 | | | |
| Faggots | 0.00 | | | |
| TOTAL | 386.03 | | | |

*Table 7–9. Pasture district: production per 100 acres in non-parliamentary enclosures in Rutland, c. 1806*

| | Values (£) | | | Net output (bushels) |
|---|---|---|---|---|
| | Gross output | Used for | | |
| | | Seed | Horses | |
| Wheat | 17.10 | 2.06 | | 34.33 |
| Barley | 15.87 | 1.50 | | 67.46 |
| Oats | 11.99 | 1.07 | 14.48 | −23.72 |
| Beans | 4.04 | 0.69 | | 13.27 |
| Subtotal | 49.00 | 5.32 | 14.48 | |
| | | | | Weight (lb.) |
| Butter | 25.62 | | | 439.25 |
| Pork | 3.87 | | | 116.20 |
| Veal | 8.61 | | | 229.73 |
| Beef | 251.37 | | | 8,618.34 |
| Mutton | 89.17 | | | 2,675.03 |
| Wool | 32.52 | | | 6,404.24 |
| Calves | −1.50 | | | |
| Foals | 3.64 | | | |
| Faggots | 5.37 | | | |
| TOTAL | 467.67 | | | |

*Table 7–10. Pasture district: production per 100 acres in parliamentary enclosures in Rutland, c. 1806*

| | Values (£) | | | Net output (bushels) |
|---|---|---|---|---|
| | Gross output | | Used for | |
| | | Seed | Horses | |
| Wheat | 44.81 | 6.71 | | 86.99 |
| Barley | 23.28 | 2.78 | | 96.24 |
| Oats | 5.99 | 1.16 | 26.82 | −146.57 |
| Beans | 13.36 | 3.07 | | 40.82 |
| Subtotal | 87.44 | 13.72 | 26.92 | |
| | | | | Weight (lb.) |
| Butter | 33.30 | | | 570.90 |
| Pork | 7.31 | | | 219.19 |
| Veal | 1.99 | | | 52.95 |
| Beef | 66.32 | | | 2,273.76 |
| Mutton | 87.63 | | | 2,628.92 |
| Wool | 25.92 | | | 5,104.11 |
| Calves | 5.38 | | | |
| Foals | 3.27 | | | |
| Faggots | 5.37 | | | |
| TOTAL | 323.93 | | | |

*Table 7–11. Pasture district: production per 100 acres in non-parliamentary enclosures in Huntingdonshire, c. 1806*

| | Values (£) | | | Net output (bushels) |
|---|---|---|---|---|
| | Gross output | | Used for | |
| | | Seed | Horses | |
| Wheat | 61.79 | 8.66 | | 121.31 |
| Barley | 39.56 | 5.30 | | 160.83 |
| Oats | 8.33 | 1.02 | 21.99 | −97.87 |
| Beans | 33.61 | 5.99 | | 109.60 |
| Subtotal | 143.29 | 20.97 | 21.99 | |
| | | | | Weight (lb.) |
| Butter | 23.60 | | | 404.65 |
| Pork | 8.08 | | | 242.36 |
| Veal | 1.41 | | | 37.47 |
| Beef | 27.21 | | | 932.93 |
| Mutton | 59.54 | | | 1,786.05 |
| Wool | 16.92 | | | 3,332.32 |
| Calves | 4.59 | | | |
| Foals | 2.84 | | | |
| Faggots | 5.37 | | | |
| TOTAL | 292.85 | | | |

*Table 7–12. Pasture district: production per 100 acres in parliamentary enclosures in Huntingdonshire, c. 1806*

| | Values (£) | | | Net output (bushels) |
|---|---|---|---|---|
| | Gross output | Used for | | |
| | | Seed | Horses | |
| Wheat | 68.43 | 9.84 | | 133.76 |
| Barley | 46.62 | 7.65 | | 182.93 |
| Oats | 3.00 | 0.41 | 21.80 | −128.07 |
| Beans | 25.59 | 5.68 | | 79.03 |
| Subtotal | 143.64 | 23.58 | 21.80 | |
| | | | | Weight (lb.) |
| Butter | 15.83 | | | 271.38 |
| Pork | 5.89 | | | 176.78 |
| Veal | 1.03 | | | 27.59 |
| Beef | 89.56 | | | 3,070.55 |
| Mutton | 70.58 | | | 2,117.27 |
| Wool | 23.57 | | | 4,642.42 |
| Calves | 2.06 | | | |
| Foals | 1.28 | | | |
| Faggots | 5.37 | | | |
| TOTAL | 358.81 | | | |

Enclosed villages in the pasture district did produce more meat and wool than open villages, but total agricultural output was frequently less. Production was highest in the non-parliamentary enclosures in Rutland (£467.67 per 100 acres), but open field villages scored second (£386.03). Villages enclosed by act in both counties produced less per 100 acres— £323.93 in Rutland and £358.81 in Huntingdon. These output declines were probably not exceptional. In his tours of England, Arthur Young collected another set of systematic information that allows the comparison of open and enclosed, arable and pasture farms. I will analyse these data later, but it is important to note here that they indicate pastoral farms usually generated less revenue per acre than arable farms in the 1760s. (See Tables 11–1 and 11–2.) Young's data thus confirm Parkinson's in showing that much enclosure involving the conversion of arable to pasture lowered agricultural production. Certainly, the available data provide no support for the optimistic Tory conclusion that enclosure led to more output in the pasture district.

## Conclusion

When I evaluate enclosure by its contribution to yields and output growth I reach a much less favourable assessment than when I assess the differential receptiveness of open and enclosed farmers to new crops and methods.

The heavy arable district is the only district where enclosure led to the adoption of a new method—hollow draining—that significantly raised yields and production. But the yield increases in the heavy arable district were only about 25 per cent and equalled only about one quarter of the growth in yields between the middle ages and the nineteenth century. Open field farmers, in other words, had accomplished about three-quarters of the advance. Enclosure in this district made a contribution to productivity growth but was only a small part of the whole story.

In the other districts, the contribution of enclosure to productivity and output growth was even less. Enclosed farmers in the light arable district were indeed more likely than their counterparts in open villages to adopt turnips, clover, and high yielding sheep, but the gains to these changes were confined to the sheep account. They made no contribution to raising yields, and, indeed, enclosed farmers produced less per acre of land than open farmers.

Likewise, enclosed farmers in the pasture district were more likely than open farmers to lay land down to grass, to grow turnips, and to practise convertible husbandry. But, again, these changes had little effect on yields. Corn output dropped sharply due to the decline in arable acreage. The main purpose of enclosure in this district was to reorientate farming from corn to livestock. Outside the old enclosures, where livestock densities were especially high, the increase in animal products was not enough to compensate for the fall in corn output, so enclosure reduced farm output.

In the light arable district and much of the pasture district, greater output was not the motive for enclosure. To understand why these districts were enclosed, it is necessary to consider the impact of enclosure on farm costs (in particular labour costs) and the level of rent.

# 8

## Employment and Labour Productivity

That this [open field] parish [Glenfield] produces more sustenance and employment for mankind, than the average of enclosed parishes in this county, of equal extent and staple of soil, I have not the least doubt; . . . enclosures are generally thrown to pasture in this county, and stocked with sheep and cattle; in which little labour is wanted, nor much attendance necessary . . .

But as enclosures have generally been a good speculation, and enable the proprietor to raise the rent, so as to pay him a good percentage, who is to prevent it, or to compel him to forego his advantage?

W. Pitt, *General View of the Agriculture of the County of Leicestershire*, 1809: 78–9

THE big split between Marxist and Tory fundamentalists is over the impact of enclosure on farm employment and labour productivity. According to Marxists, enclosure reduced agricultural employment thereby raising output per worker. In contrast, Tories contend that enclosure increased farm employment but raised production even more. The two changes together pushed up labour productivity. In this chapter I offer a resolution of this dispute. I also reassess the contribution of enclosure to the growth of labour productivity in early modern English agriculture.

My examination of the Tory and Marxist positions proceeds in three steps. The first was taken in the last chapter when I assessed the effect of enclosure on farm output in the various natural districts. The second step is to compare the effects of enclosure on farm employment in the same districts, and the third is to divide the results of the first by those of the second to compute the change in labour productivity. I find that enclosure had no effect on employment when the balance of tillage and pasture remained unchanged. When enclosure led to the conversion of arable to pasture, however, employment fell. More surprising, the impact on labour productivity was often small—either because enclosure led to little change in output or employment, or because reductions in employment were often accompanied by reductions in output. Thus, the Marxist view receives much more support than the Tory when employment is the question, but both exaggerate the contribution of enclosure to the growth in labour productivity.

Both the Marxist and Tory views have roots running back to the fifteenth century. The Marxist view derives from the critics of enclosure who claimed that it led to the conversion of arable to pasture with catastrophic effects on employment. The history of deserted villages and depopulation during the first wave of enclosures (1450–1524) strongly supports this view. For the next three hundred years, every wave of enclosures was greeted by a revival of this charge. 'Depopulation hath cast a slander on Inclosure, which because often done with it, people suspect it cannot be done without it.'[1] The Marxist position continues to reaffirm this accusation. 'As the enclosures proceeded . . . a number of men who formerly lived on and off the land were forced to leave it, and others were reduced to the status of landless rural laborers working for wages' (Wallerstein 1974: 251). 'During the seventeenth century one witnesses a wholesale transformation of formerly grain-growing areas, particularly midland England, towards animal production. The accompanying "depopulation" and freeing of labour opened the way for the rise of new industries in the neighbouring vicinities' (Brenner 1976: 311).

The simplest defence against the charge of depopulation was to counter-claim that enclosure expanded employment by creating 'new jobs'. Fitzherbert (1539) is perhaps the earliest example.

Peradventure some men would say, that this [enclosure] shuld be against the common weale, bicause the shepeherdes, heerdmen and swyneherdes shulde than be put out of wages. To that it may be answered, though those occupations be not used, there be as many newe occupations that were not used before; as gettyng of quicke settes, diching, hedging, and plashing, the which the same men may use and occupye. (Tawney and Power 1924: iii. 24)

Young repeated this argument in the eighteenth century.

Respecting open field lands, the quantity of labour in them is not comparable to that of inclosures; for, not to speak of the great numbers of men that in inclosed countries are constantly employed in winter in hedging and ditching, what comparison can there be between the open field system of one half or a third of the lands being in fallow, receiving only three ploughings; and the same portion now tilled four, five, or six times by midsummer, then sown with turnips, those hand-hoed twice, and then drawn by hand and carted to stalls for beasts; or else hurdled out in portions for

---

[1] Pseudomisus, *Considerations Concerning Common Fields*, 1654, p.39, as quoted by Beresford (1961: 40).

fatting sheep! What a scarcity of employment in one case, what a variety in the other! (Young 1774: 72)

The most influential statement of the 'new jobs' argument is Professor J. D. Chambers'.

One important factor which contributed to the stability of the agrarian population during this period [the Industrial Revolution] was the high level of employment which was maintained both in enclosed and open parishes where the improved agriculture was adopted. The explanation seems to be that new agricultural practices had developed in advance of the technical devices for dealing with them. Thus the yield of corn per acre went up (e.g. at Queniborough from 50 to 100 per cent) after enclosure but the methods of ploughing, sowing, reaping, and threshing were not substantially speeded up until the 1830s and 1840s. At the same time the spread of turnip cultivation and green fodder crops both in open and enclosed villages called for labour throughout the year in field, barn and stackyard; the maintenance of a milking herd or fat stock involved continuous field work throughout the year in pasture districts as well as in arable, except where the land was too stiff for mixed farming as in south-east Leicestershire; and the hedging and ditching of the new enclosures found winter work for casual labour to a greater extent than the open villages. (Chambers 1953: 332–3)

Hedging and ditching always appear in these lists. Since the eighteenth century, turnips have made a regular appearance.

Despite the confident statements of Professor Chambers and others that enclosure and improved agriculture increased the demand for labour, there have been few attempts to measure the impact since Arthur Young's efforts in the *Northern Tour*. The most direct attempt has been Timmer's (1969). His approach was based on an accounting by Arthur Young of the costs and profits of old and new methods. Young detailed the tasks in growing each crop—such as ploughing, harrowing, sowing, reaping, and so on—and estimated the labour cost of each. Timmer applied these figures to a model 'old farm' divided into permanent pasture and arable cultivated under the three-field system and to a 'new farm' cropped with a Norfolk rotation. Total employment was 45 per cent higher in the new farm than in the old. Timmer also estimated the total output of the two farms, and that increased by 24 per cent. Thus, both output and employment expanded, as Arthur Young predicted, but employment expanded more so labour productivity declined. This result contradicts Young's predictions about capital-intensive agriculture.

In this chapter I begin by estimating the impact of enclosure on farm employment. I use two independent methods. One is like Timmer's in that it begins by listing all the tasks in farming and estimates the labour required to do them. Instead of applying the coefficients to stylized farms as Timmer did, I apply them to villages in the three natural districts of the midlands as described by Parkinson's census. This procedure contrasts the effects of enclosure on employment in three different natural environments. Contrary to Timmer's 'optimistic' findings, I conclude that enclosure had no significant effect on employment when it left the arable orientation of agriculture unchanged, while it reduced employment when it led to the conversion of arable to pasture. I also measure the effects of enclosure on employment with an altogether different set of data—the hundreds of 'representative farms' minuted by Arthur Young in his tours of the late 1760s. Despite the great difference in character of the information, this procedure gives very similar results to the cost-accounting approach. By combining my measurements of the effects of enclosure on employment with my estimates in the last chapter of the effects on output, I estimate the effects of enclosure on the growth of labour productivity. In general, I find that enclosure did increase labour productivity, but the increment was quite small and not historically important.

## The Batchelor–Parkinson Accounting Model

The first approach I use to measure the effect of enclosure is a cost-accounting model of farm employment: I list all the tasks in farming, determine the number of hours needed for each, and value that time. Total employment is measured by total labour cost. By tailoring the calculations to the production patterns of open and enclosed farms, I can compare employment in the two systems.

Calculations of costs and profits were a common exercise for eighteenth- and nineteenth-century agronomists. Young's *Farmer's Guide in Hiring and Stocking Farms* (1770), Parkinson's *General Views* of Rutland and Huntingdon (1808, 1811), Batchelor's *General View* of Bedfordshire (1808: 70–160), and Morton's *Handbook of Farm Labour* (1868) are important examples, with Batchelor's and Morton's the most obsessive and thorough. The model presented here is based mainly on Batchelor's accounting, subject to cross-checks with the other sources.

Batchelor's accounts were prepared as part of his report to the Board of Agriculture on Bedfordshire. He took as his paradigm a farm of near average size with 150 acres of arable and 30 acres of grass 'possessing at least an average share of the conveniences of occupation which are commonly met with in this county' (Batchelor 1808: 71). He noted that 'nothing short of a very minute inquiry into every step of the business, is likely to furnish a proper basis of calculations' (Batchelor 1808: 70).

*Table 8–1.  Direct labour costs, fallow–wheat–beans*

|  | Cost per course (shillings and pence) | | |
|  | Fallow | Wheat | Beans |
|---|---|---|---|
| Ploughing | 6–0 | 4–0 | 2–0 |
| Harrowing |  | 0–9 | 0–6 |
| Rolling |  | 0–1.25 | 0–1.25 |
| Carriage of manure | 0–10 |  |  |
| Spreading, etc. manure | 0–1.25 |  |  |
| Clod breaking |  | 2–6 | 1–0 |
| Seeding |  | 0–4 | 0–4 |
| Weeding or hoeing |  | 3–6 | 3–0 |
| Reaping or mowing |  | 10–0 | 2–2 |
| Binding sheaves |  | 3–3 | 6–6 |
| Harvest carriage |  | 3–6 | 5–8 |
| Thatching ricks |  | 1–0 | 1–0 |
| Carriage to barns |  | 1–0 | 1–2 |
| Threshing and dressing |  | 12–3.5 | 7–6 |
| Marketing | 1–0 | 1–0 | 1–0 |
| Casualties | 0–6 | 0–6 | 0–6 |
| TOTAL | 8–5.25 | 43–8.75 | 32–5.25 |

Total over three years = 84s. 7.25d.
Average per year      = 28s. 2.42d.

*Source*: Batchelor (1808: 118).

Batchelor's method was to take every operation on the farm, subdivide it into as many tasks as possible, then estimate the time and cost of each. As an example of Batchelor's attention to detail, I will trace an acre of open field arable through the three years of the conventional rotation of fallow–wheat–beans. The direct labour costs are detailed in Table 8–1.

In the first year, the land was fallow. Even though no crop was grown, costs were incurred. Barnyard manure was carted to the field and spread evenly over it. 'The carriage occupies two or three drivers, average two and a half; and four men are employed in filling and spreading, which together with throwing into heaps' cost 3s. 5d. per load. 'At sixteen loads per acre, this part of the expense amounts to 4s. 8d. per acre for twenty acres, or 3s. 1¼d. each, if equally divided among the thirty acres of fallowing.' The other 10 acres of the hypothetical farm were folded by sheep. '182 sheep, in a fold of nine hurdles square, will occupy nearly one hour and an half of the shepherd's time in setting the fold, and driving backwards and forwards to the sheep-walks. The expense is consequently 3d. per head for half a year's folding.'[2] The field was ploughed three times to control weeds and to work in the manure. Table 8–1 shows a cost of marketing, since Batchelor

[2] Batchelor (1808: 94). The cost of sheepfolding is not shown in Table 8–1 since sheep expenses were included in the livestock accounts.

estimated that on a per acreage basis over the land in tillage. 'The market charges are about 3s. per day for fifty days, or 1s. per acre over 150 acres arable' (Batchelor 1808: 112).

Wheat, which succeeded the fallow, was the most expensive crop. The field was ploughed again, the land was harrowed and rolled for planting. Clods had to be broken. 'The open field farmers who sow wheat after summer fallows, seldom take any trouble to make the soil break into small pieces previous to wheat-sowing, as this is supposed to increase the May-weed, scratch-burs, &c. Consequently a dry autumn causes much expense in breaking the clods with spades, hoes, &c.' (Batchelor 1808: 105). Weeds were a recurring problem. 'The expense of weeding is extremely variable. The crops of wheat which succeed fallows could not, in some seasons, be properly cleaned for 10s. per acre; in other cases the expense is trivial' (Batchelor 1808: 105). Batchelor thought the average cost was 3s. 6d. Harvesting was the most expensive operation since it involved many labour-intensive tasks. The wheat was reaped and bound into sheaves. They were then carried to the barnyard where most were stacked in thatched ricks for storage. 'Cutting of the haulm or wheat stubble, costs about 1s. 6d. per acre, and may be included in the harvest account; it is indeed sometimes performed in the time of harvest by those who have neglected to provide straw for thatching the stacks of corn' (Batchelor 1808: 108). Over the course of the winter the corn was gradually transferred to the barn where it was threshed for sale.

The next spring the field was planted with beans. This again required ploughing, harrowing, rolling, and sowing. Clodding and weeding were less expensive than for wheat. Mowing the beans was much cheaper than reaping the wheat, but it was more expensive to bind beans into sheaves and transport them to the farm yard. Like wheat, the beans were stored in ricks and gradually threshed out over the course of the winter. All told, beans required less labour than wheat but much more than fallow.

In addition to the costs directly assignable to the various crops, Batchelor estimated the costs of grassland. He distinguished weeding, banking, spreading mole-hills, manuring dairy pastures and meadows, mowing meadows, making hay, loading, stacking, and thatching the hay, operating the dairy, folding the sheep, and tending the stock. In addition Batchelor worked out the cost of overhead activities like road maintenance, hedging, ditching, and miscellaneous expenses.

One must worry about the accuracy of Batchelor's estimates. There are three approaches to checking them. First, one must consider whether they were produced by a careful mind; the length and meticulousness of Batchelor's discussions is encouraging in this regard. Second, one can

compare Batchelor's work to that of other investigators. The best
alternative is Morton's, which was carried out sixty years later. While some
activities had changed—steam threshing was an innovation since
Batchelor's day—many were the same and Morton often gave the time and
cost of the older method to show the advantages of the newer. In the event,
Batchelor's and Morton's lists of tasks are very similar as are their cost
estimates. Third, one can compare the costs implied by Batchelor's figures
with farm survey data. I report on the results of such comparisons later, but
it should be noted now that they do substantiate Batchelor's accounting.

Using Batchelor's figures I can compute the labour cost of a farm given
the character of the soil, the land use pattern, the numbers of the various
livestock, and whether or not the sheep were folded. Table 8–2 shows the
results for the villages surveyed by Parkinson.

*Table 8–2. Enclosure and employment: Batchelor–Parkinson model (£ per 100
acres of farm land)*

|  | Open (£) | Enclosed (£) |
|---|---|---|
| Heavy arable[a] | 139.47 | 141.05 |
| Light arable[b] | 141.63 | 131.45 |
| Pasture[c]—open | 113.87 | — |
| Pasture—Rutland old | — | 76.23 |
| Pasture—Rutland parl. | — | 87.49 |
| Pasture—Hunts. old | — | 92.49 |
| Pasture—Hunts. parl. | — | 94.29 |

[a] Heavy arable: the enclosed figure is for villages enclosed by
act.
[b] Light arable: the open figure is for open, sheepfolding villages;
the enclosed figure is for non-sheepfolding villages enclosed by
act.
[c] Pasture: the open figure is for the hypothetical village
discussed in the text. For enclosed villages, employment was
computed for four cases—parliamentary and non-parliamentary
villages in Hunts. and Rutland.

The effect of enclosure on employment was usually negative and
depended on the natural district. The heavy arable district was the only one
in which employment increased and the gain was only 1 per cent.
Furthermore, this was the least important district. Only 19 per cent of the
land enclosed in the south midlands between 1750 and 1849 lay in the
heavy arable district. In the light arable district—the land of clover and
turnips—employment *declined* by 7 per cent. This district included 32 per
cent of the land enclosed 1750–1849. In the pasture district, which included
49 per cent of the land enclosed between 1750 and 1849, open field farms
offered less employment than in the arable districts because open field
farmers in the pasture district put more land down to grass. Even so,

employment was lower in the enclosed villages. Job losses ranged from 17 per cent to 33 per cent depending on the acreage withdrawn from tillage. These results strike at the heart of Tory fundamentalism. The overall effect of enclosure in the south midlands was to reduce farm employment 12 per cent.[3]

One advantage of the cost-accounting model is that I can assess in detail the Tory claim that enclosure led to new jobs that replaced old ones. Since the days of Fitzherbert, hedging and ditching has been the prime example. Batchelor (1808: 114) was to the point on its importance:

Hedges are commonly cut every twelve years: the expense of cutting, laying, ditching, with occasional repairs, and a faggot for every day's work, is about 8*d.* per pole, or 73*s.* 4*d.* for 110 poles, which is nearly 5*d.* per acre per annum over the whole farm.

5*d.* per acre per annum! Reference to Table 8–1 shows the triviality of this expense compared to most arable operations. The maintenance of hedges and ditches amounted to only a 1.5 per cent increase in farm employment according to Table 8–2.

Ever since Arthur Young's accounts of the squads of hoers, turnip culture has been the second Tory example of enclosure's raising employment. Indeed, enclosure did lead to an increase in the share of the arable planted with turnips, so it is at first surprising that enclosure in the light arable district lowered employment.

Table 8–3 shows the direct labour costs in the four courses of the Norfolk rotation. Contrasting it with Table 8–1 provides some insight into why a conventional three-field rotation gave more employment than a Norfolk rotation. Table 8–3 confirms that hoeing turnips—two times—was, indeed, expensive. Notice, however, three other features. First, turnips were generally drawn and left on the land to be eaten by sheep. This was far less expensive than the cost of harvesting grain, carting it to the rick yard then the barn, and threshing it. Threshing alone cost more than hoeing turnips. This immediately alerts us to the possibility that the employment-creating effects of turnips have been exaggerated. Second, wheat and beans both had higher labour costs than turnips. With the Norfolk rotation, only one

---

[3] I compute the average relative change in employment as $0.880 = (0.491 \times 0.798) + (0.322 \times 0.928) + (0.188 \times 1.011)$. The first term in parentheses applies to the pasture district, the second to the light arable district, and the third to the heavy arable district. The first factor in each term is the share of the land in the south midlands enclosed between 1750 and 1849 that lies in that district. These weights sum to one. The second factor in each term is the ratio of employment per acre for enclosed agriculture relative to open agriculture for that district. These ratios are computed from Table 8–2. For the pasture district, I used the average labour cost per acre for parliamentary enclosures in Rutland and Hunts. (£90.98).

*Table 8–3. Direct labour costs, turnips–barley–clover–wheat*

| | Cost per course (shillings and pence) | | | |
|---|---|---|---|---|
| | Turnip | Barley | Clover | Wheat |
| Ploughing | 7–6 | 1–6 | | 1–6 |
| Harrowing | 2–5 | 0–6.75 | 0–2.25 | 0–9 |
| Rolling | 0–3.75 | 0–1.25 | 0–1.25 | 0–2.5 |
| Carriage of manure | 0–10 | | | |
| Spreading, etc. manure | 3–1.25 | | | |
| Clod breaking | 2–0 | | | |
| Seeding | 0–4 | 0–4 | | 0–4 |
| Weeding or hoeing | 12–0 | 3–0 | 1–0 | 3–0 |
| Reaping or mowing | | 2–2 | 2–3 | 10–0 |
| Binding sheaves | | 2–8 | 2–6 | 3–3 |
| Harvest carriage | | 4–4 | 2–2.5 | 3–6 |
| Thatching ricks | | 1–0 | 0–6 | 1–0 |
| Carriage to barns | | 1–2 | | 1–0 |
| Threshing and dressing | | 12–6.25 | | 12–3.5 |
| Drawing | 3–9 | | | |
| Marketing | | 1–0 | 1–0 | 1–0 |
| Casualties | | 0–6 | 0–6 | 0–6 |
| TOTAL | 32–3 | 30–10.25 | 10–3 | 38–4 |

Total over four years = 111s.  8.25d  
Average per year    =  27s. 11.06d.

*Source*: All numbers from Batchelor (1808: 121) except for the cost of drawing turnips. Batchelor charged drawing to his sheep accout rather than the turnip account. The figure in the table reflects the modal expense per acre. It is distinctly high, however, compared to the expense of 1s. 6d. per acre that Young uses. See Timmer (1969: 390).

quarter of the arable was planted with wheat, a more expensive crop. In the conventional three-course rotation, two-thirds of the land was planted with the more costly wheat and beans. Third, although the fallow was eliminated with the Norfolk rotation, a quarter of the land was planted with clover, and its labour costs were almost as light as the fallow's. All told, enclosing land that was cropped as fallow–wheat–beans and planting it with a Norfolk rotation reduced direct labour costs from 28s. 2.42d. per acre per year to 27s. 11.06d. per acre per year.

This assessment of the effect of turnips on employment differs from Timmer's. He found that the 'new farm' which followed a Norfolk rotation employed 45 per cent more labour than the 'old farm' that followed a three-field system. Perhaps some of the difference is due to different assessments of the labour costs involved, but the most important source of the discrepancy is Timmer's peculiar assumptions about the balance of arable and grass in his old and new farms. His old farm was 72 per cent arable and 28 per cent grass, while the new farm was 96 per cent arable and only 4 per cent grass. Enclosure, in other words, led to the conversion of pasture to arable! Batchelor's cost-accounting would also show enclosure raising employment if it increased the share of land in tillage (irrespective of how it was cropped). The employment gain that Timmer attributes to turnips is better attributed to the reduction in pasture acreage on his 'new' farm. But how commonly did enclosure *reduce* the share of grass? Rarely in the midlands. Enclosure there either left the balance of arable and grass unchanged or shifted land from tillage to pasture. In the light arable district, enclosure and the adoption of the Norfolk four-course rotation did not lead to an expansion of farm employment because those changes did not reduce the share of grass.

Other job increases that can be credited to 'improved agriculture' were also small. Chambers (1953) contended that it led to higher yields which required more threshing. It was only in the heavy arable district that yields rose much, and threshing expense did increase. But the extra cost was only £2.79 per 100 acres—2 per cent of the wage bill. The cost of installing and maintaining drains was larger—£7.46 per acre—but it was still only 5 per cent of costs.

## Arthur Young's Farm Survey

The employment effect of enclosure is so important that it is desirable to check the result with a different approach. Arthur Young's survey of farms is an independent source that allows employment estimates to be made on a basis other than the task-by-task accounting that I have just undertaken. None the less, Young's data point to conclusions very similar to those I have just reached. This is encouraging corroboration from a most unexpected source.

On his English tours, Young collected information about the villages through which he passed, including, in many cases, the 'details of representative farms'. This information includes the acreage of arable and grass and the number of permanent employees broken down by age and sex. By comparing open and enclosed farms, I can determine the impact of enclosure on employment. I again measure total employment as the total cost of labour. Young distinguished four kinds of labour and each must be valued in its own way.

First, the servants (men, maids, and boys) were hired on annual contracts and provided with room and board in addition to a money payment. I multiplied the total wage (cash plus room and board) by the number of servants to compute the cost of annually hired labour on each farm.[4]

---

[4] For room and board, see Young (1771*b*: iv. 356). In the calculations I distinguished men, women, and boys. Further, Young usually reported the wage of a 'first class' servant and an 'ordinary servant'. I valued one male servant on each farm as a first class servant and the others as ordinary servants.

Second, Young recorded the number of labourers permanently hired. These were men who lived in their own cottages. Young reported the average weekly wage for each village he visited, and I took the annual earnings of a permanently hired labourer to have been fifty-two times the weekly wage.

Third, there was the labour of the farm family. In the absence of any information on the demographic composition of the households, I assumed each family contributed the labour of one man, one woman, and one boy.[5] This assumption conforms to what is known of the sexual division of labour in eighteenth-century farms (Pinchbeck 1930). I valued the family labour at the wage rates (including board) of 'first class' servants, dairymaids, and boys.

Fourth, supplementary labour was also hired during the harvest. I computed the cost of this labour by multiplying the acreages to be harvested by the appropriate mowing and reaping rates, which Young recorded. This procedure assumes that carting and rick-making were done by the family and the permanent labour force.

In addition to estimating the total labour supply, it was necessary to classify the farms as open, enclosed, or partly enclosed. Young did not always indicate this. Consequently, it was necessary to examine printed and archival sources in order to classify the farms. Large scale eighteenth-century county maps, many of which were drafted around 1770, were the most helpful printed sources, for they usually indicated, by varying the symbol identifying a road, whether it was passing through open or enclosed fields. In addition, I visited twenty-seven county record offices or comparable archives. The main aim was to examine manuscript maps, made about 1770, to ascertain the predominant field patterns. The property descriptions in glebe terriers, estate surveys, mortgages, deeds, conveyances, and leases were occasionally useful. Of the 231 farms Young described, 159 were enclosed, 27 were open, and 45 contained appreciable quantities of both sorts of land.

I use regression analysis to explore the determinants of farm employment per acre. In equation 1 of Table 8–4, employment per acre of men is regressed on a constant and dummy variables denoting enclosed and partially open farms. The corresponding dummy variable for open farms is excluded. Hence, the constant measures average male employment per acre for open farms, and the coefficients of the other dummy variables indicate

[5] Young distinguished 'gentleman' farms from 'common' farms. Following Young (1770: i. 246–80), I assumed that no family labour was utilized on gentleman farms. Gentleman farms were excluded in the analysis discussed in this chapter; however, they are included in the analysis in the next chapter.

*Table 8–4.  Employment per acre regressions (t-ratios in parentheses)*

|  | Equation | | | |
|---|---|---|---|---|
|  | 1 (men) | 2 (men) | 3 (woman) | 4 (women) |
| Constant | 0.03440 | 0.01810 | 0.01623 | 0.01244 |
|  | (12.548) | (5.321) | (7.225) | (7.082) |
| Encl | $-6.974 \times 10^{-3}$ | $-2.300 \times 10^{-4}$ | $3.117 \times 10^{-3}$ | $9.632 \times 10^{-4}$ |
|  | (−2.378) | (−0.081) | (1.297) | (0.382) |
| Part | $-8.132 \times 10^{-3}$ | $-4.579 \times 10^{-3}$ | $2.179 \times 10^{-3}$ | $1.044 \times 10^{-3}$ |
|  | (−2.359) | (−1.440) | (0.771) | (0.369) |
| Arab |  | 0.02156 |  | $-6.886 \times 10^{-3}$ |
|  |  | (7.053) |  | (−2.531) |
| $R^2$ | 0.02634 | 0.19524 | 0.00736 | 0.03349 |

*Notes:*
Arab = share of arable
Encl = 1 if the farm is enclosed, 0 otherwise
Part = 1 if the farm is partly enclosed, 0 otherwise
The dependent variable in equations 1 and 2 is the number of men employed per acre; in equations 3 and 4, it is the number of women employed per acre.

*Table 8–4.  Continued—Employment per acre regressions (t-ratios in parentheses)*

|  | Equation | | | |
|---|---|---|---|---|
|  | 5 (boys) | 6 (boys) | 7 (total labour) | 8 (total labour) |
| Constant | 0.01578 | 0.01763 | 1.08380 | 0.74385 |
|  | (6.576) | (5.387) | (13.538) | (7.129) |
| Encl | $2.259 \times 10^{-3}$ | $1.496 \times 10^{-3}$ | −0.14587 | −0.00516 |
|  | (0.880) | (0.548) | (−1.703) | (−0.059) |
| Part | $5.882 \times 10^{-4}$ | $1.860 \times 10^{-4}$ | −.17150 | −0.09735 |
|  | (0.195) | (0.061) | (−1.704) | (−0.998) |
| Arab |  | $-2.440 \times 10^{-3}$ |  | 0.44987 |
|  |  | (−0.830) |  | (4.798) |
| $R^2$ | 0.00520 | 0.00809 | 0.01379 | 0.10111 |

*Notes:*
Arab = share of arable
Encl = 1 if the farm is enclosed, 0 otherwise
Part = 1 if the farm is partly enclosed, 0 otherwise
The dependent variable in equations 5 and 6 is the number of boys employed per acre, whereas the dependent variable in equations 7 and 8 is labour cost per acre (£).

the difference between employment per acre in the categories they represent and open farms. In equation 1, those coefficients are negative and significantly different from zero by the usual tests. Thus, this equation indicates that enclosure lowered the employment of men.

Equation 2 identifies the source of this employment decline. In this equation the share of arable on the farm is added to the equation. Its coefficient is positive and highly significant, indicating that the employment of men increased with the share of the farm in tillage. Equally important, the coefficients of the dummy variables for enclosed and partly open farms become insignificantly different from zero. The lower level of employment per acre on these farms was due to their higher share of land under grass. Equation 2 shows that enclosure reduced the employment of men in so far as it led to the conversion of arable to pasture; otherwise enclosure had no effect.

Enclosure had less effect on the employment of women and boys. The employment of women was independent of enclosure status, although their employment *declined* as the share of arable increased—the reverse of the male pattern. The difference arises because women were extensively employed in dairies. The employment of boys was independent of both enclosure status and the share of arable.

Equations 7 and 8 examine total employment, which is the cost of harvest labour plus the cost of the labour of men, women, and boys. The determinants of total employment are similar to those of men. Equation 7 indicates that labour cost per acre fell with enclosure, although the statistical significance of the decline is marginal. When the share of arable is added to the regression (equation 8), the coefficients of the dummy variables for enclosed and partly open farms are driven towards zero and the share of arable is highly significant. Equations 7 and 8 indicate that farm employment decreased when enclosure led to the conversion of arable to pasture. Otherwise, enclosure had no effect.

The equations in Table 8–4, thus, point to the same conclusions about the employment effects of enclosure as the Batchelor–Parkinson accounting model. Indeed, the regression equations and the accounting model corroborate each other quite exactly. Farms in the arable district were about 80 per cent arable. If I substitute 0.8 for the share of arable in equation 8, I compute the labour cost per acre as £1.104 for open farms and £1.099 for enclosed farms. Young's data were collected in the late 1760s when a man's winter wage was about 1*s*. 3*d*. per day. At the time of Batchelor's survey, it had risen to 1*s*. 6*d*. Multiplying the computed values of labour cost by 20 per cent (to adjust for wage inflation) and by 100 gives the labour cost per 100 acres for a farm in the arable districts in *c*.1806. The result is £132.45 for an open farm and £131.83 for an enclosed farm. These numbers are very close to those computed in Table 8–2 from the Batchelor–Parkinson model. The correspondence is especially exact for light arable farms, which are the more appropriate test since Young did not report in detail on districts such as the heavy arable district.

The correspondence for farms in the pasture district is almost as good. An open field farm was 63 per cent arable, so equation 8 predicts that labour cost per 100 acres was £123.27 at *c*.1806 wage rates. This is only 8 per cent higher than the value predicted by the Batchelor–Parkinson model. (Table 8–2.) An enclosed farm with 20 per cent arable had a predicted labour cost of £99.44, which exceeds the Batchelor–Parkinson labour costs for most enclosed pastoral farms in Table 8–2 by a similar small margin. The two methods predict similar employment declines for enclosure in the pasture district during the eighteenth century: the Batchelor–Parkinson model predicts an employment fall of about 21 per cent—from £114 to £90 per acre. Young's survey data predict a fall of 19 per cent.

The Batchelor–Parkinson model and Young's survey data give results that are remarkably close considering the radical differences between the two estimation procedures. The similarity of results lends credence to both procedures. Therefore, one can be confident of the general conclusion— enclosure which did not affect the balance between tillage and pasture had negligible effect on employment even if it led to 'improved agriculture'. Enclosure that increased the share of grass reduced the level of employment.

## The Effect of Enclosing on Employment

Enclosed farming did not raise the long-run level of farm employment, but the process of enclosing did create new jobs. The parish had to be surveyed, the new fields had to be hedged and ditched, land had to be drained. Even though the jobs were temporary, their magnitude must be determined to assess the total employment impact of enclosure. I continue to measure the magnitude of the jobs created by the cost of the extra labour. I begin with the jobs created by the 'public costs' of enclosing, then proceed to those needed for hedging and ditching the allotment, and finish with those required for hollow draining.

PUBLIC ENCLOSURE COST[6]

When an enclosure was effected by parliamentary act, some costs were defrayed by a parish rate. These were the 'public costs' of the enclosure. They included legal and parliamentary fees incurred in drafting the bill and guiding it through parliament, the charges of the commissioners and their clerks, the expense of surveying the parish, the cost of fencing the tithe owner's allotment, and the cost of public roads, bridges, major drains, and so on. Until the mid–1790s, it was customary to attach a schedule of these costs to enclosure awards, and several historians have abstracted and studied them. After 1800 it is necessary to examine commissioners'

---

[6] Turner (1984*a*: 53–63) provides a good summary of the literature on public costs. My discussion is based on it.

notebooks to obtain the public costs, so knowledge of these costs is sketchier in this period.

In all the counties that have been studied, the public costs rose several fold between the 1770s and the post-Napoleonic period. In part this increase reflects wage increases; in part, it reflects procedural changes: before the 1790s roads were often built after the award, so their costs are not recorded. In the first decade of the nineteenth century, the public costs were in the order of £3 per acre.

Not all of this expense represented increased labour demand in the countryside. According to J. M. Martin (1967) and Turner (1984*a*), the legal and parliamentary fees and the expenses of the commissioners and their clerks represented about half of the cost of the enclosure. Evidently, these funds were not paid to farmers and village labourers. Surveying, fencing the tithe allotment, and building roads, etc., accounted for the other half. Much of the latter expenditure created jobs in the enclosed village. An expenditure of £1.5 per acre equalled £150 per 100 acres, which exceeded one year's agricultural wage bill.

HEDGING AND DITCHING

The cost of fencing the external boundary of the tithe allotment was usually borne as a public cost, but everyone else had to pay for hedging and ditching his or her own allotment.[7] This included subdividing the property into usable fields as well as fencing its perimeter. Even the tithe owners had to pay for internal fencing.

There are no schedules of fencing expenses, so the best way of estimating their magnitude is by costing out the work. Priest (1813: 335) describes the procedure as follows:

Ditching—New, according to the plan pursued in new enclosures, by a ditch two feet wide, and about one in depth, with quicking, that is putting in one row of white thorn; and mounding, that is, setting two rows of posts and rails to protect the fence, until it is grown to a sufficient height to protect itself [about five years].

The Board of Agriculture reporters were in close agreement that the cost per yard for a hedge and a ditch was about 1*s*. 7*d*. in the first decade of the nineteenth century.[8]

The total cost of fencing an allotment depended on the number of yards

[7] In the midlands, hedging and ditching were the universal way of fencing. See Parkinson (1808: 5–10; 1811: 87-93), Young (1813: 95–7), Pitt (1809*b*: 224), Batchelor (1808: 271–5).

[8] Pitt (1809*a*: 335) quotes a figure of 1*s*. 6¾*d*. for Leics., Pitt (1809*b*: 224) states the cost in Northants was 1*s*. 6*d*., Murray (1813: 63–4) reports 1*s*. 6*d*.–7*d*. for Warwicks., Young (1813: 95–7) reports the Oxon. cost as 1*s*. 7½*d*. Only Batchelor (1808: 271–5) was marginally higher at 1*s*. 11*d*.

of fencing per acre. That, in turn, depended on the size and shape of the fields. They were usually square or rectangular. Since the ratio of the perimeter to the area of a rectangle decreases as the area of the rectangle increases, large fields were cheaper to fence than small fields. Large farms generally had larger fields than small farms (Yelling 1977: 138). Batchelor (1808: 70–3) claimed that the average farm was 150 acres. 'It is probably that the average size of enclosed pastures is ten acres, and of enclosed arable fields is 15 acres each.' Since farms in the arable districts were about 80 per cent arable and 20 per cent grass, these figures imply an average field size of 14 acres. Supposing that fields were square (to minimize fencing cost per acre), recognizing that fences were shared between two fields, and assuming that the cost of fencing was 1*s.* 7*d.* per running yard implies that the cost of fencing was 59 shillings per acre.

Batchelor (1808) developed a slightly more realistic example with similar results. He drew the plan of a hypothetical 180-acre farm. It was 150 acres arable (83 per cent) and 30 acres grass (17 per cent), not far from the average proportion for an arable farm. The farm was rectangular—240 rods by 120 rods and subdivided into thirteen fields, the average size of which was indeed 14 acres. The smaller ones were rectangular and the larger ones, which took up most of the area, were square or virtually so. Batchelor assumed that a public road ran along 120 rods of the farm's boundary, so that the farm's owner had to bear the entire cost of that stretch of fence. Otherwise, he split the cost of fencing the perimeter with adjoining farms. Of course, the cost of internal fencing was entirely chargeable to the hypothetical farm. Still assuming that hedging and ditching came to 1*s.* 7*d.* per yard, the cost of fencing Batchelor's hypothetical farm was 64 shillings per acre—not far from the cost-minimizing figure of 59 shillings. Hence, I will assume that fencing new enclosures amounted to £3 per acre in the early nineteenth century.

HOLLOW DRAINING

In the heavy arable district, enclosure facilitated hollow draining, which was the source of the productivity growth. Drains were capital improvements that lasted in the order of ten years. They were paid for by farmers, not landowners. While I have already included the amortized cost of drains in the income statements for the heavy arable district, laying drains did have an employment-creating effect at the time of the enclosure. The jobs created are measured by the labour cost.

Vancouver (1795: 13, 43, 83, 97, 99, 100–1, 105) presented numerous estimates of the cost of draining heavy clay in Essex in the early 1790s, that is, just before the inflation during the time of the French Wars. There were three components to the cost. The first was 'Opening the hollow drains by ploughing four furrows upon each other', and this cost 3*d.* per 'score of rods', the usual unit of measure. Next came 'digging one spit with the broad spade 8 inches, and one spit with the land ditch spade, 14 inches, allowing 2 inches extra depth for the drains leading to the outfall', which was much more expensive at 3*s.* per score of rods. Finally, there was the 'value of the straw and expense of twisting it into a rope, in which form it is put into the drains'.[9] This expense amounted to 1*s.* 9*d.* per score, so the total cost of hollow draining was 5*s.* per score of rods.

The cost of draining an acre depended on the density of the drains. Vancouver assumed there were 8 score of rods (i.e. 160 rods) per acre. Since an acre is 160 square rods, Vancouver's assumption corresponds to assuming that the drains were dug in the furrows of the field, and that the strips were one rod wide, which was typical. Hence, the total cost was 40*s.* per acre—'an expense that is generally and necessarily incurred in this neighbourhood' (Vancouver 1795: 191). Gooch (1813: 242–4) and Batchelor (1808: 471) presented comparable estimates of the cost of hollow draining per score rods, and Batchelor speculated that the total cost of hollow draining 'may be about 3 pounds per acre'. The total cost exceeded Vancouver's estimate since wages had advanced.

### THE OVERALL IMPACT

Together the public and fencing costs amounted to £4.5 per acre—or £450 per 100 acres—in the early nineteenth century. That was three times or more the annual open field wage bill. In the heavy arable district, the cost of enclosing was even greater.

The expense of enclosing, of course, was not incurred all at once. Enclosures took several years to effect, and internal fencing and hollow draining could be delayed to ease the financing problems. Spreading the expenses over five years implies that they amounted to £90 per 100 acres per year in the pasture and light arable districts and £150 per year in the heavy arable district. This represents roughly a doubling of employment over open field levels. The Tory claim that the *process* of enclosing created a large number of new jobs is true.

Not all of the jobs, however, were filled by labourers in the affected village since much of the work was done by contractors from other places. Snell (1985: 182) reports that hedging and ditching were done by gangs of piece-workers who shifted from enclosure to enclosure. Hollow draining was done in the same way. Thus, Gooch (1813: 20) reported that 'Mr.

---

[9] Most of the value of straw represented the cost of labour in preparing it. I will treat the whole cost as labour, so as not to underestimate the employment-creating effects of enclosure.

Treslove, of Trumpington, is completely at home at hollow draining, and has wrought great improvement by it on his farm. He laughs at the idea of any wet land not being benefited by it, and engages (it is part of his profession) to drain effectually any land upon which he may be employed.' It is no surprise that contractors were employed. Repetition made them more efficient than inexperienced, local crews at hedging, ditching, and draining.

The construction projects that accompanied enclosure did increase the demand for labour, but the expansion was insignificant for two reasons. First, many of the jobs were performed by non-resident contractors and work crews, so local residents missed out on many of the new jobs. Second, the employment creation was ephemeral. After five years—certainly after ten—the hedging, ditching, and building were completed. From then on only the permanent effects of enclosure on farm employment mattered. In most of the south midlands, that effect was a reduction in farm employment.

## Enclosure and the Growth in Labour Productivity

The effect of enclosure on employment is important in its own right and also for its implications for the growth in labour productivity. I measure labour productivity as farm output (revenue) divided by farm labour cost. Table 8–5 contrasts labour productivity in open and enclosed villages in the three natural districts.[10]

*Table 8–5. Enclosure and labour productivity: Batchelor–Parkinson model (farm revenue divided by labour cost)*

|  | Open | Enclosed |
|---|---|---|
| Heavy arable | 3.03 | 3.36 |
| Light arable | 3.10 | 3.18 |
| Pasture—open | 3.39 | — |
| Pasture—Rutland old | — | 6.13 |
| Pasture—Rutland parl. | — | 3.70 |
| Pasture—Hunts. old | — | 3.17 |
| Pasture—Hunts. parl. | — | 3.81 |

*Note:* Labour productivitiy is measured as farm output from Tables 7–4 to 7–12 divided by labour costs from Table 8–2.

[10] Arthur Young's survey can be used to perform an analogous accounting, but I defer that exercise to Ch. 11, when I incorporate the effects of increasing farm size. This more elaborate accounting assigns only a small role to enclosure in explaining the growth of English agricultural labour productivity. The two sets of calculations, thus, corroborate each other.

The impact of enclosure on labour productivity depended on the natural district. In the heavy arable district, enclosure increased output per worker by 11 per cent. Since output grew faster than employment, the history of this district is in accord with the Tory model. In the light arable district, labour productivity increased by only 3 per cent. Both output and employment fell, with the latter falling more. In the pasture district, one can compare productivity in the hypothetical open field village with enclosed villages in Rutland and Huntingdon. Employment fell most in the old enclosures in Rutland and output increased the most as well, so labour productivity was 81 per cent higher in those villages than in the open village. In the other categories, labour productivity declined, or the gain was only 8–12 per cent.

The history of labour productivity does not accord well with either Tory or Marxist fundamentalism. To give them their due, there is one example of each theory: the heavy arable district is an example of the Tory model since employment expanded while ouput grew more, and the old enclosures in Rutland are an example of the Marxist model since employment fell while output increased. But a different pattern characterized the light arable district and the other parts of the pasture district. In these villages, employment and output both declined with employment falling the most. Enclosure led to a less intensive agriculture that boasted a small rise in labour productivity.

## Enclosure and Labour Productivity in Perspective

In most villages enclosure led to a gain of less than 10 per cent in ouput per worker. To gauge the importance of the gain one must compare it to the growth in labour productivity in early modern England and to England's productivity *vis-à-vis* other countries.

Several recent investigations have shown that agricultural labour productivity was higher in England than in the other advanced countries of Europe in the early nineteenth century. Bairoch's (1965) calculations for 1840 (shown in Table 8–6) demonstrate this clearly. UK labour productivity exceeded that in France by 50 per cent and in Belgium by 75 per cent. The British lead over central and eastern Europe was far greater.

O'Brien and Keyder (1978: 102–45) undertook a more detailed investigation of Anglo–French productivity differences. They measured agricultural output as the value of production in pounds for Britain and francs for France. French output was then revalued in sterling by using a cross-country index of agricultural product prices. O'Brien and Keyder (1978: 91) found

*Table 8–6. Agricultural labour productivity, 1840 (net output in millions of calories per male worker)*

| | |
|---|---|
| UK | 17.5 |
| France | 11.5 |
| Belgium | 10.0 |
| Germany | 7.5 |
| Sweden | 7.5 |
| Switzerland | 8.0 |
| Italy | 4.0 |
| Russia | 7.0 |

*Source*: Bairoch (1965: 1096).

that British labour productivity was more than 40 per cent greater than French after 1815.[11] Their finding is consistent with Bairoch's calculation.

Wrigley (1985) has ingeniously pursued the question of labour productivity back to 1500. For a country self-sufficient in food, a rise in labour productivity in agriculture allows the non-agricultural work-force to expand. Wrigley has computed the ratio of the total population to the agricultural population for England and France between 1500 and 1800. Table 8–7 shows his results. Labour productivity was similar in the two countries in 1500 and 1600 and grew little if at all in that century. In the seventeenth and eighteenth centuries productivity growth was slow in France but rapid in England. Between 1500 and 1800 English labour productivity increased 88 per cent.[12] By 1800, labour productivity in England was 46 per cent higher than in France. That differential agrees with the direct calculations of O'Brien and Keyder for the post–1815 period.

*Table 8–7. English and French 'agricultural labour productivity', 1500–1801*

| | England | France |
|---|---|---|
| 1500 | 1.32 | 1.38 |
| 1600 | 1.43 | 1.45 |
| 1700 | 1.82 | 1.58 |
| 1750 | 2.19 | 1.63 |
| 1801 | 2.48 | 1.70 |

*Note*: The table shows for each country the ratio of the total population to the agricultural population. The English figure shown for 1500 is properly for 1520.

*Source*: Wrigley (1985: 720).

[11] For 1781–90, O'Brien and Keyder found no difference in labour productivity between the two countries. However, this calculation is less surely founded than the calculations for the post-Napoleonic period.

[12] Clark (1987) has come to roughly the same conclusion using a different methodology.

If one considers my estimates of the effect of enclosure on labour productivity in the light of these aggregate calculations, the effect of enclosure becomes puny indeed. The 10 per cent rise in productivity that was due to enclosure in most districts was far less than either the growth in productivity that occurred between 1500 and 1800 or the gap between English and French productivity. Only the old enclosures in Rutland increased productivity by an amount approaching these differentials. And these enclosures were quite atypical.

One thing is clear—England's exceptionally high level of labour productivity in 1800 was not due to enclosure.

## Enclosure, Employment, and Labour Productivity

The evidence I have presented in this chapter suggests that neither the Tory nor the Marxist account of the impact of enclosure on labour productivity is adequate. The Tory account is flawed. While the process of enclosing did increase the demand for labour in the short term, the Tory's 'optimistic' contention that enclosure led to increased permanent employment was false everywhere, except for the heavy arable district and there the gain was only 1 per cent. In general, enclosure had no effect on employment when land remained under corn even if 'improved agriculture' was adopted (as it usually was). When arable was converted to pasture, employment declined. The Marxist account of the impact of enclosure on employment is, therefore, much closer to the facts.

While the Marxist account is closer to the truth in this regard, it is off the mark with respect to labour productivity. Often, the employment declines that followed enclosure were accompanied by some decline in output. Hence, the gain in labour productivity was rarely substantial. This conclusion gains force when one considers the effects of enclosure on labour productivity from an international or an intertemporal perspective. These comparisons show that enclosure's contribution to the rise in agricultural labour productivity was small in relation to Britain's lead over other European countries and smaller still with respect to the growth in labour productivity in early modern English agriculture. Enclosure had surprisingly little effect on output per worker.

# 9

# Rent Increases and Overall Farm Efficiency

When in particular districts, improvements are introduced which tend
to diminish the costs of production, the advantages derived from them
go immediately, upon the renewal of leases, to the landlords, as the
profits of stock must necessarily be regulated by competition, according
to the general average of the whole country. Thus the very great
agricultural improvements which have taken place in some parts of
Scotland, the north of England, and Norfolk, have raised, in a very
extraordinary manner, the rents of those districts, and left profits where
they were.

T. R. Malthus, *Principles of Political Economy*: 1836: 161.

THE case that enclosures were important for the growth in agricultural
efficiency has always rested primarily on two kinds of evidence. The first
was the observation that land use changed after enclosure and that enclosed
farmers were more likely to adopt improved methods than were open field
farmers. The evidence I presented in Chapter 6 supports this assessment, but
its significance is called into question by the evidence I presented in Chapters
7 and 8, which shows that enclosure (and the new methods that
accompanied it) made only a modest contribution to the growth of yields
and labour productivity in most parts of the south midlands. It is all the
more important, therefore, to examine the second kind of evidence for the
importance of enclosures. That evidence is the increase in rent that usually
followed them.

Table 9–1 summarizes the history of rents in the south midlands from
1450 to 1850. Throughout the period, enclosed rents per acre in the pasture
district were about double or treble open field rents. The rent gap between
open and enclosed land in the arable districts, where enclosing only got
under way in the eighteenth century, was less stable and reached a
maximum during the Napoleonic Wars.[1] The aim of this chapter is to

---

[1] None of these comparisons are exact—open field land was usually tithed and open field
farms often had access to commons whose acreage is not accounted for.

*Table 9–1. Rent in the south midlands, 1450–1849 (shillings per acre)*

| | Pasture | | Light arable | | Heavy arable | |
|---|---|---|---|---|---|---|
| | Open | Enclosed | Open | Enclosed | Open | Enclosed |
| 1450–1474 | 0.49 | 0.99 | 0.59 | | 0.71 | |
| 1475–1499 | 0.32 | | 0.59 | | 0.76 | |
| 1500–1524 | 0.37 | 1.05 | 0.42 | | 0.74 | |
| 1525–1549 | 0.45 | 1.70 | 0.65 | | 0.61 | |
| 1550–1574 | 0.60 | 3.58 | 0.49 | | 0.43 | |
| 1575–1599 | 3.38 | 7.11 | 2.93 | | 4.62 | |
| 1600–1624 | 5.53 | 13.60 | 6.53 | | 5.85 | |
| 1625–1649 | 6.71 | 15.24 | 5.29 | | | |
| 1650–1674 | 6.02 | 17.57 | 6.18 | 6.50 | 5.78 | |
| 1675–1699 | | 17.60 | 6.25 | | 5.83 | |
| 1700–1724 | 9.61 | 16.00 | | | | |
| 1725–1749 | 9.29 | 13.89 | 10.52 | 11.98 | 6.91 | |
| 1750–1774 | 8.69 | 15.70 | 11.13 | 14.38 | 8.09 | 10.32 |
| 1775–1799 | 9.33 | 19.80 | 11.77 | 16.71 | 7.61 | 14.84 |
| 1800–1824 | 15.18 | 29.73 | 19.33 | 25.82 | 9.70 | 21.16 |
| 1825–1849 | | 33.28 | 29.98 | 29.32 | | |

*Note:* The table is based on 1678 rent quotations drawn from estate surveys, rentals, and valuations—both in manuscript and as summarized in the *Victoria County Histories* and the secondary literature. The aim was to measure commercial or rack rents, so rents that appeared to be customary and non-economic were eschewed. Entry fines were amortized and added to quit rents.

It should also be noted that rents at the beginning of the period are not strictly comparable to rents at the end since early leases required tenants to pay for repairs, while landlords often paid for repairs by the end of the period.

*Source:* Allen (1988a: 43)

understand why rents rose when land was enclosed. Do the higher rents indicate an increase in farm efficiency, as Agrarian Fundamentalists usually suppose, or do the higher rents represent a redistribution of income from farmers to landlords?

Arthur Young entertained both possibilities. Not surprisingly, in the *General Report on Enclosures* he suggested that higher rents were paid from increases in output:

The great rise in rent which has taken place in nineteen cases out of twenty should alone be accepted as a sufficient proof of this point [i.e. the higher productivity of enclosed farming], for what can be the inducement to the farmer to give double or treble his former rent unless with a view to the profit of his business. And how is he to increase his profit without increasing the value of his produce? (Young 1808: 37–8, as quoted by Yelling 1977: 210)

The obvious retort is that a reduction in employment would raise a farmer's profits, thereby allowing a higher rent to be paid 'without increasing the value of his produce', but, in either case, the advance in rent indicates greater efficiency.

Young's explanation of the rent increase has a long pedigree. Fitzherbert claimed that enclosures led to convertible husbandry. As a result, 'If an acre of lande be worthe sixe pens or it be enclosed, it will be worth viii pens whan it is enclosed, by reason of the compostyng and dongyng of the cattell, that shall go and lye upon it both day and nighte' (Tawney and Power 1924: iii. 23). In the seventeenth century, Fortrey also argued that enclosure raised rents by improving agriculture:

the common fields . . . will not let for above one third part so much, as the same land would do inclosed, and always several . . . And all this by reason of the many several Interests: whence it is, that men cannot agree to employ it to its properest use, and best advantage: . . . all which inconveniences, would by inclosure be prevented. (Fortrey 1663: 16)

The idea that rising efficiency was responsible for rising rents was a common argument in the eighteenth and nineteenth centuries. Malthus, for instance, claimed:

For the very great increase of rents which has taken place in this country during nearly the last hundred years, we are mainly indebted to improvements in agriculture, as profits have rather risen than fallen, and little or nothing has been taken from the wages of families, if we include parish allowances, and the earnings of women and children. Consequently these rents must have been a creation from the skill and capital employed upon the land, and not a transfer from profits and wages, as they existed nearly a hundred years ago. (Malthus 1836: 197)

In recent years, this line of reasoning has been incorporated into neoclassical assessments of enclosure. Thus, McCloskey (1972: 33): 'The increase in rent, then . . . can be used as an estimate (although biased downwards by not including the value of the increased employment of the other, mobile factors of production) of the increase in the value of output, resulting from enclosure.'

These claims are difficult to reconcile with the evidence directly bearing on the relative efficiency of open and enclosed agriculture. In the last two chapters, I have examined that evidence systematically and concluded that efficiency gains were modest or confined to particular regions. If productivity growth was not the cause of rising rents after enclosure, the increase must have been due to a redistribution of income from farmers to landlords. That is the second possible explanation for the rent rise. Arthur Young also raised this possibility in his *General Report on Enclosures*: 'If profit be measured by a percentage on the capital employed, the old system [open field arable] might, at the old rents, exceed the profits of the new [enclosed]; and this is certainly the farmer's view of the comparison' (Young 1808: 31–

2). Like the argument that enclosure raised rents by increasing efficiency, the argument that enclosure redistributed income in favour of landlords has a long history among opponents of enclosure.

## The Ricardian Theory of Rent

The Agrarian Fundamentalist argument that rent increases were caused by increases in farm efficiency depends on the theory of rent propounded by Ricardo and elaborated by later economists. Rent and efficiency are linked by the concept of 'Ricardian surplus'.

The Ricardian surplus of a farm equals the revenues of the farm less the costs of the labour and capital supplied by the tenant:

$$sT = pQ - wL - iK \qquad \text{(Equation 9–1)}$$

Here $s$ is Ricardian surplus per acre, $T$ is the acreage of the farm, and $p$ and $Q$, $w$ and $L$, $i$ and $K$ are, respectively, the prices and quantities of output, labour, and capital. $Q$ includes all final output[2] (including any produce consumed by the farm family) and $L$ and $K$ include the labour and capital of the farm family. Since the values of the labour and capital supplied by the landlord are not subtracted from farm revenue in equation 9–1, the Ricardian surplus of the farm, as defined by that equation, includes the value of structures, roads, ditches, hedges, any other improvements, and repairs performed by the landlord, as well as the value of the land *per se*.[3]

The classical economists believed that farm rents tended to equal Ricardian surplus less rates. 'It sometimes happens that, from accidental and temporary circumstances, the farmer pays more, or less, than this; but this is the point towards which the actual rents paid are constantly gravitating' (Malthus 1836: 136). This gravitation was the result of two assumptions. The first was psychological: farmers explicitly calculated profits using equation 9–1. The second was mobility for tenancies: farmers were not attached to a farm by habit, sentiment, or tradition. They would quit a farm if the rent was set so high that they could not earn the usual return on their capital and labour. Further, they would move to another farm, if it were available on better terms.

[2] Agricultural products like hay that are produced on the farm and consumed by livestock on the farm are intermediate inputs, not final outputs, and are not included in $Q$.

[3] Ricardo drew a distinction between the rent of a farm and the rent of 'land'. He defined the latter to be 'that portion of the produce of the earth which is paid to the landlord for the use of the original and indestructible powers of the soil' (Ricardo 1817, ed. Fellner, p. 29). When a tenant leased an actual farm, his rent included payment for improvements (buildings, roads, hedges, drains, etc.) and any maintenance provided by the landlord, as well as payment for the 'original and indestructible powers of the soil'.

Calculation and mobility imply that farm rents equalled the Ricardian surplus (less rates). No farmer would pay more than this for a farm since the resulting income would be less than he and his family could earn applying their labour and capital to other activities. Every farmer would gladly pay less than the Ricardian surplus (less rates), but competition among prospective tenants ensured that the rent equalled that amount. 'If the original tenant refused [to pay that sum], some other person would be found willing to give' it (Ricardo 1817: 32).

Finally, Ricardian surplus also measures farm efficiency. In Chapters 7 and 8 I measured efficiency with output per acre and output per worker. The sense in which Ricardian surplus measures efficiency is broader than these—it is what economists call 'total factor productivity', that is, output relative to the utilization of all factors of production rather than just one as in output per acre or per worker. Equation 9–1 suggests that there is a relationship between Ricardian surplus and total factor productivity (TFP).[4] Suppose that the quantities of the inputs remain fixed and output increases —so TFP rises. In that case, farm revenues ($pQ$) rise with respect to costs ($-wL - iK$) and Ricardian surplus goes up. Alternatively, suppose that the quantity of output remains fixed and input use declines. Now surplus rises since revenues remain constant and costs fall. In either case, Ricardian surplus per acre rises if TFP rises.

While changes in Ricardian surplus per acre point in the same direction as changes in TFP,[5] Ricardian surplus overstates the magnitude of the TFP change. Suppose input use remained constant at £4 while output increased from £5 to £6 per acre. In this case TFP increases by 20 per cent, but Ricardian surplus doubles from £1 to £2: the change in Ricardian surplus per acre overstates (as a percentage) the change in TFP. The following formula provides an approximate adjustment to compute TFP changes from changes in Ricardian surplus :

$$\left\{\frac{s_e}{s_o}\right\}^a = \frac{A_e}{A_o} \qquad (Equation\ 9\text{--}2)$$

*Here* $s_e$ is Ricardian surplus per acre for enclosed farms, $s_o$ is Ricardian surplus per acre for open farms, $^a$ is the ratio of Ricardian surplus to farm revenue, and $A_e/A_o$ is the ratio of total factor productivity of enclosed farms relative to open farms.[6]

[4] Allen (1982) explores the relationship in detail.
[5] This statement is true only if the structure of agricultural prices is not changing.
[6] See Allen (1982) for a derivation of equation 9–2.

## Rent Rises and Productivity Growth in 1806

Ricardian surplus is a powerful tool for analysing rent and efficiency. By comparing the Ricardian surpluses of open and enclosed farms, I can compare their total factor productivity. This is a more encompassing test of efficiency than the comparisons of yield per acre and output per worker I undertook in Chapters 7 and 8. By comparing Ricardian surplus to rent plus taxes, I can test the Ricardian theory of rent. Finally, by verifying whether the rise in rent at enclosure matches a rise in Ricardian surplus, I can test the hypothesis of Young, Malthus, *et al.*, that the rent rise reflected a rise in farm efficiency rather than a redistribution of income from farmers to landlords.

To carry out these tests for the villages surveyed by Parkinson in 1806, it is necessary to prepare an income statement for each type of farm. Revenues and labour costs have already been computed, and I also estimate capital and incidental costs.[7] The income statements are shown in Tables 9–2 to 9–4. Clearly, enclosure had very different effects in the three districts.

Table 9–2. *Income statements, heavy arable district, c. 1806 (£ per 100 acres of farm land)*

|  | Open | Enclosed by act |
|---|---|---|
| *Revenue* | 423.26 | 474.44 |
| *Costs* | 312.00 | 312.40 |
| of which— |  |  |
| seed | 59.56 | 63.94 |
| labour | 139.47 | 141.05 |
| animals, capital costs | 33.38 | 30.84 |
| animals, other costs | 54.20 | 37.48 |
| Implements and material costs | 25.38 | 39.09 |
| *Ricardian Surplus* | 111.26 | 162.04 |
| of which— |  |  |
| rent | 58.00 | 101.38 |
| taxes | 14.33 | 21.61 |
| tithe | 23.92 | 1.27 |
| farmer's surplus | 15.02 | 37.78 |

Enclosure in the heavy arable district worked much as defenders of enclosure expected. Revenue per acre was about half a pound higher in the villages enclosed by act than in the open field villages. Costs were almost identical in the two cases. As a result, Ricardian surplus was about half a pound per acre higher in the enclosed villages. With no change in costs, the rise in output measures the rise in total factor productivity—12 per cent.[8]

---

[7]  Full details are reported in Appendix II.

[8]  This result can also be established with equation 9–2. According to Table 9–2, Ricardian

*Table 9–3. Income statements, light arable district, c. 1806 (£ per 100 acres of farm land)*

|  | Sheepfolding, open | Non-sheepfolding, enclosed by act |
|---|---|---|
| Revenue | 438.68 | 418.06 |
| Costs | 320.31 | 288.76 |
| of which— |  |  |
| seed | 60.45 | 47.18 |
| labour | 141.63 | 131.45 |
| animals, capital costs | 33.72 | 36.50 |
| animals, other costs | 59.70 | 50.78 |
| Implements and material costs | 24.80 | 22.84 |
| Ricardian Surplus | 118.37 | 129.31 |
| of which— |  |  |
| rent | 70.83 | 107.14 |
| taxes | 13.40 | 15.95 |
| tithe | 19.42 | 3.71 |
| farmer's surplus | 14.72 | 2.50 |

*Table 9–4. Income statements, pasture district, c. 1806 (£ per 100 acres of farm land)*

|  | Open field | Rutland | | Huntingdon | |
|---|---|---|---|---|---|
|  |  | non-parl. | parl. | non-parl. | parl. |
| Revenue | 386.03 | 467.67 | 323.93 | 292.85 | 358.81 |
| Costs | 283.36 | 321.31 | 233.29 | 203.28 | 255.57 |
| of which— |  |  |  |  |  |
| seed | 45.79 | 6.69 | 15.16 | 25.00 | 27.40 |
| labour | 113.87 | 76.23 | 87.49 | 92.49 | 94.29 |
| animals, capital costs | 49.46 | 195.41 | 73.59 | 35.88 | 85.09 |
| animals, other costs | 56.73 | 37.18 | 49.41 | 38.19 | 36.98 |
| Implements and material costs | 17.51 | 5.80 | 7.64 | 11.72 | 11.81 |
| Ricardian Surplus | 102.67 | 146.37 | 90.64 | 89.57 | 103.24 |
| of which— |  |  |  |  |  |
| rent | 75.90 | 123.93 | 123.18 | 104.00 | 107.22 |
| taxes | 14.00 | 11.92 | 22.58 | 14.33 | 18.46 |
| tithe | 20.00 | 8.98 | 0.00 | 15.15 | 0.00 |
| farmer's surplus | −7.23 | 1.54 | −55.12 | −43.91 | −22.44 |

*Notes:*
For the hypothetical open field farm:
Rent: £15 18s. per acre, from Table 9–1 (open land in the pasture district in 1800–24).
Taxes and tithes: approximate values for open field villages in the arable districts.
Farmer's surplus: computed as residual.

Defenders of enclosure could take less heart from the light arable district. The enclosed villages had lower revenue per acre than the open ones, but

surplus increased by a factor of 1.46 (which equals $s_e/s_o$) and the average share of surplus in revenue equals 0.3 (which equals $a$). Substitution of these values in equation 9–2 implies that TFP increased 12 per cent.

even lower costs with the result that surplus per acre increased £10.94 per 100 acres. There were efficiency gains from enclosure and improved agriculture, but the rise in TFP was only 3 per cent.

The effects of enclosure were not uniform across the pasture district. Ricardian surplus per acre was highest in non-parliamentary enclosures in Rutland. If an open field village were enclosed and turned into a village like that, then enclosure raised productivity. Output increased by about £80 per 100 acres; most land was put down to grass; costs rose due to the very high stocking rates, and surplus rose £45 per 100 acres. This represents an 11 per cent gain in TFP.

This scenario, however, did not regularly occur in the eighteenth century. Although parliamentary enclosures led to the conversion of arable to pasture, the process was not carried as far in the new enclosures as in the old, and the stocking rates in the parliamentary enclosures were not as high as in the old enclosures in Rutland. Consequently, Ricardian surplus remained constant or declined marginally. Eighteenth-century enclosures in the pasture district probably did not raise TFP.

These comparisons of total factor productivity confirm the general conclusions about the efficiency effects of enclosure that I reached by the study of yields and labour productivity in Chapters 7 and 8. In the cases where enclosure raised TFP, it is possible that the rise in efficiency accounted for the rise in rent. In other districts that conclusion is out of the question.

Thus, the efficiency gains in the heavy arable district did account for the rise in rent. Rent, taxes, tithes, and farmer's surplus were paid out of Ricardian surplus. Tithes fell to almost nothing in villages enclosed by act since the land was usually exonerated from tithe. (Some land was transferred to the tithe owners in compensation.) Taxes, however, were slightly higher in enclosed villages. The increases and decreases were close to balancing, so the rise in rent was almost equal to the rise in Ricardian surplus, which, in turn, equalled the rise in farm output. Arthur Young (1804*b*: 502) claimed of the farmers of Knapwell, Cambridgeshire, which had been enclosed twenty years before, 'They have been enabled to pay their increased rents by hollow draining'. Table 9–2 bears this out.

In the light arable district, the rent increases required a reduction in the income of farmers. Thus, rent increased by £36.31 per 100 acres, which exceeded the rise in surplus (£10.94). The decline in tithe payments by £15.71 accounted for part of the rent increase with the rest coming out of farmers' surplus, which went down by £12.22 per acre. A rent increase of £24.09 (= £36.31 − £12.22) per 100 acres would have been justified by the rise in agricultural efficiency and the exoneration of tithes. This increase would have amounted to only 20 per cent of the open field rent—a small increment.

In the pasture district, the relationship between rent change and efficiency change was mixed. Only in the old enclosures in Rutland did the rise in efficiency justify the rise in rent: rents could be raised about £55 per 100 acres given the increase in surplus and the elimination of tithes, and, in fact, rent was raised £48 per 100 acres. In other parts of the pasture district, Ricardian surplus per acre increased only trivially or actually declined. Hence, rent increases in these enclosures represented transfers of income from farmers to landlords rather than the return to rising efficiency.

In summary, enclosure significantly raised total factor productivity and justified higher rents in only two of the comparisons—in the heavy arable district and in the old enclosures in Rutland. These were the districts where enclosure led to higher yields and labour productivity. Elsewhere, the efficiency effects were small and did not justify the increases in rent that followed enclosure.

## Enclosure in the Eighteenth Century

The parliamentary enclosures in the pasture district occurred in the middle of the eighteenth century—a generation or two before Parkinson's visitation. Why were these villages enclosed at all if enclosure did not raise the efficiency of agriculture? Two factors were involved. First, the efficiency gains in the mid-eighteenth century were larger than Table 9–4 suggests. Efficiency, in the sense of TFP, depends on the structure of farm prices since it involves the difference between farm revenues and costs. In the mid-eighteenth century, corn was relatively cheap and labour relatively dear compared to their values in 1806. Since enclosure in the pasture district saved labour and reduced corn output, it increased surplus more when mid-eighteenth-century prices are used to compute Ricardian surplus than when prices *c.*1806 are used. Recomputing Table 9–4 with mid-eighteenth-century prices raises the apparent benefits of enclosure in the pasture district, but not by enough to explain the whole rise in rent.

Second, enclosures in the mid-eighteenth century also redistributed income from farmers to landlords. I can use the data that Arthur Young collected on his tours of England to substantiate this point. These data were collected in the late 1760s—right in the middle of the first spurt of parliamentary enclosures—so they provide direct evidence on their effects. Young's data confirm the implication of Parkinson's data that these enclosures mainly redistributed income instead of raising efficiency.[9]

For each of Young's sample farms, I computed Ricardian surplus per acre using procedures like those I used with Parkinson's data. I put Young's data

---

[9] This section summarizes the analysis of Allen (1982), and the reader is referred to that article for elaboration.

to two tests. The first, and simpler, was to compare rent to Ricardian surplus. For each farm, I computed farmer's surplus per acre, i.e. Ricardian surplus minus taxes minus rent. If Ricardo's theory were correct, farmer's surplus would equal zero. Further, if farmer's surplus equalled zero for both open and enclosed farms then rents equalled the value of the land, and it is likely that enclosure raised rents by raising efficiency. On the other hand, if farmer's surplus equalled zero for enclosed farms but was positive for open farms, then the rise in rents that accompanied enclosure probably indicates a redistribution of income.

Table 9–5 shows the frequency distribution of farmer's surplus per acre for enclosed, open, and partially open farms. The distribution for enclosed farms is centred near zero with a mean of £0.2351 per acre. In contrast, farmer's surplus is positive for almost every open or partially open farm and the means of those distributions are £1.2233 and £0.8847 respectively. Table 9–5 suggests that enclosure raised rents by redistributing income from farmers to landlords.

The second test was to use the estimates of Ricardian surplus to compare the total factor productivity of open and enclosed farming. This comparison is much more complicated than the corresponding comparison with Parkinson's data since Young's farms were scattered over the whole of England. Hence, the comparisons are affected by regional differences in prices, differences in the fertility of the soil, and large variations in the amount of common pasture attached to the farms. The adjustment for these differences is complicated, and I have discussed it elsewhere (Allen 1982).

*Table 9–5. Farmer's surplus per Acre*

| Farmer's surplus per acre | Number of farms | | |
|---|---|---|---|
| | Enclosed | Open | Partly open |
| less than −1.5 | 4 | — | — |
| −1.0 to −1.5 | 4 | — | — |
| −0.5 to −1 | 13 | 1 | 3 |
| 0 to −0.5 | 46 | 2 | 6 |
| 0.5 to 0 | 38 | 3 | 11 |
| 1.0 to 0.5 | 31 | 5 | 6 |
| 1.5 to 1.0 | 8 | 7 | 7 |
| 2.0 to 1.5 | 5 | 5 | 6 |
| 2.5 to 2.0 | 5 | 1 | 1 |
| 3.0 to 2.5 | 4 | 1 | 3 |
| 3.0+ | 1 | 2 | 2 |
| Number | 159 | 27 | 45 |
| Mean farmer's surplus per acre | 0.2351 | 1.2233 | 0.8847 |

*Note:* farmer's surplus equals a farm's surplus minus taxes and tithes paid minus the rent actually paid.

The conclusion is important here: there was no systematic difference in efficiency between open and enclosed farms in Young's sample.

The data collected by Arthur Young in his tours of England therefore support two conclusions. First, only half of the surplus generated by open field farms accrued to the landlord as rent and to the church and state as tithes and rates. Second, enclosure did not raise efficiency. Together these findings indicate that the major economic consequence of the enclosure of open field arable in the eighteenth century was to redistribute the existing agricultural income, not to create additional income by raising efficiency. And it was this redistribution of income that made the mid-eighteenth-century enclosures profitable to landlords.

## The Ricardian Theory of Rent: An Assessment

The comparisons I have presented between rents and Ricardian surpluses show that there were many exceptions to Ricardo's theory of rent— sometimes rent exceeded the economic value of the land by a large measure, and sometimes rent fell short. While the differentials probably persisted for many years, they were not permanent. Thus, the gap between rent and surplus in open field farms disappeared between Young's tours and Parkinson's census. In the long run, English landowners succeeded in capturing the full Ricardian value of the soil, so, over very long periods, Ricardo's theory applies. In a shorter time frame, which may have lasted a generation, there were significant discrepancies. For that reason, the rent increases at enclosure often included an important redistributive element.

This conclusion immediately raises the question: Why did competition fail to equate rents with surpluses in accord with Ricardo's analysis? The answer is that both postulates of the theory—that farmers used equation 9–1 to value land and that they moved readily in search of a better deal— corresponded badly to the facts of the eighteenth century.

The evidence regarding mobility is clear. Certainly some farmers shopped around for farms, but most stayed where they were born. As Batchelor (1808: 43) noted, 'the farms of Bedfordshire generally descend from father to son through a long series of years, and perhaps as frequently change their owners as their occupiers'. Thus, rents were not set by competition—they were determined administratively or bargained between lord and tenant. It was therefore possible for rents to diverge from Ricardian surplus.

Indeed, if we examine how rents were set on large estates, it is not surprising that they were often out of line with the commercial value of the land. A sense of social responsibility meant that many large landowners were reluctant to charge their tenants the full value of their farms. Young railed against this behaviour: 'The landlords, who, through a false pride, will not raise [rents], when they easily might, do an inconceivable prejudice to their country' since 'high rents are an undoubted spur to industry; the farmer who pays much for his land, knows that he must be diligent, or starve.' Hence, 'the man who doubles his rental, benefits the state more than himself' (Young 1771*a*: iv. 343–5). The social prestige that came from being a 'good landlord', however, sometimes outweighed a fat rent roll. In *Pride and Prejudice*, a turning point in Elizabeth Bennett's attitude to Darcy was the observation of his housekeeper that 'he is the best landlord and the best master . . . that ever lived', and that he was 'affable to the poor' (Austen 1813: 208, 207). While the gentry and aristocracy were often eager for more cash, many of them also valued the prestige that money could not buy. The rewards to Darcy were substantial.

Even when landlords sought to rack rent their farms they had the problem of deciding what the Ricardian surplus was. The problem was intractable, for no one—landlord, farmer, or surveyor—could use equation 9–1 to perform the calculation. This is clear from Philip Tuckett's prize essay 'On Land Valuing' published in the *Journal of the Royal Agricultural Society of England* in 1863—a late date by which time one might have thought that land appraisal would have been established on a firm, Ricardian basis. Tuckett did, indeed, allow that Ricardian surplus governed rent in the long run: 'there can be no doubt that the difference between the produce and the expenses incurred must, in the end, regulate the rent that a farmer can afford to pay.' However, he rejected Ricardian calculations as a practical basis for land valuation. 'Indeed, my chief object in submitting this paper to the Society is to point out the fallacy of expecting reliable practical results from this source.'[10]

There were three reasons why neither farmers nor landowners could put Ricardo's theory directly into practice. First, it was impossible to predict accurately revenues and costs over the term of a lease. 'Many men can calculate to a nicety the yield of a crop before harvest; but who can accurately estimate the produce of a farm during a series of years? The same remarks, though in a less degree, apply to the estimated expenses' (Tuckett 1863: 6). Second, when changes in farm operation were anticipated, it was also necessary to estimate how the quantities of inputs and outputs would change. How much would hollow draining boost yields? What stocking rates would a new pasture in Huntingdonshire support over the long term? Today, farmers can consult the results of controlled experiments for help in framing their expectations, but controlled experiments were not undertaken

---

[10] Both quotations from Tuckett (1863: 6).

in the eighteenth century. Probability theory was in its infancy and statistical inference not yet dreamed of, so the problem of distinguishing a temporary improvement from an underlying change could be solved only with enormous practical experience. Innovation was, therefore, guesswork. As a result, as Adam Smith (1776: 154) remarked: 'Such comparisons, however, between the profit and expense of new projects, are commonly very fallacious; and in nothing more so than in agriculture.' These problems were, of course, particularly acute when enclosures were undertaken, since they often led to changes in land use.

In addition to the problems of forecasting and experimental design, there was a third reason that people in the eighteenth century could not use equation 9–1 to value land—they could not do the requisite accounting. In his survey of business accounting during the Industrial Revolution, Pollard (1965: 285) concluded that 'The evidence presented here appears to be largely negative. Accountancy in its wider sense was used only minimally to guide businessmen in their business decisions, and where it was so used the guidance was often unreliable.' If this was true of the leading industrial concerns, what of the average farm? Capital was a particular problem. 'Contemporaries did not attempt any calculations of the profit rate on capital in the modern sense.' (Pollard 1965: 274). The concept of depreciation was only imperfectly comprehended and frequently ignored. Further, eighteenth-century entrepreneurs typically allowed 5 per cent interest on the investment and only computed the rate of profit on the excess earnings. In *The Farmer's Guide in Hiring and Stocking Farms*, Young (1770) followed all of these conventions in his voluminous accounts. He ignored depreciation on farm capital, allowed 5 per cent interest, and computed the rate of profit on the residual. His object was to determine how much a farm was worth, but his calculations ignored most of the capital costs and so overstated the value. Further, his comparisons of the profitability of different farming systems—especially when they differ in capital intensity—were incorrect and misleading. If farmers made calculations like Young's when they bid on a farm, then they often made mistakes.

What did ordinary farmers do? Most, of course, kept no accounts. Hueckel (1976) has used the surviving accounts of eight farms from the period of the Napoleonic Wars to compute the rate of return to farm capital. He had to rectify numerous errors in accounting to arrive at profits —the accounts typically ignored farm produce consumed by the farm family, ignored payments in kind to employees, failed to value the labour of family members, and left out depreciation. In most cases, there were no capital accounts at all, so Hueckel had to estimate farm capital in order to compute its rate of return. The range of realized returns vary considerably between farms and over time, indicating that the adjustment of rents to land values was imperfect and slow.[11]

There were thus three reasons why it was difficult, if not impossible, for either farmers or landlords to value land with equation 9–1. How was a value then to be determined? One procedure was to auction farms to the highest bidder. Tuckett immediately rejected that possibility: in an auction for farms, the top bids would come from prospective tenants aiming to rent the farm for only a few years and cream off a high income by robbing the land of its fertility. 'Not that the fair rent is the highest sum that some reckless speculator can be induced to give for it, who will probably get all he can out of it for a few years, and then leave it in deteriorated condition.' From the landlord's point of view, the right rent was 'that amount of rent which can readily be obtained from a substantial, respectable tenant during a series of years' (Tuckett 1863: 6).

Other writers also denounced auctions as a method of letting farms. Marshall doubted that the system would produce bids that equalled the Ricardian value of the land.

A single instance will serve to show . . . the impropriety of letting farms by *public biddings*. Three of the principal farms of an estate were let, some years ago, in that way. One of them was let twenty percent too dear. The consequence was—the tenant, after running his farm to the lowest state of ruin, left it in that condition; and it remained untenanted, two years: the proprietor losing, in rents, taxes, repairs, and unprofitable management, several hundred pounds. The other two were let twenty to thirty percent too cheap!!—and the tenants are now enjoying them, on lease, at rents which are not much less than fifty percent beneath their *present* values! (Marshall 1804: 389)

Tuckett's preferred solution to the problem of valuing land was to use the rents received from new lettings as a guide to the Ricardian surplus. 'Yet every experienced land-valuer can readily compare in his mind almost any description of land with farms of similar quality in different localities, which he has previously valued and let, where there has been sufficient competition on the one hand, or difficulty in finding a suitable tenant on the other, to form a test of value.' Tuckett claims that 'such is at present the basis on which the rental value of land is usually calculated, and I do not see how it is possible to obtain any better or more practical one.'[12]

The procedure was a practical one, but it did not guarantee that rent

---

[11] Hueckel's estimates overstate the rate of return in the sample farms for two reasons—he omits depreciation and he values the labour of women and children in the farm family at zero. Including these costs would probably make many farms unprofitable.

[12] Quotations in this paragraph from Tuckett (1863: 7).

equalled Ricardian surplus. Marshall (1804: 388), for instance, doubted that even experienced appraisers would agree.

But let three or four surveyors, or land valuers, (all of them noted for being great judges of land) go over a farm, separately, and their several valuations will differ very materially; especially if they go over it, in different seasons. Instead of ten percent . . . I have known twenty, thirty, or even fifty percent, and in one instance (a difficult subject of valuation) nearly centpercent, different, in the estimates of men who stand well in their profession.

Tuckett accepted that substantial discrepancies between valuations and actual values probably existed.

I believe the rent which can be obtained for poor thin-skinned clays, low as it may seem, is yet higher in proportion to their true value than that which can be obtained for really good land, high as that may appear. But if farmers in general underrate the difference, it is beyond the power of the land-valuer to alter their views . . . there is great danger of valuing the best soils considerably below, and the inferior ones above, their real value. (Tuckett 1863: 7)

The figures for enclosed farms in Tables 9–2 to 9–4 support Tuckett's thesis, for the rent was always close to £1 per acre, while the Ricardian surpluses varied widely. The implication of Tuckett's remarks is that farmers—like landlords and surveyors—had trouble directly applying equation 9–1 and fell back on convention in determining their assessments of the value of farm land.

The problem of land valuation in the eighteenth and nineteenth centuries was an exercise in what H. A. Simon (1982) has called 'bounded rationality'. Application of equation 9–1, which would have been 'fully rational', was beyond the computational abilities of contemporaries. (Indeed, I have been able to apply the equation only because of microcomputers, modern methods of data analysis, and because I am investigating a simpler question: instead of trying to predict future land values, I am only asking how reasonable rents were in hindsight.) With optimality unobtainable, farmers, landlords, and surveyors fell back on rules of thumb that provided satisfactory, if not perfect, solutions. The result was a circle of convention and experience. In the eighteenth century, open field land was often let at 10s. per acre and enclosed land at 20s.—not because of close calculations but as rules of thumb. Since farmers paid these rents, surveyors then used them as guides for valuation, and experience reinforced convention. Eventually experience might show a rent to be out of line so an adjustment was made, but that process might well be effected by changes in productivity that would ratify or undermine the conventional numbers.

The use of conventional numbers to value land explains some of the disequilibrium in Tables 9–2 to 9–5. Thus, by the beginning of the eighteenth century, open field land came to be valued at about 10s. per acre and enclosed land, which was mainly like the high quality old enclosures in Rutland, at closer to 20s. In the next half century, open field farms were amalgamated, and their efficiency increased as they shed labour. (I will discuss this process in detail in Part III.) Larger size raised their value. When the pasture district was enclosed in the mid-eighteenth century, rents were successfully raised to levels like the old enclosures in Rutland. Farmers paid the higher rents, not because their pastures were more productive—they were not—but because their farms were bigger and employed less labour. Adjusting the rents to the Ricardian surpluses took decades, however, because of the failures of accounting on capital-intensive pasture farms.

In the arable districts, the conventional rents did not apply as well, and this led to changes. The heavy arable district was so much less fertile than the pasture district that conventional open field rents never reached 10s. per acre, but when the heavy arable district was enclosed during the French Wars, rents were raised there, too, to 20s. per acre. In this case, farmers could pay them by hollow draining. In the light arable district, open field land did rent at 10s. per acre in the early eighteenth century, and rents were raised to 20s. per acre on new enclosures during the French Wars. However, there was really no efficiency gain to enclosures in this district. Farm amalgamation allowed enclosed farmers to pay the higher rent, but so could large-scale open field farmers. In the light arable district, the rent gap between open and enclosed farms disappeared as rents in the remaining open fields were raised to enclosed levels during the nineteenth century.

How rents were actually set warrants further research. The calculations and the considerations advanced here, however, indicate that the Ricardian model is an oversimplification. Rents did tend to follow the value of land in the long run, but that was often very long indeed. In many cases, rents were set at conventional levels, which reflected productivity differences in some districts. In other districts, the use of these conventional rents meant that enclosure resulted in a considerable redistribution of income. Further, the rent changes that followed enclosure in those districts reflected conventional expectations rather than actual changes in efficiency.

## Conclusion

For two centuries, defenders of enclosure have pointed to the resulting rent increases as an indicator of the large productivity gains that were supposedly caused by the enclosures. The evidence reviewed here indicates that the rent increases bore little connection to productivity increases given the way in which rents were set in the eighteenth century. Far from establishing that enclosure raised productivity, the study of rents supports the conclusions advanced in the previous three chapters; namely, that enclosures in most cases made only modest contributions to the growth in productivity. Had English agriculture not been enclosed, its land use and production patterns would have been different, but its overall level of efficiency would not have been much lower.

# PART III

# Capitalist Agriculture and Productivity Growth

# 10

# Yeomen, Capitalist Farmers, and the Growth in Yields

A considerable farmer, with a greater proportional wealth than the smaller occupier, is able to work great improvements in his business, and experience tells us, that this is constantly the case; he can build, hedge, ditch, plant, plough, harrow, drain, manure, hoe, weed, and, in a word, execute every operation of his business, better and more effectually than a little farmer . . . He also employs better cattle and uses better implements; he purchases more manures, and adopts more improvements; all very important objects in making the soil yield its utmost produce. The raising of great crops of every sort, so far increases the solid publick wealth of the kingdom; himself, his landlord, and the nation are the richer for the size of his farm; his wealth is raised by these improvements which are most of them wrought by an increase of labour; he employs more hands in proportion than the little tenant, consequently he promotes population more powerfully.

Arthur Young, *Political Arithmetic*, 1774: 287–8

IF enclosure does not explain the growth in efficiency in English farming between 1500 and 1800, what does? Many factors were involved, but, so far as the institutional basis is concerned, there are two candidates. The first is emphasized by the Agrarian Fundamentalists—namely, the replacement of small owner-occupied family farms by large-scale capitalist enterprises. This change proceeded rapidly in open field villages in the eighteenth century and so may account for their high level of efficiency. The second is, of course, the yeomen themselves. In other times and places, peasant agriculture has been the social basis of agricultural progress. It is time to reassess its role in England too. I concentrate on crop yields, output per worker, and total factor productivity as the main indicators of performance, although I also give some attention to changes in method.

The theory of the 'capital-intensive farmer' was the eighteenth century's contribution to agrarian thought. The quotation at the head of this chapter indicates the five main propositions in Arthur Young's formulation. First, farm operations depended on the personal wealth of the farmer. Second, large-scale farmers were proportionately wealthier than small-scale farmers. Third, as a result, large farmers used more capital-intensive methods and, fourth, realized higher yields than small farmers. Fifth, large farms employed more people per acre than small farms. All of these ideas have been incorporated into Agrarian Fundamentalism except for the last, which is accepted by Tories but rejected by Marxists. In this chapter I will focus on the claims dealing with farm capital and yields.

In developing his analysis, Young probably drew both on his own difficulties in financing his farms (Gazley 1973: 26–8) and on Quesnay's theories about agricultural development. Indeed, the distinction between the *grande culture* and the *petite culture* prefigured Young's distinction between the large-scale 'capital' farmers and the small-scale farmers of the open fields. But Young did not simply repeat Quesnay's views. Instead, Young aimed 'to treat the subject in a new way, by presenting facts' (Young 1771a: iv. 192). Admirable intention! No one before—or since—has measured the effect of farm size on indicators of performance such as yield. The systematic nature of Young's inquiry has given his opinions a unique authority that has lasted for two centuries.

What were Young's facts? In the late 1760s he toured England and recorded voluminous details about the practice of agriculture. He noted average yields, seed rates, livestock productivity, wages, and prices in the villages he visited. In many cases, he also recorded the 'details of representative farms'. For these sample farms, he noted the size, rent, cropping pattern, as well as the number of each kind of animal and the permanent labour force. In the *Northern Tour*, he used these data to compute the relationship between farm size and yield. He claimed to find a positive relationship, and that is the most important evidence that has ever been presented for the productivity-raising effects of large farms.

However, the case for the indispensability of the large farm is a flimsy one. First, it rests principally on the researches of Arthur Young and the claims of eighteenth-century apologists for the landed interest. Twentieth-century historians have produced no new evidence to verify the putative links between farm size, capitalization, and yield. Second, the history of other countries has so frequently shown that peasant farmers adopt new methods and secure rising yields that the views of Quesnay and Young have become increasingly anomalous. We need to re-examine English agricultural history and ask whether the large farm was really as essential for improvement as Agrarian Fundamentalists claim.

## *Farm Size and Capital c.1770*

I begin the reassessment by using Young's data to test his views on farm credit. In the eighteenth century, the provision of agricultural capital was divided between landlord and tenant. Their economic roles reflected their legal relationship: the landlord provided improvements like buildings, roads, fences, etc. that became part of the freehold of the property; the tenant provided the capital that remained part of his personal estate, including livestock, implements, and the payment of seed, wages, rent, taxes, etc. in advance of the harvest. Loosely speaking, the landlord provided the fixed capital and the tenant provided the working capital, although, it should be noted, some of the tenant's capital—implements and some livestock—lasted for years and was in that sense fixed. There was also a grey area that included the expense of marling and hollow draining. While these were clearly improvements to the freehold, landlords expected tenants to bear the cost.

Both landlord and tenant contributed substantial amounts of capital but at different tempos. Feinstein (1978: 50) estimated that the buildings and improvements provided by landlords had a capitalized value in 1760 of about £4 per acre—admittedly a highly approximate figure.[1] These assets had very long lives, and the stock was built up by a modest investment rate over many years. In the 1760s landlords were investing about 6 per cent of the gross rental of the country—£0.04 per acre per year.[2] This sum equalled 1 per cent of gross farm income, also about £4 per acre per year.

The capital provided by tenants was of the same magnitude as that supplied by landlords—about £3–£4 per acre per year. For tenants, however, the annual investment was also of this order since most of this capital consisted of the advance payment of wages, rents, and taxes or the purchase of livestock that were bought for fattening and sold within the year. Tenants invested at a higher rate than landlords, but the assets acquired were far more ephemeral.

Young's theory of the capital-intensive farmer relates to the provision of capital by tenants. To see whether small-scale, open field farmers could adequately stock their farms, I have analysed Young's own data. For a sample of 241 farms, I can determine capital and size. The farms span the spectrum from peasant holdings of 20 acres to mammoth enterprises of hundreds of acres.

There are three relevant definitions of 'capital'. The first is 'finance'. It

[1] This equals £125 million worth of buildings and permanent improvements divided by 31 million acres of land.

[2] Dividing the £1.2 million of investment by 31 million acres of farm land implies a rate of about £0.04 per acre per year (Feinstein 1978: 49).

equals the cash outlay needed to stock a farm and operate it for a year, that is, the cost of buying the animals and implements plus one year's expenditure for labour, seed, rent, and taxes. This is the concept of capital that Young uses in the *Farmer's Guide in Hiring and Stocking Farms*, and it is consistent with his views on rural credit. The classical economists used the same definition.[3]

The second definition of 'capital' is a modern one—capital cost. It equals the depreciation plus the interest of livestock and implements, that is, the assets with lives longer than one year. Including interest presumes that the funds invested in these assets could have been invested otherwise—or borrowed—at that rate. This definition thus presumes that there were rural credit markets. Young was not very consistent in these matters, and in his representative farm accounts he often charges farmers 5 per cent for their capital. I will do likewise.

The third definition of 'capital' is livestock density. Following Yelling (1977: 159), I aggregate animals in terms of their feed requirements. There are two reasons for this. First, it allows comparison with the probate inventory data, which I will introduce later. Second, the consumption of feed is directly related to the production of manure. Feed consumption is, therefore, the appropriate capital aggregate to test the claim that higher livestock densities led to higher yields.

Arthur Young's data support three important generalizations about farm

Table 10–1. *Capital per acre (£ per acre)*

| Farm size (acres) | Arable farms | | | Pasture farms | | |
|---|---|---|---|---|---|---|
| | Finance | Capital | Animal density | Finance | Capital | Animal density |
| 0–50 | 4.6 | 0.80 | 0.28 | 4.8 | 1.84 | 0.46 |
| 50–100 | 4.6 | 1.03 | 0.28 | 4.0 | 1.41 | 0.37 |
| 100–150 | 3.9 | 0.85 | 0.25 | 3.6 | 1.25 | 0.33 |
| 150–200 | 3.9 | 0.79 | 0.21 | 3.4 | 1.29 | 0.31 |
| 200–250 | 3.7 | 1.40 | 0.35 | 3.6 | 1.12 | 0.31 |
| 250–300 | 3.5 | 0.75 | 0.20 | 2.6 | 0.96 | 0.22 |
| 300–350 | 4.3 | 1.24 | 0.27 | 2.1 | 0.49 | 0.17 |
| 350–400 | 3.3 | 0.69 | 0.18 | 2.7 | 0.80 | 0.24 |
| 400–450 | 3.8 | 1.27 | 0.24 | 3.1 | 0.97 | 0.26 |
| 450–500 | 3.9 | 1.08 | 0.21 | — | — | — |
| 500–550 | — | — | — | 3.1 | 2.19 | 0.36 |
| 550–600 | 3.4 | 1.08 | 0.21 | 3.0 | 1.91 | 0.32 |
| 600–650 | 2.8 | 1.15 | 0.19 | — | — | — |
| 650–700 | 3.3 | 1.04 | 0.22 | 2.6 | 0.67 | 0.14 |

[3] See Eagly (1974: 4). Feinstein (1978: 70) defined circulating capital to include the cost of horses, livestock, and the value of harvested and standing crops. The latter resolves into the wage expense, seed cost, taxes, and rent that I have included in my definition of 'finance'.

capital in the eighteenth century. First, capital per acre declined with farm size. Second, when finance is the measure of capital, arable farms were more capital-intensive than pastoral farms. This result is due to the fact that finance includes the advance payment of wages, and arable farms were more labour-intensive than pastoral farms—a theme to be pursued in the next chapter. Third, when capital is measured by either capital cost or animal density, pastoral farms used more capital per acre than arable farms. These conclusions are supported both by inspection of Table 10–1 and by the regressions in Table 10–2. It is remarkable that Young's data contradict his belief that large-scale farmers practised a more capital-intensive agriculture than small-scale farmers.

*Table 10–2. Capital per acre regressions (t-ratios in parentheses)*

| Dependent variable | Constant | Share arable | Acres | Acres squared | $R^2$ |
|---|---|---|---|---|---|
| 1. Finance | 4.52 | 0.929 | $-0.896 \times 10^{-2}$ | $0.980 \times 10^{-5}$ | 0.18 |
| | (19.6) | (3.36) | (−5.12) | (3.59) | |
| 2. Capital cost | 1.87 | −0.885 | $-0.280 \times 10^{-2}$ | $0.399 \times 10^{-5}$ | 0.10 |
| | (11.0) | (−4.39) | (−2.19) | (2.00) | |
| 3. Animal density | 0.479 | −0.189 | $-0.774 \times 10^{-3}$ | $0.877 \times 10^{-6}$ | 0.17 |
| | (15.3) | (−5.05) | (−3.26) | (2.37) | |

## The Capital Intensity of Midlands Agriculture, c.1600–c.1800

The regressions in Table 10–2 allow me to simulate the capital intensity of south midlands agriculture over the early modern period. The results are in Table 10–3. In this table, 'capital' means 'capital cost'. Panel A shows estimates of capital per acre for three subdivisions of the south midlands. Notice that capital per acre was higher in enclosed pastoral farms than in arable farms, and that capital per acre declined over time in the arable sector as farm size increased. Panel B shows the implied totals.

Panel C is an estimate of the annual capital cost of enclosing. Most of this was borne by landlords. Following the discussion in Chapter 8, the public cost of a parliamentary enclosure in the early nineteenth century was about £3 per acre and the cost of hedging and ditching the allotment amounted to another £3 per acre. Perhaps one-third of the public costs were transfer payments associated with the parliamentary process rather than additional physical capital, so I assess the total resource cost of enclosing at £5 per acre (McCloskey 1972; Turner 1984a: 60–3). Interest on this at 5 per cent was £0.25 per acre. Hedgerows did not disappear if they were maintained, so I assess depreciation as the annual cost of hedging and ditching—5d. per acre (£0.02)—again following the discussion in Chapter 8. The total annual cost of enclosing, thus, comes to £0.27 per acre in early nineteenth-century

Table 10–3. The capital intensity of midlands agriculture, c. 1600–c. 1800 (£)

|  | c. 1600 (£) | c. 1700 (£) | c. 1800 (£) |
|---|---|---|---|
| A. Farmers' capital per acre |  |  |  |
| Enclosed pasture | 1.308 | 1.308 | 1.308 |
| Open pasture | 1.085 | 1.048 | 0.933 |
| Light and heavy arable | 0.935 | 0.897 | 0.783 |
| B. Farmers' total capital |  |  |  |
| Enclosed pasture | 577,345 | 927,874 | 2,000,085 |
| Open pasture | 1,231,490 | 908,642 | 44,123 |
| Light and heavy arable | 822,815 | 789,374 | 689,053 |
| Overall | 2,631,650 | 2,625,890 | 2,733,261 |
| C. Landlords' capital cost of enclosures |  |  |  |
| Overall | 117,901 | 187,758 | 488,562 |
| D. Farmers' cost of hollow draining |  |  |  |
| Heavy arable | 0 | 0 | 97,354 |
| E. Total capital |  |  |  |
| TOTAL | 2,749,551 | 2,813,648 | 3,319,177 |

Sources:
Panel A:1 computed capital per acre for enclosed farms in the pasture district, open farms in the pasture district, and open farms in the arable districts using equation 2 in Table 10–2. These calculations require the share of arable, and I assumed it to have been 0.2, 0.63, and 0.8, respectively. (See Ch. 6.) Capital per acre also depended on farm size. I used the distributions in Table 4–4. For each size category, I computed capital per acre with the regression equation, and then computed capital per acre as a weighted average. For enclosed farms in the pasture district, I used the farm size distribution for enclosed farms c. 1800 for all years since the sample size was so small before c. 1800. For all other farms I used the open distributions for early 17th century, early 18th century, and c.1800. (As a check I computed capital per acre with the enclosed distribution c.1800 for the arable districts, but it gave virtually identical results.)
Panel B: To compute capital per acre in the south midlands as a whole, I combined these estimates using the proportions of farm land in each district and estimates of the proportions of open and enclosed land in the pasture district at each date. (See Fig. 2–1.)
Panel C and D: See text.
Panel E: sum of Panels B, C, And D.

prices. Young's figures are in prices of c.1770, so I reduced £0.27 to £0.22 to reflect wage inflation of 25 per cent.

Panel D shows the additional cost of hollow draining borne by tenants in the heavy arable district. Again following Chapter 8, I assume that cost was £3 per acre in the early nineteenth century. These investments only last ten years, so the annual cost was £0.45 allowing 5 per cent for interest and 10 per cent for depreciation. Allowing for wage inflation makes the cost £0.36 per acre in 1770 prices. I assess this charge on the enclosed acreage in the heavy arable district. I assume hollow draining was not done c.1600 or c.1700.

The total capital requirements of south midlands agriculture are shown in panel E. Capital requirements increased 2 per cent in the seventeenth century and 18 per cent in the eighteenth. Virtually all of the increase was due to the investment by landlords in enclosing. Capital supplied by tenants remained fairly constant due to offsetting trends: enclosure and the conversion of arable to pasture increased their capital requirements as livestock densities rose, while the increase in farm size economized on capital.

These conclusions rest on backward extrapolations from eighteenth- and nineteenth-century information—inevitably a hazardous procedure. One can, however, test the results against other information, in particular, the history of livestock densities. They are of considerable interest in their own right since livestock was a large share of a farmer's capital costs and since supposed increases in livestock densities are often imagined to have been the cause of higher corn yields.

Table 10–4 compares stocking rates from the mid-sixteenth century to the early nineteenth. For the period 1550–1727, I compute livestock densities from a sample of ninety Oxfordshire probate inventories that I will discuss shortly. These inventories describe open field farms mainly in the light arable district and so shed no light on the impact of enclosure. For the late 1760s I again use Young's figures. For the early nineteenth century, I rely on Parkinson's censuses of Rutland and Huntingdon.[4]

Table 10–4 teaches two important lessons. First, within arable agriculture, livestock densities remained roughly constant. Since the probate inventories derive mainly from farms in the light arable district, the most pertinent comparison is between the inventory results and those for Parkinson's open farms in that district. That comparison shows little change —perhaps a slight decline—in livestock numbers. Animal densities in the heavy arable district were less than in the light arable district, which is consistent with the poor pasture potential of the heavy arable lands. The constancy of livestock numbers between the mid-sixteenth century and the early eighteenth century has been noticed before by students of probate inventories (e.g. Yelling 1977: 158; Mingay 1984: 102). Table 10–4 extends that finding into the nineteenth century, and confirms the absence of evidence for any increase (by farmers) of the capital intensity of arable agriculture.

Second, enclosure that led to the conversion of arable to grass often

---

[4] The data sources are not strictly comparable, so precise comparisons are impossible. The probate inventories recorded cropped acreage, so it was necessary to pick a value for the ratio of cropped to total acreage in order to compute the livestock density of the farm as a whole. I chose 0.5, since the three-field system was the norm in most villages in the sample. About 75 per cent of the village lands were in the open fields, and the rest was meadow and pasture (Havinden 1961a). Hence, half the land (0.67 × 0.75 ) was cropped. Young's data exclude the acreage of commons, and so overstate livestock densities for the farms as a whole. Parkinson's data are the best, although he likely included in his tallies animals such as the cottagers' cows, so his figures overstate livestock densities on large farms.

*Table 10–4. Livestock densities, 1550–1806*

| | Arable | Pasture |
|---|---|---|
| *Probate inventories:*[a] | | |
| 1550–1574 | 0.28 | — |
| 1575–1599 | 0.24 | — |
| 1600–1624 | 0.22 | — |
| 1625–1649 | 0.26 | — |
| 1650–1674 | 0.19 | — |
| 1675–1699 | 0.25 | — |
| 1700–1727 | 0.13 | — |
| *Young's data, late 1760s*[b] | 0.25 | 0.29 |
| *Parkinson's data (1806):*[c] | | |
| Light arable district, open | 0.21 | — |
| Light arable district, parl. enclosed | 0.15 | — |
| Heavy arable district, open | 0.17 | — |
| Heavy arable, parl. enclosed | 0.14 | — |
| Pasture district, open | 0.19 | — |
| Pasture district, Hunts, parl. enclosed | — | 0.19 |
| Pasture district, Hunts, old enclosed | — | 0.17 |
| Pasture district, Rutland, parl. enclosed | — | 0.22 |
| Pasture district, Rutland, old enclosed | — | 0.31 |

*Note*: Livestock densities equal a weighted sum of animals divided by the acreage of the farm, as estimated. Weights are from Yelling (1977: 159). Horses and foals equal 1.0, beef cattle and stores equal 1.2, cows and calves equal 0.8, sheep, lambs, and swine equal 0.1, sucklers equal 0.
[a] Probate inventories: I first computed animal density per cropped acre by dividing the weighted sum of animals by the cropped acreage. Then I computed animal density for the whole farm by dividing that figure by 2 on the assumption that farms were 75% arable and farmed on the three-field system (so two-thirds of the arable was cropped).
[b] Young: I weighted livestock density in Table 10–1 by the acreage of farm land in each size group as indicated by Table 4–4 for enclosed farms. The match-up of size groups is not exact.
[c] Parkinson: computed from Tables 6–3, 6–5, 6–8, and the assumptions I made for livestock densities on open farms in the pasture district in Ch. 7.

resulted in an increase in livestock. Young's data show this unquestionably, as does the comparison between the pre-parliamentary enclosures in Rutland and the stocking density for the (hypothetical) open field village in the pasture district. The other enclosures in the pasture district, especially those in Huntingdon, did not result in increases in livestock, but those pastures were not especially productive.

Table 10–4 presents a more nuanced picture of farm capital requirements than Table 10–3 but confirms the major conclusions of the simulations: the capital supplied by farmers did not go up much over the early modern period, the shift to large farms economized on capital in arable farming, and the conversion of land to pasture raised capital requirements if the grass was of high quality.

## The Farm Capital Problem in the Eighteenth Century

The farm capital problem in the eighteenth century was not what Young imagined—it was *not* the case that large-scale farmers used a more capital-intensive technology than small-scale farmers, so Young was wrong to argue that agriculture was held back by the small farm sector. Nor was it the case that farm capital was rising over the century, and there were problems in financing that increase. The farm capital problem in the eighteenth century related to the distribution of capital across farms rather than to the total quantity of agricultural capital.

There appeared to be a farm capital problem in the eighteenth century since the traditional ways of financing farmers did not work smoothly during the landlords' agricultural revolution. For centuries, most farms were initially stocked through inheritance and kept going with retained earnings and the breeding of replacement animals on the farm. Such additional credit as may have been needed from time to time was secured from relatives, corn merchants,[5] or even landlords.[6] What was essentially a system of self-financing was adequate as long as agriculture simply reproduced itself. In the eighteenth century, however, enclosure followed by the conversion of arable to pasture and the growth in farm size required the declining number of successful farmers to increase their capital rapidly. 'The avarice of hiring a large quantity of land' meant 'that farms are every day hired with much smaller sums of money than the most considerate people would allot for the purpose . . . even farmers themselves will often own that a larger sum of money is really necessary than often possessed upon the hiring of a farm' (Young 1770: i. 49–50). The result was a tendency to gamble. A farmer might take a big farm and strap himself for cash in the hope that high prices or bumper crops in the first years of his tenancy would give him the capital to stock the farm for the long term.[7] In the meantime, he could pay his harvesters by selling the standing grain to a corn merchant, neglect maintenance, and scrounge money from relatives. Under this scenario, two things would be true. First, some farmers would lose the gamble and fail. Landlords would complain about the difficulty of attracting sufficiently solvent tenants. Second, the farmers who won would have the money to stock their farms.

My view of farm size and capitalization is the reverse of Young's. For Young, large farms were part of the solution to the agricultural problem, for

---

[5] See Defoe (1727: ii, pt II, p. 36) and Chartres (1985: 471–3).

[6] Batchelor (1808: 476) discusses landlords' loaning tenants money to install hollow drains.

[7] Young (1770: i. 108–10) describes behaviour very much like this. Holdnerness' (1975–6: 103–4) finding that probate inventories show farmers' making scarcely any loans at times of low prices but appreciable loans when prices were high is consistent with this conjecture.

they would conjure up a supply of unusually rich tenants who would raise the capital intensity of English farming. Where would these tenants come from? We are never told. Did they exist? The data say no. For me, large farms created a pseudo-problem. The large farm did not require more capital per acre to stock, but it did require that the newly created large-scale farmers amass the livestock of their less fortunate neighbours who went out of business. That was, indeed, a financing problem, and it might have led to a reduction in the capital intensity of English farming. In the event, it did not. But that is no reason to regard the large farm as progressive.

## Farm Size and 'Improvement'

Farm capital was important to Young for what it implied about farm productivity. Young was adamant that large farmers would undertake more 'improvements' than small farmers, since the large farmers had more cash on hand. 'Great farmers are generally rich farmers; and it requires no great skill in agriculture to know that they who have most money in their pockets, will, upon an average, cultivate the soil in the most complete manner' (Young 1771a: iv. 253). Like his views on farm capital, this theory can also be tested with Young's own data.

'Improvement' encompassed many changes in method, and I can investigate only one—the adoption of clover and turnips. These crops were diffusing across England at the time of Young's tour, and he was an enthusiast for both. His reports are skimpy in regions where turnip culture was inappropriate—the boulder clays of the south midlands are an example —and detailed where they could be grown. In such places he could laud the innovators and chastise the laggards. Since my abstraction of Young's information relies on the more complete reports, the data set is confined primarily to places that practised turnip husbandry or where it was well suited. Since Young reported the cropping pattern of each farm, the data set is reasonably well constructed to test the effect of farm size on cropping.

Table 10–5 reports regressions of the shares of arable devoted to wheat, barley, oats, peas, beans, turnips, clover, and fallow. Farm size was invariably insignificant. Sometimes the arable orientation of farming influenced the choice of crop—the greater the share of arable, the more it was planted with barley, peas, and beans, and the less with oats. One might find large farmers more innovative than small farmers with a different data set or by another measure. But so far as the cultivation of clover and turnips is concerned, Young's data show small farmers adopting the new crops as rapidly as large farmers.

*Table 10–5. Crop choice and farm size (t-ratios in parentheses)*

| Dependent variable | Constant | Share arable | Acres | Acres squared | $R^2$ |
|---|---|---|---|---|---|
| 1. Wheat | 0.223 | 0.035 | $0.403 \times 10^{-4}$ | $-0.159 \times 10^{-6}$ | 0.02 |
| | (12.1) | (1.66) | (0.316) | (−0.808) | |
| 2. Barley | 0.136 | 0.066 | $-0.141 \times 10^{-3}$ | $0.343 \times 10^{-6}$ | 0.04 |
| | (5.59) | (2.36) | (−0.840) | (1.33) | |
| 3. Oats | 0.336 | −0.243 | $-0.247 \times 10^{-3}$ | $0.206 \times 10^{-6}$ | 0.23 |
| | (12.3) | (−7.83) | (−1.32) | (0.711) | |
| 4. Peas | 0.008 | 0.054 | $0.113 \times 10^{-3}$ | $-0.133 \times 10^{-6}$ | 0.04 |
| | (0.451) | (2.64) | (0.917) | (−0.702) | |
| 5. Beans | 0.007 | 0.053 | $0.150 \times 10^{-3}$ | $-0.273 \times 10^{-6}$ | 0.03 |
| | (0.364) | (2.27) | (1.06) | (−1.25) | |
| 6. Turnips | 0.082 | 0.012 | $-0.942 \times 10^{-4}$ | $0.279 \times 10^{-6}$ | 0.02 |
| | (3.85) | (0.481) | (−0.643) | (1.24) | |
| 7. Clover | 0.055 | 0.044 | $-0.980 \times 10^{-4}$ | $0.191 \times 10^{-6}$ | 0.02 |
| | (2.47) | (1.73) | (−0.636) | (0.802) | |
| 8. Fallow | 0.153 | −0.020 | $0.277 \times 10^{-3}$ | $-0.455 \times 10^{-6}$ | 0.01 |
| | (4.45) | (−0.500) | (1.17) | (−1.24) | |

## *Young's Mismeasurement of the Yield–Farm Size Correlation*

The relationship between size and farm capital or cropping was less significant than the effect of farm size on yield. Young accepted that. In the last volume of the *Northern Tour*, he analysed the particulars of representative farms and concluded that they showed that large farms realized higher yields than small farms. This investigation demands careful scrutiny since it is the sole empirical basis for the belief that large farms raised corn yields in early modern England.

Young's statistical investigation is precocious, but the data do not support his conclusion. The basic problem is that the survey was not well designed to explore the relationship between size and yield. Young did not obtain the individual yields of his sample farms: instead he recorded the average yields of the village where each farm was located. As a result he had to apply the same village yield to every farm in the village irrespective of its size, or he had to compute an average farm size for each village and correlate that with yield. He chose the second approach. The procedure was poor—so was the alternative—and Young knew it:

There is something of uncertainty in this article: the Products [yields] are the average of each neighbourhood, and the size of the farms is also the same average; consequently neither of them are drawn from particular farms; and as the average product is general, it includes that of all sizes; so that the result can only shew any general tendency of countries that are pretty strongly marked by large or small farms. (Young 1771*a*: iv. 255)

Moreover, Young's decision to match average village yield with average farm size in the village raised a further problem—how to measure average farm size. Young had rarely enquired about it. The only relevant question he asked regularly was the minimum and maximum rent of farms in each village. He used their average as his proxy for average farm size. There are two obvious problems with that procedure. The first is that rent is a less direct measure than acreage. In analysing his employment data, which were collected farm by farm and which could consequently be correlated with farm acreage, Young had considered measuring farm size by rent but rejected that approach: 'In the scale of this comparison, I think it will be stating it with more precision, to be guided by the acres rather than rent; the latter is a capricious circumstance, varying according to favour and other extraneous causes; whereas the former always is decisive of the size of the farm' (Young 1771*a*: iv. 207). The second is that the average of the minimum and maximum rent is an unreliable indicator of the average in the village. We will see an example shortly.

In spite of these problems, Young used his data to measure the association between farm size and yield. Table 10–6 shows the results. As farm size increased from a rent of less than £50 per year to £300 per year, the average yield increased from 27 to 29 bushels.[8] The average yield jumped to 34 bushels for villages with average farm rental above £300. Young (1771*a*: iv. 265) was impressed by the regularity of the increase and the high level for the largest group. He emphasized the latter in his 'general recapitulation'. 'Farms of above 300 l. a year yield a PRODUCT of corn and pulse superior to smaller ones, as 8½ to $6\frac{127}{136}$,' (Young 1771*a*: iv. 267). (The practice of reporting many insignificant digits in statistical work evidently predates the computer.) This was Young's principal evidence for concluding that the movement to large farms raised corn yields.

But that is a misleading conclusion to draw from Table 10–6. Notice, first, that, at the average rent of 10*s*. an acre that Young assigned to England, the largest farm size group was over 600 acres, which exceeded almost all farms in the country. The move to large farms in the eighteenth century was an adjustment within the three smallest groups, for which Table 10–6 shows essentially equal yields. Thus, Young's results imply that this reorganization had no impact on yields. Second, the average yield in the smallest group (27 bushels) greatly exceeded the comparable number for medieval farms (12 bushels).[9] Young's data show that even the smallest eighteenth-century farms had greatly surpassed medieval levels. Con-

---

[8]  This average is across all grains, beans, and peas.

[9]  12 is the average of the yields of wheat (10.7 bushels per acre), barley (16.8), oats (11.7), and beans (10). These are the medieval baseline values discussed in Ch. 7.

*Table 10–6. Arthur Young's results on farm size and yield*

| Size (£ rent) | Average crop yield (bushels) | Number of villages |
|---|---|---|
| less than £50 | 27 | 14 |
| £50–£100 | 27 | 19 |
| £100–£200 | 28 | 16 |
| £200–£300 | 29 | 12 |
| £300 and over | 34 | 2 |

*Source*: Young (1771a: iv. 264).

sequently, Table 10–6 is evidence against the historical myth that the shift to large farms explains the rise in corn yields in early modern England.

Indeed, Young's case is even shakier since the high yield for farms over £300 is itself doubtful. It is based on only two villages—Bendsworth, Gloucestershire, and Fenton, Northumberland. Bendsworth's assignment to this group is almost certainly a misclassification. It was put in the largest group since the range of farm sizes was reported to have been £40 to £1,000, with an average of £540. Since the average rent in that village was 21s., £540 corresponds to 514 acres. Young (1771a: iii. 318) also reported, however, that the parish included 1,500 acres of agricultural land and fourteen farms, so the average farm size was, in fact, only 107 acres. The broad range was the result of one 850-acre farm. A discrepancy of this magnitude, however, is exactly what one would expect when 'average farm size' is measured with such a dubious proxy as the mean of the minimum and the maximum rent.

The classification of Bendsworth is extremely important, however, for the conclusion that large farms had high yields rests entirely on its performance. The average yield for Fenton was 28 bushels, while Bendsworth was 40. Reassigning Bendsworth to another group destroys the positive correlation of size and yield. It is instructive that the evidential base for the tradition that large farms had higher yields than small farms depends on a single erroneous classification.

The problems with average rent as a measure of farm size can be avoided by imputing village yields to farms and examining the correlation of yield and size. Regressions of yield on size and size squared showed no significant relationship. These results establish that Young's data, when sensibly analysed, do not show a positive correlation between size and yield. But the data are not really up to that exercise since they do not match farms with their respective yields. We need other evidence to explore that issue.

## The Yeomen's Agricultural Revolution

To measure the correlation between farm size and yield, I have assembled another data set from probate inventories. These documents were routinely drawn up when someone died in early modern England. They listed and valued the deceased's personal property; this included livestock and standing grain. Overton (1979) was the first to suggest that dividing the value per acre of grain by its price might retrieve the appraiser's estimate of the yield. When allowance is made for tithes and harvesting costs, one can, indeed, compute crop yields that can then be correlated with livestock densities, cropping patterns, and cultivated acreage.[10] In this way, probate inventories allow further tests of capital-intensive agriculture.

I have worked out yields for a sample of probate inventories from Oxfordshire.[11] The earliest was prepared in 1550 and the latest in 1727. Most relate to the portion of the county between the Cotswolds and the Chilterns. The soil was light and well adapted to turnip cultivation in the nineteenth century. Almost all the corn land was in open fields, so the results are not muddied by any yield-boosting effects attributable to enclosure. The full sample consists of ninety inventories.

The biggest change in farm methods between 1550 and 1727 was in cropping. The cultivation of rye, barley, and oats decreased, while the acreage devoted to wheat and pulses expanded. The share of the cropped land planted with peas, beans, and vetches increased from 15 per cent to 34 per cent over the period. Recently Chorley (1981) has argued that the adoption of nitrogen-fixing crops like these explains the rise in yields in much of Europe. Inventories can be used to test this theory for England. Average farm size and livestock density remained approximately constant.

Table 10–7 reports regressions that test Young's views about farm capital and innovativeness. In the first regression, animal density, the measure of capital per acre, is correlated with farm size and year raised to the third power. (In almost all regressions reported here, year cubed was more significant than year to the first or second power.) The time trend was insignificant. Again animal density *declined* significantly with size. This result is the reverse of what Young contended.

[10] See Allen (1988*b*) for a more detailed discussion of the method. See also Glennie (1988*b*, 1989) and Overton (1989) for discussion of the method.

[11] Havinden (1961*a*) read through the Oxfordshire probate inventories and extracted those that distinguished the acreages of the various crops. I have read the inventories for the people that he lists and recorded the acreages of the crops, their valuations, and the livestock owned by the farmers. (The inventories are now deposited at the Oxfordshire County Record Office, and I am grateful to its staff for expediting my access to them). 16th-cent. inventories were taken from the printed collection, Havinden (1965). The data sets analysed in this paper generally consist of all the relevant, usable inventories obtained in this way.

*Table 10–7. Animal density and cropping (t-ratios in parentheses)*

| Dependent variable | Constant | Size | Year cubed | $R^2$ |
|---|---|---|---|---|
| 1. Animal density | 0.937 (0.592) | $-0.477 \times 10^{-2}$ (−2.09) | $-0.691 \times 10^{-10}$ (−0.506) | 0.11 |
| 2. Share wheat | −0.466 (−2.057) | $0.139 \times 10^{-3}$ (0.159) | $0.161 \times 10^{-9}$ (3.072) | 0.19 |
| 3. Share rye | 0.571 (4.797) | $0.237 \times 10^{-4}$ (0.052) | $-0.118 \times 10^{-9}$ (−4.285) | 0.31 |
| 4. Share barley | 1.232 (5.468) | $0.339 \times 10^{-3}$ (0.390) | $-0.182 \times 10^{-9}$ (−3.505) | 0.22 |
| 5. Share oats | 0.00538 (0.059) | $0.420 \times 10^{-4}$ (0.118) | $0.377 \times 10^{-11}$ (0.178) | 0.00 |
| 6. Share legumes | −0.343 (−2.18) | $-0.543 \times 10^{-3}$ (−0.896) | $0.136 \times 10^{-9}$ (3.743) | 0.25 |

The other regressions in Table 10–7 show that farm size had no impact on crop choice. Again this result fails to accord with Young's views.

Was the variation in farm size, crop mix, or capital intensity related to the growth of yields between 1550 and 1727? The regressions in Table 10–8 provide some evidence. For both wheat and barley, size was insignificant—the Oxfordshire inventories provide no evidence to support Young's view that large farms had high yields. The share of land planted with beans, peas, pulses, and vetches was majestically insignificant throughout. Moreover, it always appears with a negative sign. Chorley's hypothesis receives no support from these data.

The regressions do provide some support for the hypothesis that high animal densities raised yields. This is especially true for barley, where the coefficient of animal density is always substantial and significant. For wheat the coefficient of animal density in equation 4 is also substantial but marginally insignificant by the usual criteria. It is not surprising, however, that animal density influenced barley yields more forcefully than wheat yields. Fallow was usually folded with sheep in preparation for wheat, while barnyard manure was usually applied in advance of barley. Most of the livestock on these farms were cattle and horses that produced barnyard manure. Not many sheep were kept.

The Oxfordshire inventories do support the agronomic observation underlying the capital-intensive farmer theory. More livestock per cropped acre raised yields. The inventories, however, refute the economic assumptions embodied in that theory. Large farms did not keep more livestock per acre than did small farms: in fact, fewer animals were kept on big farms. Large farms did not reap higher yields than small farms. Moreover, the rise in yields in seventeenth-century Oxfordshire cannot be explained by rising animal densities since, on average, densities did not rise.

*Table 10–8. Crop yields, legume cultivation, and animal density (t-ratios in parentheses)*

| | Constant | Anim | Spul | Year cubed | Size | $R^2$ |
|---|---|---|---|---|---|---|
| Part A: Dependent variable is wheat yield | | | | | | |
| 1. | −28.05 | | | $0.970 \times 10^{-8}$ | 0.3 | 0.32 |
| | (−2.03) | | | (3.26) | (0.80) | |
| 2. | 16.36 | 4.92 | | | | 0.05 |
| | (8.31) | (1.14) | | | | |
| 3. | 19.13 | | −3.82 | | | 0.00 |
| | (6.27) | | (−0.36) | | | |
| 4. | −35.75 | 6.18 | −13.06 | $0.157 \times 10^{-7}$ | 0.034 | 0.45 |
| | (−2.69) | (1.61) | (−1.37) | (3.80) | (0.884) | |
| Part B: Dependent variable is barley yield | | | | | | |
| 5. | −14.44 | | | $0.714 \times 10^{-8}$ | 0.002 | 0.24 |
| | (−1.41) | | | (3.14) | (0.06) | |
| 6. | 14.99 | 7.51 | | | | 0.11 |
| | (8.79) | (2.04) | | | | |
| 7. | 17.92 | | 0.55 | | | 0.00 |
| | (6.78) | | (−0.06) | | | |
| 8. | −22.86 | 8.06 | −11.83 | $0.885 \times 10^{-8}$ | 0.017 | 0.42 |
| | (−2.36) | (2.34) | (−1.24) | (3.71) | (0.489) | |

*Notes:*
'Anim' is the livestock density of the farm.
'Spul' is the share of cropped land planted with pulses, i.e. peas, beans, vetches, and pulse.
'Size' is cropped acreage.
There are 28 observations in the wheat regressions and 35 observations in the barley regressions.

## New Seeds and Rising Yields

The time trend is the most important variable in explaining the rise in yields in Table 10–8. It represents changes in farm practices that are not included in the regressions. In many places rising yields have been the result of new varieties of seed, and better seeds were probably responsible for rising yields in Oxfordshire as well.

The improvements in seed were probably the result of the development of a broad market in seed after the Restoration. Earlier, most marketed seed in England was imported from the continent, but a rising demand for vegetables after 1660 led to the proliferation of the domestic breeding and marketing of garden seed (Thick 1985). As farmers became accustomed to buy seed for clover, turnips, and produce, they also began buying corn seed. Thus, Chartres (1985: 478) reports that 'cereal farmers' in the period 1640–1750 'tended to buy in seed from well outside their immediate locality, seeking special qualities in the yield or the straw'. Diffusion of the better qualities of seed across the country increased the average yield. In addition, the transference of seed between geographical regions was itself sufficient to boost yields. Thus, Bacon in his prize-winning *Report on the Agriculture of Norfolk* (1844: 24) reports that

The reason why new wheats are found to be so much more productive than the old stock, making a fair allowance for their naturally superior productiveness, depends very much upon the change of soil to which they have been subjected. A new wheat from the alluvial soils of Cambridgeshire would find perhaps more suitable elements in the good mixed silicious soils of West Norfolk, or the vegetable moulds and fine loams of the East.

The developing trade in seed facilitated the spread of the best seeds in the kingdom and the matching of seeds to the soils in which they grew best.

The extension of the market for seed had another effect—it increased the profitability of selecting and breeding seed for sale. Plot (1677: 151) described how farmers were developing improved varieties of wheat by culling and propagating the most productive grains and selling the seed in other regions. The first variety he mentions 'was first propagated from some few *ears* of it pickt out of many *Acres*, by one *Pepart* near *Dunstable*, about fifty years ago'. It was 'sowed by it self till it amounted to a quantity, and then proving *Mercatable*, is now become one of the commonest *grains* of this *County*, especially about *Oxford*'. This corn resisted smut, but was being replaced by an improved variety at the time Plot wrote. The new wheat, which was 'first advanced like the former from some few *ears*', was 'found to yield considerably better than most other *wheat, viz.* sometimes *twenty for one*'. Consequently, 'it is now become the most eligible *Corn*, all along the *Vale* under the *Chiltern* Hills'. This is the region where most of the inventories in my sample originated. A return of 'sometimes twenty for one' on the seed corresponds to a yield per acre of 'sometimes' 45 bushels. The spread of varieties that productive would explain the rise in yields I measured.

## The History of Corn Yields

The regression analysis of the Oxfordshire probate inventories undermines the theory that the amalgamation of farms was responsible for the rise in yields since the data show no correlation between yield and farm size. The data also undermine the theory in a second respect since they indicate that most of the growth in yields between the middle ages and the nineteenth century was accomplished before the elimination of peasant agriculture in the eighteenth century.

Table 10–9 shows predicted values for wheat and barley yields over the sixteenth and seventeenth centuries. These calculations presuppose a 65-

Table 10–9. *Predicted crop yields in Oxfordshire (bushels per acre)*

| Period | Wheat | Barley |
|--------|-------|--------|
| 1550 | 9.0 | 12.2 |
| 1600 | 12.7 | 14.9 |
| 1650 | 16.5 | 17.7 |
| 1700 | 20.6 | 20.7 |

*Note*: The yields were predicted with equations 1 and 5 in Table 10–8. The calculations presume that the farm has 32.5 cropped acres, which correspond to a 65-acre farm if that has 75% of its land arable and that is cropped on a three-field system, so two-thirds of the arable is under crops.

acre farm, which was the average for open field villages before the shift to capitalist agriculture in the eighteenth century. (Table 4–4.) For wheat, the predicted yield rose from 9.0 bushels in 1550 to 20.6 bushels in 1700. The barley yield increased from 12.2 to 20.7 bushels over the same period. Since wheat yields in enclosed villages in the light arable district *c.*1800 averaged 19.7 bushels and barley yields averaged 29.3 bushels (Table 7–2), yeoman farmers had accomplished all of the advance in wheat yields and half of the advance in barley yields realized by enclosed, capitalist farmers in the early nineteenth century. These calculations thus reinforce the conclusion of the regression analysis—the yeoman farming system of seventeenth-century England produced a revolution in corn yields.

## Conclusion

There was a biological revolution in English agriculture between the middle ages and the nineteenth century—corn yields approximately doubled. That revolution was not the result of large-scale, capital-intensive farming. Also, it did not require enclosure, for the Oxfordshire inventories analysed here described open field farms almost exclusively. The unique institutional feature of the sixteenth and seventeenth centuries was tenurial: many of the farmers held their land for several lives or for long terms of years on copyholds or beneficial leases. Their forefathers in the middle ages had not had such security of tenure, nor did their descendants in the eighteenth century who farmed as tenants at will. It was only the yeomen of early modern England who had a long-term interest in the soil and who thereby benefited from the rise in land value caused by a rise in productivity.

Eighteenth-century discussions of leases highlight the importance of these considerations. Most commentators believed that tenancy at will reduced the farmers' incentive to raise productivity. The problem was not so much that farmers were arbitrarily evicted, since they rarely moved. Rather the danger was that landlords would expropriate the benefits of improvement by raising rents.[12] Batchelor (1808: 43) remarked: 'The large farmers, therefore, are secure in their possessions, and the benefit of their improvements, if not overburdened with increased rents, descends to their children.' The qualifying clause is to the point, for the Board of Agriculture reports are full of examples and warnings that rents were frequently increased.

Not far from Olney [Bucks.], a farmer, whose character is that of being a very honest and industrious man, had banked [removed the anthills from] 40 acres, which in the course of three years cost him 200 l. [H]e had no lease: just as this improvement was finished, the person of whom he hired his farm, observing (or perhaps informed of) the state to which the farmer had brought the land, sent a surveyor to examine it. The surveyor, deaf to any observations of the farmer, and intent only upon giving the real value, to which the farmer had increased the land, put 9s. per acre upon it in addition to the former rent. Add to this the interest of the capital (200 l.) sunk upon the 40 acres, and a serious lesson is taught every farmer, not to lay out one shilling upon the improvement of an estate without a lease. (Priest 1813: 91)

Trust in the good intentions of one's landlord was not enough:

A farmer cannot exercise his skill and industry, with that spirit which is necessary in all important undertakings, without some probable security for the enjoyment of the fruits of his labour; death may perhaps take from him a landlord on whom he could depend, and whose word was equal to his bond, and the estate devolve to another, who regardless of the engagements of his predecessor, may give him notice to quit or may raise his rent. He may be so unhappy as to differ with him in politics; or his dog may unfortunately kill a hare, which has been bred on the farm; the consequences of such slight offences are well known. (Pitt 1809a: 50–1)

The conventional solution was long leases. Parkinson (1811: 45), for instance, 'recommend[ed] leases for twenty-one years with proper restrictions on all *large farms*, as one great means of their being essentially improved, by thus giving the tenant a security for expending money, and a proper scope to exert his abilities'. (See also Young 1813: 64–8 and Gooch 1813: 39–40).

There is a good deal of irony in the proposal for long leases, for they would have only restored the security of tenure that farmers had enjoyed

[12] Mokyr (1985: 86–7, 104–9) provides an insightful discussion of such 'predatory' behaviour.

when they held their land on copyholds and beneficial leases. Kent (1794: 35–6) observed:

The ancient feudal tenures had undoubtedly a strong tendency to enslave mankind, by subjecting tenants to the controul and power of an arbitrary lord; but like all other things, there were some advantages to be found in the system. Every man who held land, had a certainty in it, as the tenant generally held possession for life.

Pitt (1809a: 51) took the argument for leases to its logical —heretical— conclusion that 'the cultivation of small or moderate sized farms by their owners, is generally productive of the best and most improved modes of agriculture, as the farmer finds himself doubly encouraged by interest, and the security of enjoying the fruits of his labour'. This, of course, was the situation that had prevailed before the yeoman mode of production was superseded by the capitalist. Is it really so surprising that the yeomen made an agricultural revolution?

# 11

# Labour Shedding and the Growth in Productivity

> Improvements in agriculture are of two kinds: those which increase the
> productive powers of the land, and those which enable us to obtain its
> produce with less labour.
>
> David Ricardo, *On the Principles of Political Economy and
> Taxation*, 1817: 80

LARGE farms had no advantage over small using corn yields as the measure
of performance. But there are other indicators of efficiency. In this chapter I
will show that the advantage of large farms lay in labour—rather than land—
productivity.

This result explains the most distinctive characteristic of agricultural
productivity growth in early modern England—the rapid and exceptional
increase in output per worker. Early in the nineteenth century, output per
worker in England was 50 per cent higher than in France and two and a half
times the Russian level.

Over the early modern period, and especially the eighteenth century,
average farm size in England increased considerably. Both Tory and
Marxist fundamentalists argue that such a change should have increased
labour productivity. The Tories contend that the large farms employed
more people per acre than the small farms, but that output per acre grew
even more, so output per worker rose. This argument is doubtful, however,
in view of the finding that large farms did not reap greater yields than small
farms. The Marxist view is more plausible; it contends that the large farms
reduced employment per acre. That would certainly have raised output per
worker, if it were not offset by a fall in production.

My finding in the last chapter that small farmers raised yields in the
seventeenth century suggests a third hypothesis to account for the growth in
labour productivity. Since the labour requirements of most tasks in

agriculture depended on the acreage rather than the yield,[1] a rise in yield alone would raise output per worker. Perhaps England's rapid growth in labour productivity was entirely the result of the yeomen's efforts and had nothing to do with enclosures or the growth in average farm size?

I will analyse these possibilities by decomposing output per worker into two terms—output per acre and acres per worker. I shall show that the increase in English labour productivity between 1600 and 1800 had two causes. The first was the rise in crop yields; this cause operated across northwestern Europe causing labour productivity to rise throughout the region. The second was the shift to large, capitalist farms. This reorganization increased labour productivity by reducing employment per acre. Moreover, that increment to efficiency was about equal to the Anglo–French productivity gap. The English superiority noted by Wrigley, Bairoch, and O'Brien and Keyder was due to England's peculiar agrarian institutions—in particular, large farms operated with wage labour.

## Farm Size and Performance

To determine the significance of the eighteenth-century growth in farm size, it is necessary to compare the operations of small and large farms. The most systematic source for these comparisons is the survey of farms collected by Arthur Young in his tours of England in the late 1760s.[2] These data show that small and large farms differed in several important ways.

Both Tory and Marxist fundamentalists contend that overall efficiency increased with farm size. Ricardian surplus, the difference between farm revenue and the opportunity cost of all inputs other than land, is a convenient indicator of efficiency or total factor productivity. (See Chapter 9.) Tables 11–1 and 11–2 show how surplus varied with size for arable and pastoral farms. Expansion in size was accompanied by a continuous rise in Ricardian surplus but at a diminishing rate and reaching a peak at about 200 acres. Further increases of surplus with size look to have been unlikely for arable farms. For pastoral farms, the data indicate decreasing returns to

[1] See Tables 8–1, 8–3, and Appendix II. See Parker and Klein (1966: 528) for parallel results for American agriculture, and Grantham (1989) for French agriculture.

[2] The data are described in Chs. 8 and 9 and in Allen (1982). In Ch. 9, as in Allen (1982), farm-specific input and output prices were used, but in Allen (1982) it was shown that the variation in those prices explained little variation in surplus. In this chapter, revenue, cost, and surplus are computed using average prices. Also harvest labour is estimated as the cost of mowing and reaping rather than as a proportion of the wage payments of labourers, which was the method Young (1771a: iv. 356–7) used. The results are similar on average, but the procedure adopted here seems appropriate since we are investigating the effects of farm size on employment and since the mix of servants and labourers varied with size.

*Table 11–1. Revenue and cost in arable farms (£ per acre)*

| Farm size (acres) | Total revenue per acre | Total cost per acre | Labour cost per acre | Ricardian surplus per acre | Number of farms |
|---|---|---|---|---|---|
| 0–50 | 4.0399 | 3.0615 | 1.5016 | 0.9783 | 8 |
| 50–100 | 4.3094 | 3.1146 | 1.3023 | 1.1948 | 45 |
| 100–150 | 3.7694 | 2.4290 | 1.0215 | 1.3404 | 16 |
| 150–200 | 4.4774 | 2.3931 | 0.9598 | 2.0842 | 22 |
| 200–250 | 4.0605 | 2.5283 | 0.6228 | 1.5322 | 4 |
| 250–300 | 3.5078 | 2.0145 | 0.7641 | 1.4934 | 12 |
| 300–350 | 5.0522 | 2.7088 | 0.9025 | 2.3434 | 4 |
| 350–400 | 3.2252 | 1.8165 | 0.6191 | 1.4087 | 2 |
| 400–450 | 4.9065 | 2.4322 | 0.6422 | 2.4743 | 2 |
| 450–500 | 5.3128 | 2.3392 | 0.6781 | 2.9736 | 3 |
| 500–550 | — | — | — | — | — |
| 550–600 | 4.8578 | 2.2600 | 0.6199 | 2.5478 | 6 |
| 600–650 | 3.3538 | 2.0028 | 0.4878 | 1.3510 | 1 |
| 650–700 | 4.0183 | 2.1921 | 0.5910 | 1.9162 | 3 |

Note: Arable farms are more than 45% arable and pastoral farms less than 45% arable. Initially a 50% division was used, but inspection of the data showed that there was a more natural division at 45%. Farms operated by gentlemen are excluded in this and all other calculations as are farms of more than 700 acres.
Source: This table was computed from the sample of farms surveyed by Arthur Young c.1770 and analysed in Allen (1982). As noted in the text, the analysis in this paper introduces two changes—a uniform set of prices is used in the computations and harvest labour is estimated by piece rates.

*Table 11–2. Revenue and cost in pasture farms (£ per acre)*

| Farm size (acres) | Total revenue per acre | Total cost per acre | Labour cost per acre | Ricardian surplus per acre | Number of farms |
|---|---|---|---|---|---|
| 0–50 | 3.8464 | 3.5715 | 1.4518 | 0.2749 | 16 |
| 50–100 | 3.3481 | 2.6911 | 0.9687 | 0.6570 | 35 |
| 100–150 | 3.2639 | 2.2647 | 0.7394 | 0.9992 | 19 |
| 150–200 | 3.1823 | 2.1576 | 0.5989 | 1.0246 | 23 |
| 200–250 | 2.8764 | 2.1280 | 0.7463 | 0.7484 | 1 |
| 250–300 | 2.4017 | 1.6609 | 0.4958 | 0.7408 | 8 |
| 300–350 | 1.8683 | 1.2689 | 0.5396 | 0.5993 | 1 |
| 350–400 | 2.1593 | 1.5232 | 0.5178 | 0.6361 | 5 |
| 400–450 | 2.4385 | 1.7417 | 0.4639 | 0.6968 | 2 |
| 450–500 | — | — | — | — | — |
| 500–550 | 3.4157 | 2.5452 | 0.2789 | 0.8705 | 1 |
| 550–600 | 2.9644 | 2.3241 | 0.2863 | 0.6403 | 1 |
| 600–650 | — | — | — | — | — |
| 650–700 | 2.8003 | 1.6181 | 0.5526 | 1.1822 | 1 |

Source: As for Table 11–1.

scale above 200 acres. As a rule of thumb, 200 acres was the rent-maximizing size of a farm, although it is difficult to be precise given the random fluctuations around the trend.

To test for increasing returns to scale, I regressed Ricardian surplus per acre on a quadratic function of farm size. The results are shown in Table

11–3. All the equations show surplus increasing with size (the positive coefficient of acres) but at a diminishing rate (the negative coefficient of acres squared). Equations 2 and 4 include a variable to represent the stint of farms with commons.[3] The coefficient of commons is positive and significant, an important result in its own right since it establishes that the management of commons was effective enough to prevent rent dissipation. The inclusion of commons improves the estimation of the coefficients of acres and acres squared. The low $R^2$ show that many things besides size influenced farm efficiency.

*Table 11–3. Scale economies in farming (t-ratios in parentheses)*

|  | Constant | Acres | Acres squared | Commons | $R^2$ |
|---|---|---|---|---|---|
| 1. Arable | 0.77309 | 0.00592 | $-0.57053 \times 10^{-5}$ |  | 0.09 |
|  | (2.587) | (2.319) | (−1.504) |  |  |
| 2. Arable | 0.51296 | 0.00621 | $-0.67250 \times 10^{-5}$ | 0.51788 | 0.18 |
|  | (1.756) | (2.560) | (−1.860) | (3.795) |  |
| 3. Pasture | 0.45569 | 0.00292 | $-0.41591 \times 10^{-5}$ |  | 0.04 |
|  | (2.903) | (1.929) | (−1.586) |  |  |
| 4. Pasture | 0.09571 | 0.00485 | $-0.64839 \times 10^{-5}$ | 0.23579 | 0.28 |
|  | (0.646) | (3.601) | (−2.817) | (6.138) |  |

*Note*: The dependent variable in the regressions is Ricardian surplus per acre.

The equations show that estates organized in large farms generated more income than those divided into small farms. Equation 1 predicts for arable farms that an increase in size from 50 to 200 acres would raise Ricardian surplus per acre 64 per cent. Amalgamating twenty 50-acre farms into five 200-acre farms would increase the annual value of a 1,000-acre estate by that proportion. For pasture farms, regression 3 predicts a 48 per cent rise in Ricardian surplus per acre with an increase in size from 50 to 200 acres. Financial considerations of this sort prompted the extinction of small-scale farming.

The advantage of large farms lay in lower costs, not greater output. There were economies of scale in implements and draught animals which probably reflected each farmer's wanting his own complement of equipment and horses so as not to be dependent on renting these crucial inputs. The savings on these inputs, however, were small compared to those arising from economies in labour.

---

[3] For farms without common rights, the commons variable was assigned a value of zero. For farms with common rights, the number of sheep on the farm was taken as the measure of commons. See Allen (1982).

## Labour Saving and The Economies of Large-Scale Farming

Most of the advantage of large farms came from lower labour costs. As Table 11–4 shows, employment per acre declined with size for men and, especially, for women and boys. Savings were effected in both family labour and hired labour. Table 11–5 incorporates farm size into the regression analysis of Table 8–4 that measured the impact of enclosure and pastoralism on employment. So far as enclosure and the conversion to grass are concerned, the story remains the same—enclosure itself had no effect on employment, while laying land down to grass reduced overall employment and the employment of men, while slightly increasing the employment of women. The employment of children was unaffected. The inclusion of farm size in the regressions confirms that employment per acre declined with size for all categories of workers and especially for women and children.

*Table 11–4. Employment per acre (workers per acre)*

| Farm size (acres) | Arable farms | | | Pasture farms | | |
|---|---|---|---|---|---|---|
| | Men | Maids | Boys | Men | Maids | Boys |
| 0–50 | 0.0364 | 0.0329 | 0.0392 | 0.3300 | 0.0407 | 0.0420 |
| 50–100 | 0.0398 | 0.0233 | 0.0225 | 0.0245 | 0.0252 | 0.0215 |
| 100–150 | 0.0309 | 0.0179 | 0.0175 | 0.0210 | 0.0170 | 0.0142 |
| 150–200 | 0.0316 | 0.0142 | 0.0119 | 0.0179 | 0.0139 | 0.0095 |
| 200–250 | 0.0197 | 0.0105 | 0.0095 | 0.0250 | 0.0125 | 0.0125 |
| 250–300 | 0.0268 | 0.0096 | 0.0082 | 0.0160 | 0.0084 | 0.0084 |
| 300–350 | 0.0328 | 0.0091 | 0.0105 | 0.0187 | 0.0094 | 0.0062 |
| 350–400 | 0.0182 | 0.0079 | 0.0104 | 0.0180 | 0.0095 | 0.0065 |
| 400–450 | 0.0199 | 0.0079 | 0.0092 | 0.0167 | 0.0044 | 0.0067 |
| 450–500 | 0.0228 | 0.0065 | 0.0071 | — | — | — |
| 500–550 | — | — | — | 0.0091 | 0.0055 | 0.0055 |
| 550–600 | 0.0225 | 0.0048 | 0.0048 | 0.0117 | 0.0033 | 0.0033 |
| 600–650 | 0.0185 | 0.0031 | 0.0077 | — | — | — |
| 650–700 | 0.0211 | 0.0063 | 0.0068 | 0.0214 | 0.0071 | 0.0043 |

The rapid decline with size in the employment of women and boys compared to men meant that the eighteenth-century shift to large farms changed the sex balance of rural employment. Earlier I showed that farms of less than 50 acres were family farms in the sense that a family could operate the farm without much, if any, hired labour. Table 11–4 shows a second sense in which these small farms were family farms—they employed men, women, and boys in about equal numbers, so they offered reasonably full employment to all family members. The shift to large farms meant that only the husbands in labourers' families were employed in agriculture.

Enclosure and farm amalgamation interacted in reducing employment. In districts that remained in tillage, enclosure had no effect on employment.

*Table 11–5. Employment per acre regressions (t-ratios in parentheses)*

| | Equation | | | |
|---|---|---|---|---|
| | 1 (Men) | 2 (Women) | 3 (Boys) | 4 (Total labour) |
| Constant | 0.02929 | 0.03838 | 0.03592 | 1.30405 |
| | (8.668) | (17.728) | (14.754) | (16.623) |
| Acres | $-9.256 \times 10^{-5}$ | $-1.458 \times 10^{-4}$ | $-1.642 \times 10^{-4}$ | $-4.793 \times 10^{-3}$ |
| | (−5.728) | (−14.083) | (−14.105) | (−12.775) |
| Acres sq | $9.399 \times 10^{-8}$ | $1.578 \times 10^{-7}$ | $1.889 \times 10^{-7}$ | $5.139 \times 10^{-6}$ |
| | (3.725) | (9.757) | (10.389) | (8.772) |
| Arab | 0.02370 | −0.00375 | 0.00082 | 0.55392 |
| | (8.688) | (−2.149) | (0.417) | (8.748) |
| Encl | $-9.541 \times 10^{-4}$ | $1.290 \times 10^{-4}$ | $9.099 \times 10^{-4}$ | −0.03406 |
| | (−0.377) | (0.080) | (−0.498) | (−0.579) |
| Part | $-5.495 \times 10^{-3}$ | $-9.814 \times 10^{-5}$ | $-7.522 \times 10^{-4}$ | −0.13636 |
| | (−1.938) | (−0.054) | (−0.368) | (−2.072) |
| $R^2$ | 0.37084 | 0.60821 | 0.56482 | 0.59728 |

*Notes:*
Acres sq = acres squared
Arab = share of arable
Encl = 1 if the farm is enclosed, 0 otherwise
Part = 1 if the farm is partly enclosed, 0 otherwise
The dependent variable in equations 1–3 is the number of people whereas the dependent variable in equation 4 is labour cost in £.

However, the growth in farm size led to employment declines. In districts that were converted to pasture, both enclosure and large farms reduced employment. All districts, therefore, suffered job losses from 'improved agriculture' with the declines being the fiercest in the pasture district where enclosure and farm amalgamation worked in tandem. A consideration of farm amalgamation and enclosure together leads to a decisive rejection of the optimism of Tory fundamentalists like Professor Chambers.

My findings also differ from those of Snell (1985), a recent critic of Chambers. Snell studied the pattern of seasonal unemployment, as measured by the date of application for poor relief, rather than labour demand as I have done. He found that enclosure in regions where there was considerable conversion to pasture (e.g. Leicestershire and Nottinghamshire) had little impact on male relief applications, while enclosure led to an increase in male winter unemployment relative to male summer unemployment in regions that continued to specialize in grain. This result suggests that enclosure had the biggest seasonal impact on the labour market where it had the smallest effect on farming. However, the male labour demand equation in Table 11–5 establishes that enclosure reduced employment most in places where it led to widespread conversion to pasture.

Snell (1985: 15–66, 155–8) discovered that women's wages declined relative to men's after the middle of the eighteenth century. He attributed the decline to enclosure. My labour demand functions, however, show that enclosure actually improved the employment prospects of women when it led to a shift to pasture, a situation Snell (1985: 40–6) also noted. What reduced female labour demand and led to declining women's wages was the pronounced reduction in female employment that resulted from the transition to large farms. This reorganization occurred during the eighteenth century and accounts for the trends in Snell's wage data.

## Arthur Young's Analysis of Farm Size and Population

Young collected his data in order to measure the effect of farm amalgamation on performance indicators, including employment. His work is the most systematic statistical attempt to gauge the relationship, so it remains the fundamental evidence in favour of Tory fundamentalism. However, Young's data collection was in advance of his statistical abilities: his demonstration that large farms raised employment is flawed and unconvincing.

The problem with Young's analysis is that he did not distinguish between employment and the population supported by agriculture. He regarded them as equivalent. 'In every branch of industry *employment is the soul of population*' (Young 1774: 278). He began his analysis by tabulating the average number of servants, maids, boys, and labourers according to farm size (Young 1771a: iv. 246). There is nothing peculiar about this table and if it is analysed as I have analysed Young's data it yields the same answer— employment declined with farm size. Young obtained his contrary result by estimating the population supported by agricultural employment. He assumed that 90 per cent of the labourers were married and had families of five people on average, while five-sixths of the farmers were married and had families of four. (Servants, maids, and boys were unmarried and so had families only of one.) He then computed the total population, including dependants, supported by farms of each size. Since larger farms employed a higher ratio of labourers to servants than small farms, the population supported by agriculture increased with farm size.[4] So he concluded 'that the farms most advantageous to population, without exceptions, are those from five hundred acres upwards; and of such, those above a thousand acres are the superior; those under five hundred acres much inferior' (Young 1771a: iv. 254). Whatever the truth of this conclusion for population, it has nothing directly to do with employment.

---

[4] See Kussmaul (1981) for a discussion of farm service and especially pp. 18–20 for a discussion of Young's data.

Not only does Young's account shift too readily between population and employment, it shifts back and forth between ideal and reality. Young should have been suspicious of his theory of employment on the basis of the characteristics of the 'average' farm he observed on the northern tour. In discussing his data, he remarked that 'the article of labour is much below what it ought to be'. There were too few employees to cultivate the average farm 'in a complete manner, or, indeed, upon the improved system of several counties'. He lamented that 'very few farmers employ the hands they ought' (Young 1771*a*: iv. 204, 195). When Young extolled the large, rich farmer who employed many workers, he was not describing 'best practice' —let alone 'average practice' as minuted on his tours—but instead an ideal that was rarely, if ever, realized. Why Young confused his ideal farmer with the ones he met on his tours is a tantalizing question.

## Farm Size and the Management of Labour

Why did large farms employ less labour per acre than small farms? The answers are simplest for boys and women. The immediate reason is that their total employment increased more slowly than farm size. Thus, the average arable farm of less than 50 acres employed on average 1.4 maids and 1.7 boys, including family labour, but even very large farms on average did not employ more than three of each. As a result, employment per acre of maids and boys declined with size.

The failure to expand the use of women and boys proportionately to increases in farm size is not mysterious. Even though boys were cheap, they were difficult to supervise, so their use was limited on large farms. The employment of maids did not increase much with size because farmers did not increase the size of their dairies a great deal. Dairy cows yielded little profit (Young 1771*a*: iv. 167) and were kept only because they provided a steadier source of cash income than corn. Large farmers made do with little more 'insurance' than small farmers. Thus arable farms of less than 50 acres had on average five cows while farms of 150–200 acres had twelve cows. Larger farms had only ten to twelve cows. Consequently, employment per acre of women declined with farm size.

The employment of men did expand continuously with farm size, but economies were effected on larger farms that reduced cost per acre. Specialization and division of labour were the source of the savings. On small farms without much hired labour, the farmer had to do all the jobs— ploughing, sowing, harvesting, threshing, carting, hedging, ditching, and tending the sheep, horses, and cattle. Large farms employed specialists to do some of these tasks. This reduced costs in three ways.

First, specialization allowed the employment of less skilled labour in place of the farmer's more valuable time. The possibilities are suggested by a Rutland wage assessment of 1563. It distinguished three kinds of farm servants:

A chief servant of husbandry of the best sort, which can eire [plough], sow, mow, thrash, make a rick, thatch, and hedge the same, and can kill and dress a hog, sheep, and calf, may have in wages by the year 40s., and for his livery 6s.

A common servant in husbandry, which can mow, sow, thrash, and load a cart, and cannot make a rick, hedge and thatch the same, and cannot kill and dress hog, sheep, and calf, may have in wages by the year 33s. 4d. and for his livery 5s.

A mean servant in husbandry which can drive the plough, pitch the cart, and thrash, and cannot expertly sow, mow, nor make a rick, nor thatch the same, may have for his wages by the year 24s. and for his livery 5s. (Rogers 1866–1892: iv (1882), p. 120)

A farmer of 50 acres approximated a chief servant in husbandry since he had to perform all those tasks himself. A farm with more employees could hire less skilled—and cheaper—people to perform the simpler tasks. On nineteenth-century farms 'plough men are the staple class of labourers'.

His principal skill consists in the management of the plough; and his other and aggregate skill must comprehend a practical knowledge of everything connected with the care of his horses, and with the uses and working of the implements they draw. (J. Wilson 1847: 234, 874)

The important point about this description is how limited the skills were— the nineteenth-century ploughman was comparable to the 'mean servants' in the 1563 assessment, and their substitution for the more skilled yeoman was one reason for the lower costs of larger farms. Furthermore, reducing the number of small farms and reducing their occupiers to ploughmen amounted to the deskilling of a considerable fraction of the rural population.

The cow-herd on the nineteenth-century farm was another example of substituting cheap labour for expensive labour.

The cow-herd or cattle-man is generally a man past middle life, and sometimes an old and almost superannuated ploughman. He receives considerably lower wages than a ploughman. (Wilson 1847: 234)

Farms had to be large, however, for there to be enough work for this substitution to be feasible. Savings of this sort were imperfectly incorporated into the analysis of Young's data by valuing the farmer and the first servant hired as 'first class' servants and other servants as 'ordinary' servants.

*Capitalist Agriculture and Growth*

Second, even though the yeoman had to perform a variety of tasks, he was likely to have been 'a jack of all trades and a master of none'. Large farms could employ men specialized in single lines of work who could perform it more effectively than the family farmer. Thus large midlands farms employed shepherds to manage the sheep. Hedging was a task performed by specialists on large farms.

A superior hedger ranks with a master-ploughman, requires to be a person of considerable intelligence, and has charge of planting, pruning, plashing, and repairing hedges, of pruning orchard and forest trees, and of superintending all the other shrubby and dendritic plants of the farm . . . Only a farm of large extent . . . employs a superior hedger. (Wilson 1847: 234)

The employment of specialists allowed reductions in employment per acre.

Third, the large farm could employ gangs of men to perform tasks at high efficiency.[5] The corn harvest was often done by gangs. In this case, small farms were not at a disadvantage when mobile gangs of reapers contracted to do that work. Such contractors, however, were not commonly available for mowing hay, which was also done by gangs. In his *General View of . . . Bedford*, Batchelor (1808: 110) reported that 'Cocking and dragging [required] five men to ten acres. Carrying, ten men at six acres per day . . . ' Indeed, any activity involving the transporting of bulky materials like hay, grain, or manure was done by teams. Thus, of the carriage of corn, Young (1805: 423) observed:

In a farm-yard where there are teams enough, carting the wheat crops requires three wagons: one loading in the field, one unloading, and one upon the road going backwards and forwards: five or six horses are sufficient for them, and two men to pitch, two to load, one to drive, and two to unload; in all seven: which make good dispatch.

Batchelor (1808: 109) made a similar point about building ricks:

A field of fifteen acres, when laid on one stack, appears to occupy two men about one day in covering it securely with thatch, and these, together with the four *yelmers* and *servers* cost about 20s. per day.

Later in the winter when the corn was taken from the rick yard to the barn for threshing, Batchelor (1808: 111) observed that 'six men are employed in this business'. Similarly, 'the carriage [of manure] occupies either two or three drivers . . . and four men are employed in filling and spreading' (Batchelor 1808: 106).

---

[5] Fogel and Engerman (1974: esp. i. 203–9) is the classic discussion of the efficiency-raising effects of gang labour in agriculture.

Marx was aware of this reasoning and appealed to it as an explanation for the advantages of large-scale production. He quoted from John Arbuthnot's *Inquiry into the Connection Between the Present Price of Provisions, and the Size of Farms* (1773):

'There is also' (when the same number of men are employed by one farmer on 300 acres, instead of by ten farmers with 30 acres a piece) 'an advantage in the proportion of servants, which will not so easily be understood but by practical men; for it is natural to say, as 1 is to 4, so are 3 to 12: but this will not hold good in practice; for in harvest time and many other operations which require that kind of despatch by the throwing many hands together, the work is better and more expeditiously done: f.i. in harvest, 2 drivers, 2 loaders, 2 pitchers, 2 rakers, and the rest at the rick, or in the barn, will despatch double the work that the same number of hands would do if divided into different gangs on different farms.'[6]

While Arbuthnot concluded that large farms did not actually depopulate, his analysis of gangs implies that large farms employed fewer men per acre than did small.

These quotations point to the widespread utilization of gangs on large farms. There was obviously latitude in the size of the work crews, but the numbers mentioned in these examples are not arbitrary. Six to ten men are common, and that was the range of men employed full-time on large farms in the eighteenth century. The 'specialists' like the ploughmen spent much of the year doing their peculiar task and then worked together in a team during hay-making, corn harvesting, and when produce and manure had to be moved about the farm.

Further confirmation of the efficiency-raising effects of the gang system comes from an unlikely source—Arthur Young's *Political Arithmetic*. In the course of arguing that large farms increased employment, he considered the counter-arguments. He may have had Arbuthnot's tract in mind since it was written the year before, and since the close of Young's argument runs over the ground covered by Arbuthnot. In any event, Young had to admit some merit:

It is said, that large farms are in fact machines in agriculture, which enable the cultivators of the soil to do that with few hands which before they did with many; resembling a stocking-loom, for instance, which enables the master manufacturer to turn off half his hands, and yet make more stockings than ever. A lively argument but false in almost every particular; indeed the resemblance holds no further than the capacity of performing in some operations much more with ten men in one farm,

---

[6] As quoted by Marx (1867: 358 n. 2). The quotation is reasonably accurate and the inserted material appropriate. See Arbuthnot (1773: 7–8). I am indebted to Greg Clark for very generously bringing Marx's quotation to my attention.

than with the same number divided among five farms; of which there can be no doubt: But I appeal to all persons conversant in husbandry, if this holds true through one-tenth of the labour of a farm; witness ploughing, harrowing, sowing, digging, mowing, reaping, threshing, hedging, ditching, and an hundred other articles in which one man, separately taken, performs the full tenth of ten men collected. The saving of labour is but in few articles, such as carting hay or corn; carting dung or marle; keeping sheep, &c. (Young 1774: 294)

The economies of scale in carrying bulk commodities are conceded. It should be noted that some of the items, like hedging and ditching, where there were not economies from the gang system, were subject to economies through the employment of specialists. The issue is how much labour, in total, was saved by large farms from all of these sources. As Table 11–5 shows, it was much more substantial than Young was willing to concede.

So in addition to being an agronomist, a large-scale farmer had to administer a complex system of employment. This is a recurring theme in the early nineteenth-century agricultural handbooks. Thus, Loudon (1831: 548) says

The grand point to be aimed at by . . . the occupier of a large farm, is to hit on the proper number of sub-managers; and to assign each his distinct province, so that the one may never interfere with the other. Having attained this, the next thing is to keep the whole machine in regular action; to keep every man, from the lowest operator to the highest, strictly to his duty.

Orwin and Whetham (1964: 82), in describing employment conditions in the middle of the nineteenth century, observed that

On large farms the head horseman, the head cowman and the shepherd held positions of real responsibility, often stayed on the same farm for most their lives, and identified themselves completely with its fortunes. Their knowledge of and pride in their charges were such that they would refer to '*my* sheep,' '*my* horses' and so on.

The effectiveness of specialization and the work crew system in raising labour productivity depended on the foremen being this committed. In addition, the contemporary farm management literature anticipated twentieth-century scientific management in urging farmers to keep time books in which every employee's exact whereabouts and activities were carefully monitored. These records allowed the deployment of labour to be rationally planned (Loudon 1831: 550; Morton 1855). By exploiting the possibilities of an internal division of labour, large farms economized on adult male labour and generated more Ricardian surplus per acre.

## Farm Amalgamation and the Growth of Labour Productivity

I can now confront the question: Why did labour productivity rise in early modern English agriculture? Two developments were working in the same direction. First, English farm sizes increased substantially in the eighteenth century, and larger farms employed less labour per acre than smaller farms. Second, English grain yields increased in the early modern period, especially in the seventeenth century. Since the labour requirements of most tasks in cereal production depended on the acreage involved and not on the volume of grain harvested, an increase in yield also caused labour productivity to rise. I can use the details of Young's representative farms to determine their relative performance. I distinguish arable from pasture farms. I estimate how output per acre and labour per acre varied with changing crop yields and farm size and then combine those results to simulate labour productivity.

Consider, first, output per acre, that is, total farm revenue divided by total farm acreage. Table 11–6 shows the results of computing output per acre using average medieval yields, open field yields and enclosed yields c.1800. The values in Table 11–6 are averages across all farm sizes and thus incorporate the assumption that changes in the farm size structure did not affect yields. Further, the weighting of the changes in yields reflects the cropping patterns and prices c.1770. No allowance is made for changes in livestock efficiency.[7] Between the middle ages and the nineteenth century, output per acre increased 54 per cent in arable farms with most of the gain occurring before the eighteenth century. The gain in output per acre was less than the increase in crop yields since the farms also produced animal products.

Next consider labour per acre. Table 11–7 uses a variant of equation 4 in Table 11–5 to estimate average employment per acre for the distributions of farm size drawn from estate surveys and shown in Tables 4–4 and 4–5. Employment per acre was computed for each range of farm size using the acreage of the midpoint of the range and an overall average computed by weighting these values with the acreage of land in each range. Table 11–7 shows a continuous decline in employment per acre in open, arable farms with most of the decline occurring in the eighteenth century when average farm size increased dramatically. The calculations for enclosed, pasture farms were erratic due to the small number of such farms tabulated for the early seventeenth and early eighteenth centuries.

Table 11–8 combines results from Tables 11–6 and 11–7 to estimate

[7] For this reason output per acre rises considerably more on arable farms than on pasture farms.

*Table 11–6. Output per acre with various corn yields (£ per acre)*

|  | Medieval yields | c.1800 | |
|---|---|---|---|
|  |  | Open yields | Enclosed yields |
| Pasture farms | 2.6654 | 3.1112 | 3.1649 |
| Arable farms | 2.5451 | 3.7452 | 3.9147 |

*Note*: This table is based on the following yields:

|  | Medieval | c.1800 | |
|---|---|---|---|
|  |  | Open | Enclosed |
| Wheat | 10.7 | 20.9 | 21.9 |
| Barley | 16.8 | 28.0 | 32.2 |
| Oats | 11.7 | 36.9 | 38.1 |
| Beans/peas | 10.0 | 22.4 | 23.4 |

*Notes*:
The medieval yields are discussed in Ch. 7 and the 1800 yields are from Table 7–2 for the pasture district. These yields are close to Turner's (1982*a* 500) yields for the grains.

Output per acre equals revenue per acre for arable and pasture farms computed by substituting the assumed yields for the actual yields in the computer programme that generated Tables 11–1 and 11–2. These calculations thus embody the production patterns and price structures of the arable and pasture farms in Young's sample.

*Table 11–7. Labour per acre with various farm size distributions (£ per acre)*

|  | c.1600 | c.1700 | c.1800 |
|---|---|---|---|
| Arable (open) | 1.241021 | 1.174268 | 0.911618 |
| Pasture (enclosed) | 0.421973 | 0.739729 | 0.584714 |

*Note*:
Employment was computed with the following equation:

$$\text{employment} = 1.25928 + 0.54416\,S - 0.00479\,A + 0.51590 \times 10^{-5}\,A^2$$

where $S$ is the share of arable and $A$ is farm size. Arable farms were presumed to be 75% arable and pasture farms 20% arable. The distribution of the acreage of open farms in Table 4–4 was interpreted as the distribution of the acreage of arable farms. Likewise, the distribution of enclosed farms in Table 4–4 was interpreted as the distribution of pasture farms. Labour per acre was computed for the midpoint of each size category in Table 4–4. For farms of less than 30 acres, which lay outside the range of farm sizes for which the employment equation was estimated, employment per acre was set to the value for farms of 30–60 acres. The value of employment per acre for farms of 400–500 acres was also assigned to larger farms since the employment equation began increasingly above 500 acres, a result that seems spurious on inspecting the data.

labour productivity in arable farming over the early modern period. Comparing open field farms *c*.1600 with enclosed farms *c*.1800 shows that labour productivity increased by a factor of 2.1. This is the same order of magnitude as Wrigley's estimate (1.9) of the growth in labour productivity in English farming over the same period. Moreover, Table 11–8 shows that just over half the proportional increase occurred in the seventeenth century. That growth in labour productivity was almost entirely due to the increase

*Table 11–8. Labour productivity in arable farming*

|  | *c.*1600 open | *c.*1700 open | *c.*1800 open | *c.*1800 enclosed |
|---|---|---|---|---|
| Output per acre (£ per acre) | 2.55 | 3.75 | 3.75 | 3.91 |
| Labour per acre (£ per acre) | 1.24 | 1.17 | 0.91 | 0.91 |
| Output per worker (£ of output per £ of labour cost) | 2.05 | 3.21 | 4.12 | 4.30 |
| Index | 1.00 | 1.56 | 2.01 | 2.10 |

*Source:*
Row 1: Table 11–6. Values for arable farms are used. Using the *c.*1800 open value for *c.*1700 presumes that open field farmers had accomplished the 1800 yields by 1700.
Row 2: Table 11–7. Values for arable farms are used. This presumes that the labour requirements for grain growing were the same in open and enclosed villages. This assumption is reasonable in view of equation 4 in Table 11–5.
Row 3: Row 1 divided by Row 2.
Row 4: Row 3 divided by 2.05

in grain yields since farm size increased little. Of the rise in labour productivity in the eighteenth century, most was due to the amalgamation of farms.

Table 11–8 indicates that enclosure led to only a 4 per cent increase in output per worker. In Chapter 8, I used Parkinson's census and Batchelor's cost-accounting to compute the effect of enclosure on labour productivity in each natural district. In the heavy arable district, enclosure raised output per worker 11 per cent and in the light arable district 3 per cent. These measurements are in reasonable agreement with the estimates of this chapter, although the estimates in Chapter 8 are more certain since they are 'fine tuned' to the possibilities of each natural district, while the results using Young's survey encompass a variety of soils and climates. Once again, however, Young's survey and Batchelor's cost-accounting corroborate each other by giving similar results.

Table 11–8 embodies many assumptions. The most controversial is probably the assumption that all of the increase in grain yields that occurred in open field villages between the middle ages and the nineteenth century occurred in the seventeenth century. Some may have occurred earlier, in which case the first value in Table 11–8 applies to the late middle ages. If some of the yield increase occurred in the eighteenth century, the estimate of labour productivity *c.*1700 should be reduced with the result that more of the growth in labour productivity should be assigned to open field farms in the eighteenth century. This change would not assign any greater importance to enclosure.

Table 11–9 extends the calculation of labour productivity to the south midlands as a whole.[8] Output per acre increased 33 per cent between

[8] This table embodies four more assumptions than the previous table. First, on the basis of

*c*.1600 and *c*.1800, although it fell slightly in the eighteenth century: corn yields were rising, but the conversion of arable to pasture swamped the gain. Putting land down to grass also lowered employment per acre. The implied evolution of labour productivity in the south midlands as a whole was not very different from the experience of arable farms. The overall increase in productivity was, again, similar to Wrigley's estimate for English farming as a whole. Productivity increased in about equal proportional amounts in both the seventeenth and eighteenth centuries.

*Table 11–9.  Labour productivity in south midlands farming*

|  | *c*.1600 | *c*.1700 | *c*.1800 |
|---|---|---|---|
| Output per acre (£ per acre) | 2.57 | 3.51 | 3.42 |
| Labour per acre (£ per acre) | 1.13 | 0.96 | 0.71 |
| Output per worker (£ of | | | |
|    output per £ of labour cost) | 2.27 | 3.66 | 4.82 |
| Index | 1.00 | 1.61 | 2.12 |

*Source*:
Row 1: The *c*.1600 value is a weighted average of the values shown in Table 11–6 where the weights are estimates of the shares of land open and enclosed. The *c*.1700 value is similar except that it uses the *c*.1800 open values. This calculation presumes, therefore, that open field farms had realized the 1800 yields by 1700. The *c*.1800 value is a weighted average of the *c*.1800 open, enclosed, and pastoral values in Table 11–6, where the weights are estimated shares of open field land, enclosed land devoted mainly to arable farming, and enclosed land converted to pasture.
Row 2: Each entry is a weighted average of the corresponding value of labour per acre in open fields shown in Table 11–7 and 0.584714, and *c*.1800 value of labour per acre for enclosed farms, which was assumed to apply to *c*.1600 and *c*.1700 for which the underlying samples were quite small and erratic. (See Table 4–5.) The weights are estimated proportions of open field and enclosed land in 1600 and 1700. In 1800 the open field labour value is weighted by the estimated share of open field land plus enclosed land devoted mainly to arable farming. The 1800 enclosed labour per acre value is weighted by the estimated share of land that was enclosed and converted to pasture.
Row 3: Row 1 divided by Row 2.
Row 4: Row 3 divided by 2.27.

Table 11–8 suggests a reconciliation between English and continental history. Farmers in Belgium and northern France realized crop yields *c*.1800 of the same order as English farmers, but farms in Belgium and France were much smaller than English farms. Table 11–8 indicates that if the size distribution of continental farms was similar to the distribution of English

subsidiary estimates of enclosure chronology, rough estimates of the acreage of arable and pasture in the south midlands were made for 1600, 1700, and 1800. These estimates presumed that enclosure before 1800 was to convert arable to pasture, while enclosure after 1800 was to improve arable farming. (See Ch. 2.) Second, employment per acre in pasture farming was assumed to equal the 1800 value shown in Table 11–7 for all years. Third, the three values shown in Table 11–6 for output per acre in pasture farming were assumed to apply to 1600, 1700, and 1800. Fourth, output per acre in arable farming was taken from Table 11–8, and the 1800 values were weighted roughly in proportion to the acreages of open and enclosed arable land at that date.

farms *c.*1700—before the transition to capitalist agriculture—then Belgian and French labour productivity should have been two-thirds of English productivity early in the nineteenth century. And that was indeed the size of the gap. (See Table 8–6.) Moreover, Russian farmers in the nineteenth century reaped yields like English medieval yields and had small, peasant farms, so one might expect labour productivity in nineteenth-century Russian agriculture to have been of the same order as English medieval productivity. Indeed, Bairoch (1965) found Russian productivity to have been 40 per cent of English productivity (Table 8–6), while Table 11–8 indicates that medieval English labour productivity was 48 per cent of nineteenth-century English productivity. These rough calculations suggest that my analysis of the growth in English agricultural labour productivity also explains the international differences in productivity at the beginning of the nineteenth century.

## *The Growth in Total Factor Productivity*

Labour productivity is an important indicator of farm efficiency, but it is not the only indicator. Total factor productivity—the ratio of output to all inputs—is the most complete measure of efficiency. Total factor productivity provides a framework for comparing the importance of the yield gains realized by seventeenth-century yeomen, the employment reductions of large farms and grass enclosures, and the improving effects of capital-intensive agriculture. This perspective shows that labour shedding was not as progressive a development as the history of labour productivity suggests.

I will simulate the growth in total factor productivity from the late middle ages to *c.*1800 by combining my estimates of yields, employment, and capital over the period. The simulation requires a production function, and I use the Cobb–Douglas specification:

$$Q = AK^aL^bT^cM^d \qquad \text{(Equation 11–1)}$$

*Q* is real farm output, *K* is farm capital, *L* is labour, *T* is land, and *M* is other miscellaneous inputs, i.e. seed and fodder. The exponents *a*, *b*, *c*, and *d* are the cost shares of the inputs. *A* is the index of total factor productivity, which I want to simulate.

Assuming that *a*, *b*, *c*, and *d* are constants, while *A*, *K*, *L*, *T*, and *M* change over time, the growth in total factor productivity can be computed as

$$\frac{A_2}{A_1} = \frac{q_2/q_1}{(k_2/k_1)^a(l_2/l_1)^b(m_2/m_1)^d} \qquad \text{(Equation 11–2)}$$

*where* $q = Q/T$, $k = K/T$, $l = L/T$, and $m = M/T$. The subscripts *1* and *2* denote the two time periods being compared. This formulation is convenient since it is in 'per acre terms'.

In computing efficiency changes, I calculated the cost shares from my reworking of Arthur Young's national farm accounts for *c*.1770. (See Appendix IV.) The shares of capital, labour, land, and miscellaneous expenses were 0.13, 0.27, 0.40, and 0.20, respectively. Since the farmer received the return to capital and miscellaneous expenses, these shares are not far off the traditional equal division of output among farmers, landlords, and labourers.

Equation 11–2 also requires output per acre, labour per acre, capital per acre, and miscellaneous inputs per acre. I took the first two from Table 11–9 and the third from Table 10–3. I estimated miscellaneous inputs (seed and fodder for draft animals) per acre as follows. Since these expenses were incurred in farming arable land, I used the share of arable land in the south midlands as an estimate of the trend in seed and fodder per acre of farm land in the region as a whole.[9] Table 11–10 summarizes the evolution of inputs per acre and shows the implied levels of total factor productivity in south midlands farming.

*Table 11–10. Total factor productivity in south midlands farming*

|  | Late medieval | *c*.1700 | *c*.1800 |
|---|---|---|---|
| Output per acre | £2.57 | £3.51 | £3.42 |
| Labour per acre | £1.13 | £0.96 | £0.71 |
| Capital per acre | £1.12 | £1.15 | £1.35 |
| Miscellaneous per acre | 0.61 | 0.57 | 0.42 |
| TFP | 1.00 | 1.44 | 1.58 |

*Note*: In this table I interpret values *c*.1600 as late medieval.

*Sources*:
Row 1: Table 11–9, row 1.
Row 2: Table 11–9, row 2.
Row 3: Table 10–3. I divided the total capital in south midlands farming by the acreage farmed.
Row 4: An estimate of the share of arable in south midlands farm land. On the assumption that miscellaneous inputs per acre were constant in arable husbandry and zero in pastoral husbandry, this proportion measures the trend in miscellaneous inputs per acre overall in £ *c*.1770.
Row 5: computed with equation 11–2 using the exponents discussed in the text.

The character of productivity change differed considerably in the seventeenth and eighteenth centuries. In the seventeenth century, the most significant change was the rise in output per acre: the increase in capital per acre and the declines in employment and miscellaneous inputs per acre were modest in comparison. The cumulative effect of these changes was to raise total factor productivity by 44 per cent.

[9] This calculation also embodies the assumption that seed and fodder rates per acre remained constant over the early modern period.

In the eighteenth century, productivity growth was much more modest. The conversion of arable to pasture led to a reduction in output per acre. While capital per acre increased somewhat, employment and miscellaneous inputs per acre declined. Total factor productivity increased by only 26 per cent. Table 11–10, thus, reiterates the character of agrarian change that emerged in the more detailed studies of enclosure in earlier chapters. Production was often constant or declining, employment fell when land was converted to pasture, capital rose moderately as herds increased. The overall growth in efficiency was small.

Table 11–10 also has important implications for Young's theory of the capital-intensive farmer. The important postulate of the theory—that greater output is caused by greater capital intensity—is not borne out by the table. In the eighteenth century, when capital intensity went up the most, output per acre fell. In the seventeenth century, both capital per acre and output per acre increased, but the rise in capital intensity was not enough to explain the rise in output. The contribution of rising capital intensity to output growth equals the relative rise in capital per acre raised to a power equal to capital's share in costs: $(1.15 \div 1.12)^{0.13}$. The implied increase in output is only 0.3 per cent, which is far less than the 37 per cent growth that actually occurred. Technical progress, not the greater intensity of farming, was the cause of rising output in the seventeenth century. This calculation thus buttresses the argument of the last chapter that improved seeds were the basis of progress at that time.

## Productivity Growth and the History of Rents

Ricardo's theory of rent implies that a rise in total factor productivity translates into a rise in rent according to the formula:

$$\frac{s_2}{s_1} = \left\{ \frac{A_2}{A_1} \right\}^{1/c} \qquad (Equation\ 11\text{--}3)$$

Here $s_2$ and $s_1$ are Ricardian surplus per acre in the two situations compared and $A_2$ and $A_1$ are total factor productivity. Since $c$ is a positive fraction (the share of land in costs), $1/c$ is greater than one: Ricardian surplus rises by a greater proportion than productivity since the extra income generated by greater efficiency accrues only to land. In Chapter 9, I showed that there could be considerable divergences between Ricardian surplus and rent in the short term given the way rents were set in early modern English agriculture. Now I propose to test how well the theory predicts rent changes from the late middle ages to the nineteenth century.

Figure 11–1 plots the history of real rent in the south midlands from 1450/74 to 1800/24. In constructing the series, I first computed average rent as a weighted average of open and enclosed rents in the various natural districts, where the weights were the acreages of land of each type.[10] Then I deflated the series with an index of the prices of farm outputs and inputs. I distinguished nine products (wheat, barley, oats, beans, wool, beef, mutton, pork, and cheese), labour, and the rental prices of six kinds of farm capital (implements, horses, cows, beef cattle, sheep, and pigs). The price series were aggregated using a Laspeyres index with *c*.1770 weights derived from Young's national farm accounts. (The details are in Appendix IV.) These prices all rose over the early modern period and that inflation was a factor in driving up rents. Deflating nominal rents with an index of these prices shows what rents would have been without inflation. It is those 'real' rents that equation 11–3 implies would rise as efficiency rises.

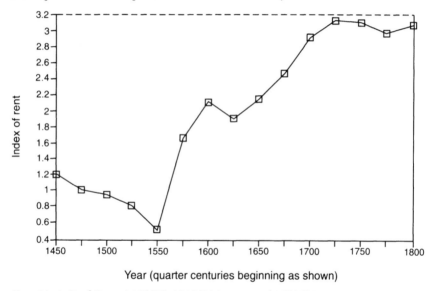

Fig. 11–1 Real Rent, 1450/74–1800/24 (average of 1450/74 to 1575/99 = 1.00)

Figure 11–1 shows several important features of the history of rent. First, real rent rose over the early modern period, in accord with productivity growth; it rose during the seventeenth century and stagnated during the eighteenth. Second, real rent also declined, in particular, during the sixteenth century. This fall does not mean agriculture was becoming less efficient. Rather, rents were not keeping up with inflation during the price

[10] The rents are shown in Table 9–1. Allen (1988*a*: 43) shows the average rent.

revolution. This is the same sort of discrepancy between rent and Ricardian surplus that I found in the open fields in the mid-eighteenth century.

To test the Ricardian theory of rent, it is necessary to compare the course of rents implied by the growth of TFP with the actual history of real rents. Substituting the values of TFP in Table 11–10 into equation 11–3 implies that real rent increased from 1.00 in the late medieval period to 2.49 in the early eighteenth century and 3.14 in the early nineteenth century. Comparison with the history of rents in Figure 11–1 is difficult due to rent lag in the sixteenth century. If I take the average level of rents from 1450 to 1599 as representative of 'late medieval' and assign it an index value of one, then the rent index is 2.89 in 1700/24 and 3.07 in 1800/25. These numbers are not far off the values predicted by equation 11–3. Over the very long term, Ricardo's theory predicts the rise in land values reasonably well. However, rents adjusted slowly and often lagged behind the values predicted by Ricardian theory for a generation or more.

## Two Agricultural Revolutions: Yeoman and Landlord

My findings emphasize the importance of distinguishing the yeomen's agricultural revolution from the landlords'. The yeomen's revolution was primarily output-expanding; its main achievement was the doubling of corn yields. As a concomitant, labour productivity increased over 50 per cent. Total factor productivity grew almost as much. In contrast, the landlords' revolution was primarily labour-shedding. This reduction in employment pushed output per worker in English agriculture to unparalleled levels. The rise in TFP, however, was much more moderate.

# PART IV

# *Agrarian Change and Industrialization*

# 12

## The Failure of Protoindustry and the Origins of the Surplus Labour Economy, 1676–1831

The Farmers' Wives can get no Dairy-Maids . . . truly the Wenches Answer, they won't go to Service at 12*d*. or 18*d*. a week, while they can get 7*s*. to 8*s*. a Week at Spinning.

Daniel Defoe, *The Behaviour of Servants in England*, 1724: 84 (Pinchbeck 1930: 140)

QUESTION: 'Have you any and what employment for Women and Children?'

ANSWER: 'None for Women; Children pick a few stones off the land.'

Question 11, Rural Queries, *Report on the Poor Laws*, 1834, and the answer of William C. Carver, Overseer, Melbourn, Cambs.[1]

THE major productivity-raising effect of enclosure and farm amalgamation was to reduce the agricultural work-force. From the early 1700s to the early 1800s, the employment of men, women, and boys declined 24 per cent, 29 per cent, and 36 per cent. These reductions exceeded the seventeenth-century declines of 8 per cent, 12 per cent, and 14 per cent. (Table 12–1.) Marxist Agrarian Fundamentalists argue that these declines contributed to the growth in manufacturing by supplying industry with cheap labour. In this chapter I will show, however, that agriculture's release of labour did not lead to industrial growth.

Falling farm employment was only one factor pushing labour into industry. Population growth was another. After a century of stasis, the fertility rate rose in the middle of the eighteenth century, and the English population once again began to grow. Between 1676 and 1831, it increased

---

[1] *Report from His Majesty's Commissioners for Inquiring into the Administration and Practical Operation of the Poor Laws*, Appendix (B.1.), *Answers to Rural Queries*, Part I, BPP 1834: xxx. 2a and 60a.

*Table 12–1. Agricultural employment in the south midlands*

| Approximate period | Men | Women | Boys |
|---|---|---|---|
| Early 1600s | 79,135 | 52,148 | 50,210 |
| Early 1700s | 72,801 | 46,144 | 43,128 |
| Early 1800s | 54,976 | 32,902 | 27,507 |

*Source:*

This table was constructed in steps. First, the total acreage of land in the arable and pasture districts was multiplied by the ratio of agricultural land to the total land. On the basis of Table 6–7 I assumed this was 0.9 for the pasture district and 0.8 for the arable districts.

For the arable districts, I assumed that 80% of the agricultural land was arable whether open or enclosed. (See Tables 6–1 and 6–4.) I then computed employment per acre using the regressions in Table 11–5 and the farm size distributions for open villages in Table 4–4. (Using the distribution for enclosed villages *c.*1800 gives almost identical results.) Total employment was obtained by multiplying average employment per acre by the total agricultural acreage.

For the pasture district, I divided the total acreage into open and enclosed acreage based on the enclosure chronology described in Ch. 2. I assumed that enclosed agricultural land was 20% arable while open agricultural land in this district was 63% arable. (See Ch. 6.) For the open land, I estimated employment per acre and total employment in the manner described above for arable farms. A similar procedure was used for enclosed land except that I used the *c.*1800 distribution of farm sizes for all years. See Ch. 11 for a discussion of this issue.

by a factor of 2.65 (Wrigley and Schofield 1981: 208–9). Over the same period, the population of the south midlands doubled.[2] Tory Agrarian Fundamentalists see the population explosion as a consequence of the yield increases they believe followed the shift to large farms. My finding that large farms did not reap higher yields than small farms undermines this argument.

Recent demographic research ties the rise in fertility to changes in the social relations of production. Goldstone (1986) has shown that the relationships between fertility, the percentage of people marrying, the average age of marriage, and the wage rate all changed after 1700: the dramatic rise in fertility during the eighteenth century was not a response (of the sort envisaged by Malthus) to a previous high wage. Instead, Tilly (1981) and Levine (1987: 68–93) regard proletarianization as the cause of the rise in the birth rate. Levine has shown that protoindustrial workers had high fertility rates, so the spread of protoindustry raised the total number of births. The more general was proletarianization in the eighteenth century,

---

[2] Table 12–2, which is based on Deane and Cole's (1969) population estimates for 9 counties that closely overlap the region discussed in this book, shows that the population grew by a factor of 1.9 between 1701 and 1801. For 1,130 villages in the south midlands, I could estimate the population in 1676 and 1831. For that sample, population grew by a factor of 2.1 between 1676 and 1831.

the more credible does it become as an explanation for the rise in England's fertility. Many historians, under the influence of Mingay and Thompson, have doubted that there was a major extension of capitalist relations in agriculture in the eighteenth century. However, as I argued in Part I, landlessness increased dramatically during that century. If yeomen farmers delayed fertility until they came into their property, that reason for restraint disappeared for many in the eighteenth century. While the issue needs much more investigation, it is plausible that the revolution in landownership caused the explosion in population.

In any event, declining farm employment and a soaring rural population created a burgeoning supply of labour that might have built cities and operated factories. The lost opportunity of the eighteenth and early nineteenth centuries was that this never happened. English agriculture released labour, but English industry failed to absorb it. The south midlands became glutted with surplus labour; unemployment in rural villages soared. For the few who found work in manufacturing, competition was intense, hours were long, and earnings a pittance. Agrarian reorganization and population growth led to immiserization—not industrialization.

## Theories of Labour Release and Labour Absorption

The theories that have been proposed to explain the transfer of labour from agriculture to industry can be divided into two groups. The first group, which includes the seventeenth- and eighteenth-century theories, are supply-side theories that focus on *labour release*. Implicit in these theories is the idea that once labour is expelled from agriculture, it will automatically be absorbed by industry. Since 1950, demand-side theories have been developed. They concentrate on the problem of labour *absorption*. While none of these theories fully accounts for what happened in the midlands during the eighteenth and nineteenth centuries, each throws light on some aspect of the problem.

Fortrey's (1663) theory of enclosure was the first supply-side theory of manufacturing growth. He accepted that enclosure led to depopulation: 'by experience it is found that many towns, which when their lands were in tillage had many families, now they are inclosed have not so many inhabitants in them'. The people who were forced out entered the wool industry. Since clothiers 'do not always finde it most convenient for them to live, just on the place where the wooll groweth' the new spinners and weavers moved to the wool towns. Moreover, 'as many or more families may be maintained and employed in the manufacture of the wooll that may arise out of one hundred acres of pasture', enclosure contributed to the growth of England's population as well as her industry. 'Cities and great towns are peopled, nothing to the prejudice of the kingdom.' According to Fortrey, there was no problem of labour absorption—everyone who left agriculture became re-employed in spinning and weaving wool. The growth of the manufacturing sector was determined by the decline of farm employment.

The second supply-side theory was built around population growth, and it was developed by Arthur Young. He was familiar with the work of Fortrey[3] but disagreed with it, for he believed that enclosure and large farms increased agricultural employment. Where then did the manufacturing labour force come from? Young's answer was that food production increased even faster than farm employment. The population expanded in proportion to the food supply thereby enlarging the manufacturing work-force. J. D. Chambers' (1953) restatement of this theory has given it wide currency in recent decades, and it is a standard textbook interpretation.

All supply-side theories of industrialization have certain features in common. Whether the demand for agricultural labour falls (Fortrey/Marx) or the work-force expands (Young/Chambers), the result is an increase in the supply of labour to the manufacturing sector. None of these theorists believes there is a problem of unemployment or labour absorption: industry willingly accepts all the labour that rural society gives up. If there is a poverty problem, it is one of long hours and low wages since the wage has to fall enough for manufacturing firms to sell at a profit all that their workers produce.

At the same time that Chambers was restating Arthur Young's theory, Sir Arthur Lewis was developing a radically different theory of the growth of the manufacturing sector. This was a demand-side theory since the growth of industry depended on its demand for labour.

Lewis put surplus labour, not population growth, at the centre of his theory. According to Lewis, 'an unlimited supply of labour may be said to exist in those countries where population is so large relatively to capital and natural resources, that there are large sectors of the economy where the marginal productivity of labour is negligible, zero, or even negative'. Where would such surplus labour be found? In the tradition of Young, Lewis denied that technical progress in agriculture (or elsewhere) was an important source. Instead he listed several sectors of the economy as likely to harbour labour of zero marginal product. Agriculture is a prime possibility when 'the family holding is so small that if some members of the

---

[3] In *Political Arithmetic*, Young (1774: 363–6) reprinted the pertinent passage by Fortrey as well as Hartlib's (1652) earlier but clumsier insightful analysis.

family obtained other employment the remaining members could cultivate the holding just as well'. Other possibilities include 'petty retail trading', domestic service, and women generally since 'the transfer of women's work from the household to commercial employment is one of the most notable features of economic development' (Lewis 1954: 141–3).

Lewis assumed that anyone can receive a conventional, probably subsistence, income in traditional agriculture, petty trading, domestic service, and so on. That income determined the wage throughout the economy; in particular, 'new industries can be created, or old industries expanded without limit at the existing wage' (p. 142). With the supply of labour unlimited, industrial employment depends on industrial demand, which Lewis, in turn, linked to the rate of capital formation.

Lewis's theory differs from Chambers's in one critical way. In Chambers's model, unemployment did not exist because wages fell to clear the labour market. In Lewis's model, the wage is fixed at a traditional level. Hence, there is no guarantee that the labour market clears. Unemployment is the result.

## Did Southern Labour Go into Manufacturing?

If either of the supply-side theories were correct, one would expect to see people born in the south midlands flowing into industry. Throughout the early modern period, the major manufacturing centres were outside the region. In the seventeenth century, there is evidence which suggests that people were moving from the south midlands to London, which was growing rapidly. Otherwise—and particularly during the Industrial Revolution of the eighteenth and nineteenth centuries—there is little evidence for a movement of labour to manufacturing districts.

Enclosure had its most catastrophic impact on population before 1525, but it is very unlikely that the farmers who were forced to move became manufacturing workers, for two reasons. First, only a small fraction of villages were enclosed and depopulated. Since population was very light, there was a labour shortage rather than a land shortage. Hence, the displaced farmers probably relocated in neighbouring open field villages and continued in agriculture. Second, there was hardly any urban industry for them to engage in anyway. London, the biggest city in the kingdom, had only 40,000 residents. The next largest cities—Bristol, Exeter, Newcastle, and Norwich—had populations of only 10,000 each. Oxford, Cambridge, and Leicester, the biggest cities in the south midlands, had populations of only 4,000 or 5,000.[4] Overall, Wrigley (1985: 700) has estimated that only 24 per cent of the English population was engaged in non-agricultural activity. At the end of the fifteenth century, England was still an overwhelmingly agricultural society—there was not much of an industrial proletariat for the midlands enclosures to have created.

The situation changed dramatically in the century before Fortrey wrote. The second wave of enclosures (1575–1674) forced many people to migrate. There was still no significant urbanization in the south midlands, but wool manufacturing was growing rapidly elsewhere in the kingdom. Norwich was the centre of this new trade, and so was presumably the centre Fortrey had in mind when he formulated his theory. However, there was negligible migration from the midlands to Norwich. Patten (1976: 122), who studied the origins of 5,835 apprentices indentured in Norwich from 1510 to 1699, found that only fifty-eight came from the ten south midlands counties discussed in this book.

Much evidence suggests—but does not prove—that many of the 'freed' labourers were going to London. Its population was exploding: it reached 200,000 in 1600 and 575,000 in 1700. This growth depended on massive immigration from country districts. Finlay (1981: 9) has shown for the period 1550–1650 that 'London was . . . absorbing the natural increase of . . . about half the English national population', and Wrigley (1967) earlier established a parallel result for 1650–1750. The south midlands were an important source of these migrants: almost one third of the apprentices to London companies between 1570 and 1640 came from the midlands.[5] Moreover, rates of migration from the midlands were far higher than for other districts. Dividing the number of apprentices by an estimate of the population of each sending county in 1630 gives a rate of 35.2 for the south midlands and 25.8 for the north midlands. The rate for the home counties was only 13.7 and all other regions of England had even lower rates of migration to London. Thus, between the middle of the sixteenth century and the Compton Census of 1676, when enclosed villages in the south midlands were forcing out the natural increase of their populations,[6] the region as a whole was sending migrants to London at an extremely high

---

[4] All population figures are from de Vries (1984: 270). For 1500, de Vries gives the population of Leicester as 4,000 and that of Oxford as 5,000. He puts Cambridge in the category of less than 10,000. The first year he gives a population for Cambridge is 1650 when its population was 9,000. At the same time Oxford's population was also 9,000. Wrigley (1985: 686) gives the population of Cambridge as 5,000 in c.1600. He indicates that the Oxford population was also 5,000 at the same time.

[5] I refer to Finlay's (1981: 64) 'south midlands' (Beds., Berks., Bucks., Northants, Oxon.) and 'north midlands' (Derbys., Leics., Notts., Staffs., and Warwicks.).

[6] See Ch. 3.

rate. Furthermore, the inflow from the midlands was important in sustaining the continued growth of the Metropolis. To be certain, interregional migration needs more study, but the presumption must be that the Elizabethan and Stuart enclosures in the midlands were making an important contribution to the growth of London's population.

During the eighteenth century, migration from the south midlands declined, although the evidence—none of which is very good—does not concur as to exactly when the change set in. Finlay's (1981: 65) calculations suggest that the turning point was in the late seventeenth century. For the period 1674–90, he found that London drew a much higher share of her immigrants from counties close by[7] and that the midlands' emigration rates were no longer exceptionally high. However, Deane and Cole's work (1969) dates the decline from the middle of the eighteenth century. Table 12–2, which summarizes their estimates, indicates that 73 per cent of the natural increase of the south midlands emigrated between 1701 and 1751, but that the percentage fell sharply thereafter.[8] After 1750, the south midlands developed into a self-contained labour market, no longer linked by migration to the rest of the country.

*Table 12–2. Migration from the south midlands, 1701–1831*

|  | Population | Natural increase | Net emigration | Share of natural increase emigrating (%) |
|---|---|---|---|---|
| 1701 | 608,367 |  |  |  |
| 1751 | 630,491 | 80,589 | 58,465 | 73 |
| 1801 | 820,147 | 282,945 | 93,289 | 33 |
| 1831 | 1,145,041 | 435,184 | 110,290 | 25 |

*Note*: The natural increase equals the excess of births over deaths in the period between the year for which it is entered and the preceding year. The net emigration is the number of emigrants minus the number of immigrants in the period between the year for which it is entered and the preceding year. Thus, 630,491 = 608,367 + 80,589 − 58,465. Share of the natural increase emigrating equals the net emigration divided by the natural increase, e.g. 73% = 58,465 ÷ 80,589.

*Source*: Deane and Cole (1969: 103, 108–9). The table includes the counties of Beds., Berks., Bucks., Cambs., Hunts., Oxon., Rutland, Leics., and Northants.

Wage patterns show that the south midlands was an insulated labour market at the beginning of the nineteenth century. Nominal wages in the midlands were considerably lower than in London: reporters for the Board of Agriculture reported steep declines within short distances of the capital. Mavor (1813: 415) noted that winter wages for men in Berkshire varied

---

[7] Wareing (1980: 243) confirms the growing importance of the midlands as a source of London migrants during the 16th cent. and shows a decline in their importance in the 18th, but the decline is more modest than Finlay's numbers indicate. See also Wareing (1981).

[8] See Hobsbawm and Rudé (1968: 43) for a similar table.

from 9 to 12 shillings per week depending on the distance from London. Young (1804c: 221) noted for Hertfordshire that 'its vicinity to the Metropolis secures higher wages, and more constant employment, than in many other districts'. High wages did not extend into Bedfordshire, however. At the other end of the midlands, Murray (1813: 167) remarked: 'Warwickshire being a great manufacturing county, the price of labour may be in some degree governed by the state of trade, and in the vicinity of the large manufacturing towns, labourers must be unsettled, and always changing from one master to another.' He also reported exceptionally high wages.

As Mavor and Young suggested, the impact of growing urban economies on the southern labour market was confined to the periphery of the region. Within the south midlands, wages were low and remained low. In part, low nominal wages reflected the lower cost of living in country districts and the better quality of life (particularly longer life expectancies) outside of the large cities. Williamson (1987: 641–60) has measured these effects carefully. For northern England he found that they fully explain the rural–urban gap in wages, but for southern England they do not. The southern labour market was self-contained and not well integrated with the rest of the country.

By the middle of the nineteenth century, net emigration from the south midlands had almost ceased, as shown by the 1851 census, which cross-tabulated place of birth and place of residence. There were 1,254,495 males living in England and Wales who were born in the counties of Buckingham, Oxford, Northampton, Huntingdon, Bedford, Cambridge, Leicester, and Rutland; 993,751 of them were still living in those counties in 1851. In addition 175,242 males born elsewhere resided in those eight counties so their total male population was 1,168,993. Thus, net emigration was 85,502 or 6.8 per cent. This is a small number and on an annual basis was negligible.

Why did migration from the south midlands decline in the eighteenth century? The question is puzzling since the economic benefits to migrating were increasing over the period. The settlement provisions of the Poor Law have often been suggested as a hindrance, for instance by Adam Smith, but other contemporaries (e.g. Sir Frederick Eden) discounted their import-ance.[9] Redford (1926: 95) ended his survey of the debate with the conclusion that the 'mental obtuseness of the English peasantry' was most to blame. 'Peasants, especially uneducated peasants, are immobile except under the spur of extreme necessity.' Indeed, one might argue that the

---

[9] Boyer (1986b) has carefully considered this issue and come to a similar conclusion.

dependence on poor relief was a result, rather than a cause, of the decline in migration.

The overpopulation of the south midlands must be seen, in part at least, as a result of the disappearance of peasant agriculture and the rise of the great estate. In the eighteenth century, many small farmers lost their land (as was discussed in Chapter 5), and the income of labouring families generally declined (as will be discussed in Chapter 14). Before the eighteenth century, yeomen had helped their children emigrate by paying for apprenticeships and establishing them in trade (M. Campbell 1942: 162–288). By 1800, their descendants were too poor to offer such help. An accumulation of population in midlands villages was the result (Howell 1983: 268–70).

An indubitable consequence of the decline in emigration from the south midlands was that the region did not supply a work-force to the modern industries of the Industrial Revolution. The new factories were located mainly in the north of England, but southern labour did not flow there. The growing number of men and women, superfluous to the needs of agriculture, remained in the region.

## Protoindustrialization

If agrarian reorganization or population growth in the south midlands was encouraging manufacturing growth, it was doing so in the region itself by fostering 'protoindustry'. That was a system of production for national and international markets in which manufacturing was carried out by artisans in their own homes or in small workshops. Frequently this work was organized in the 'putting out' system. Protoindustrial districts were scattered across England and indeed most of Europe.[10]

There was some growth in protoindustrial employment in the south midlands—at least for men—but it was limited and concentrated in Leicestershire, Northamptonshire, and northern Oxfordshire. The rest of the south midlands was devoid of male manufacturing employment. Our best guide to the employment of men in protoindustries is the 1831 census, which tallied it approximately. (See Table 12–3.) With the exception of woollen weaving, all industries were near their peak levels of employment.[11]

Framework knitting was the biggest employer by far. In the nineteenth century, it was mainly located in Leicestershire, Nottinghamshire, and Derbyshire, although there were some knitters in Northamptonshire as

---

[10] Berg (1985) provides a good overview of protoindustry in 18th cent. England and of the literature on the subject.

[11] Production of sacks and mats were too scattered and the preparation of straw for plaiting too minor to merit attention.

*Table 12–3. Male protoindustrial employment in the South Midlands in 1831*

| | |
|---|---:|
| Framework knitting | 10,200 |
| Shoes | 2,100 |
| Weaving | |
| silk plush and girth | 845 |
| blankets | 271 |
| carpets | 120 |
| woollen cloth | 50 |
| linen | 40 |
| Mats and sacks | 267 |
| Gloves | 100 |
| Machines and metals | 91 |
| Straw preparation | 12 |
| TOTAL | 14,096 |

*Source*: 1831 census, summaries of manufacturing employment by county. The summaries are rough and non-exclusive. This table includes 2,100 shoemakers originally returned in retail and handicraft, and an allowance for 100 glovemakers in the Woodstock region. The summaries indicate that there were some plush and girth weavers in Bloxham, Adderbury, and other villages near Banbury. I assigned all men returned in manufacturing in such villages to the plush and girth industry.

well.[12] These knitters produced stockings, gloves, and other garments on a frame propelled by the operator. The knitting frame was invented by William Lee at the end of the sixteenth century. By the mid-seventeenth century, the industry was established in London producing high quality silk stockings. By the mid-eighteenth century, the number of frames had increased over twentyfold, and the industry had relocated to the midlands. Derbyshire specialized in silks, Nottingham in cottons, and Leicester in cotton and worsted. The number of frames continued to grow through 1844. Output grew even faster since the wide frame, introduced in the early nineteenth century, especially in the cotton sector, produced more goods per week than the traditional 'narrow' frames. Their productivity was rising as well, however, as the working day lengthened during the nineteenth century.[13]

Two other protoindustries were smaller but grew appreciably over the eighteenth century. The first was the Northampton boot and shoe industry. While most English towns had a cobbler serving the local market, what

---

[12] Table 12–3 excludes Notts. and Derbys.
[13] Histories of the framework knitting industry include Thirsk (1973), Felkin (1867), Wells (1935), Head (1961–2), Chapman (1972, 1974), Osterud (1986), and *VCH, Leics.*, iii. 2–23.

developed in Northampton was a true protoindustry supplying a national market. As early as 1642, Northampton producers were making boots for army contracts (*VCH, Northants*, ii. 319). Employment grew to about 600 in 1777, 1,000 in 1794, and 2,100 in 1831.[14] Second, Banbury, Oxfordshire, developed a weaving industry making webbing and girth for horse harnesses, and plush or shag. (Webbing and girth were made of wool, while plush or shag were fancy cloths made of worsted, silk, hair, or cotton in the cheaper varieties.) (Beckinsale 1963.) Growth may have accelerated when the canal to Coventry was opened in 1778 (*VCH, Oxon*. ii. 245; x. 63–5). About 1820 the production of silk plush was also undertaken in the neighbouring Northamptonshire towns of Kettering, Rothwell, and Desborough (*VCH, Northants*, ii. 334).

Since the time of James I, blankets were woven in Witney, Oxfordshire (*VCH, Oxon*. ii. 248). Many were sold to the Hudson's Bay Company for use in Canada. Production remained remarkably constant. Contemporary descriptions report that 5,000 packs of wool were woven in 1677, 7,000 in 1767, and 7,000 again in 1807, 5,000 in the late 1830s, and 6,000 in 1852. The introduction of the spring loom *c.*1800 had saved the industry from the competition of the West Riding by raising labour productivity (Young 1813: 325). With constant output and rising productivity, employment in the industry declined between 1677 and 1852.[15]

The last protoindustry in the south midlands that employed men was the worsted cloth industry in Northamptonshire.[16] At the end of the seventeenth century, it began a phase of rapid growth. In 1741 Cowper (1741: 39) recalled 'I remember, in my Time' that '*Kettering*, a little Market Town in *Northamptonshire*, from manufacturing 20 or 30 pieces of Dy'd Serges Weekly, fell into making Shalloons . . . and sent to *London* upwards of 1000 Pieces *per* Week.' Employment reached about 2,000 weavers by 1777 (Hatley 1973). Prosperity continued until the French Wars when the industry collapsed. In 1821, half of Kettering was in receipt of poor relief, and in 1830 the poor rate reached 20 shillings in the pound (Greenall 1973: 254). An 1824 directory printed the industry's obituary: 'There was formerly a very considerable trade carried on here in the woollen manufacture of serges, tammies, etc. This has become very nearly extinct'

---

[14] Hatley (1973), Donaldson (1794: 10), *Abstract of the Population Returns of Great Britain, 1831* (BPP 1833).

[15] For output, see Plot (1677) as quoted by Plummer and Early (1969: 10–11), Young (1768: 100), Young (1813: 325–6); *Reports from Assistant Hand-Loom Weavers' Commissioners*, pt. v, BPP 1840: xxiv, report by W.A. Miles, p. 547; and *VCH, Oxon.*, ii. 251. These sources also contained scattered employment figures that indicate a downward trend during the Industrial Revolution.

[16] See James (1968) for a history of the worsted industry.

(quoted by Greenall 1973: 253). In 1831, there were only fifty men weaving wool in Northamptonshire. Some of the displaced weavers had moved into plush, but most had vanished.

The overall performance of these industries was disappointing so far as employment creation is concerned. The industries that employed women, however, were a greater disaster. The three main female protoindustries were spinning, pillow lace making, and straw plaiting. The employment history of women cannot be traced with as much definitiveness as can the corresponding history for men, but it is clear that most of the south midlands was deindustrialized in so far as women's work was concerned.

Spinning yarn is an essential process in all textile manufacture, and it was performed domestically by women before the use of spinning machinery in the late eighteenth century. It was widespread in Leicestershire, Northamptonshire, and northern Oxfordshire, where it was the preliminary stage in knitting hose and weaving shalloons and blankets. The employment levels of women followed those of men, generally increasing over the century. Spinning may also have been practised in other parts of the south midlands.[17]

The rise in spinning employment was reversed after the 1790s, however, when domestic spinning became the first casualty of factory competition.[18] Hargrave's jenny, Arkwright's frame, and Crompton's mule were perfected in the 1770s and 1780s and initially applied to spinning cotton. By the end of the eighteenth century, factory-made worsted yarn was eliminating domestic spinning in the midlands. Within thirty years, if not earlier, no significant numbers of women were spinning domestically in the south midlands.

The manufacture of pillow lace was introduced into England from the continent at the end of the sixteenth century.[19] In the next fifty years, its production spread through Bedfordshire and Buckinghamshire, partly at the initiative of Poor Law overseers anxious to increase employment. By the end of the seventeenth century, it was intensively fabricated in Bedfordshire (especially the northern part), Buckinghamshire, and in adjoining portions of Northamptonshire and Oxfordshire. Devon was also an important centre of the trade until the eighteenth century, but thereafter production became concentrated in the south midlands. Batchelor (1808: 594), usually a careful

---

[17] Young (1804c: 221) noted that 'About Stevenage, spinning has given place to plaiting straw, by which they earn three or four times as much.' Batchelor (1808: 593) reported that in Beds. 'Spinning of hemp and cotton is almost entirely laid aside' in preference to lace making or straw plaiting.

[18] See Chapman (1965) for an account of the early shift to factory spinning of worsted in the midlands.

[19] See Spenceley (1973) for a history of the trade.

observer, estimated that lace making 'probably employs three-fourths of the female population [of Bedfordshire], with the exception of servants, &c.' Given the population increase, this estimate implies a considerable increase in employment during the eighteenth century. Further growth and the extension of production throughout Northamptonshire probably occurred in the first third of the nineteenth century.[20] By then, however, expansion was checked by the development of factory lace making. In Byfield, Northamptonshire, 'Lace-making [by hand] has been much superseded by machinery lace' (1834 Poor Law Report, BPP 1844: xxx. 332*a*). Fancy goods continued to be made by hand for many years, but the industry was subjected to intense competition.

Straw plaiting was carried on in Buckinghamshire, southern Bedfordshire, Hertfordshire, and Essex. The scale was small in the early eighteenth century when the trade was noticed by Defoe (1724–6: 427–8) and when it petitioned parliament for protection against foreign straw hats. With the disruption of commerce with Italy during the French Wars, the demand for English straw hats shot up. Demand for English straw plait surged, wages rose, and female labour was diverted from agriculture and lace into plaiting. With the return of peace, however, Italian imports rebounded. The industry avoided extinction by importing Italian straw, which was superior to English wheat straw for hats, but the employment of English plaiters declined.

In summary, the job prospects for women in manufacturing were favourable in the eighteenth century. Spinning was widespread in Leicestershire, Northamptonshire and northern Oxfordshire and probably elsewhere in the south midlands. Likewise, pillow lace was widespread in Buckinghamshire, Bedfordshire, and parts of Oxfordshire and Northamptonshire. Employment in these activities probably increased in step with the population, if not faster due to the rise in real income and the growth in textile production. During the French Wars the employment prospects for women were exceptional due to the boom in straw plait, which drew even more women out of farming into manufacturing in Bedfordshire and Buckinghamshire.

After Waterloo, however, this favourable situation was reversed as the domestic industries declined. The rural queries of the 1834 Poor Law Report show the extent of the collapse.[21] Spinning was an industry of the

[20] Donaldson (1794: 10) reported that women made lace near Wellingborough and in the southwest of Northants, while the rural queries of the 1834 Poor Law Report indicate that women throughout the county made lace.
[21] The rural queries include a survey of employment opportunities for women and children in 114 villages in the south midlands. The responses are not all reliable, in particular, those

Agrarian Change and Industrialization

past: 'Formerly spinning was the work for women in villages, but now there is scarcely any done' (BPP 1834: xxx. 368a, Banbury). Seven parishes in Bedford, Buckingham, and Huntingdon returned that there was employment in straw plaiting, but most parishes in these counties made no such indication. The only activity with important employment opportunities was lace knitting. Virtually all the responding parishes in Bedford, Buckingham, and Northamptonshire, as well as a couple in Oxford and Huntingdon, indicated that poor women could find work knitting lace. A constant refrain is that wages in the trade were very low: 'Lace-making, the remuneration for which is greatly reduced in consequence of the introduction of machinery and foreign lace' (BPP 1834: xxx. 31a, Beaconsfield). Moreover, outside of the core area of the lace industry, job opportunities for poor women and their children were almost non-existent.

## Labour Release and Employment Growth

How did the growth in population over the eighteenth century compare in magnitude to the release of labour by agriculture? If manufacturing employment was not growing, what were people doing instead? These questions can be answered by comparing the structure of employment in 1676 and 1831. I begin with men, for whom the 1831 census provides detailed statistics, and work backwards. The numbers are in Table 12–4.

In 1831, half of the male work-force[22] in the south midlands[23] was still agricultural. What is particularly surprising is that the number of farmers and labourers returned in the census is about twice the number of full-time male jobs needed for farming. The total number of men in agriculture is about equal to the peak labour demand at the harvest. Half the agricultural labourers were unemployed for most of the year.

The table puts the protoindustrial revolution into perspective. Only 15,950 men (8 per cent of the work-force) were employed in manufacturing —proto or factory.

pertaining to agricultural employment where the answers are often nonsensical. Thus, 55 per cent of the parishes indicated there was agricultural employment for women. Results reported earlier lead us to doubt that such jobs could have been of any consequence. This doubt is reinforced by the pattern of responses themselves. Thus, only 2 of the 15 Leics. parishes report agricultural employment opportunities for women, while all 13 of the adjacent Warwicks. parishes did so. It looks as though the official line in Warwick was that since it was an agricultural county, there must be agricultural jobs for women.

[22] Table 12–4 does not include anyone classified as 'other' since that category comprised the retired and infirm. See Wrigley (1986: 304) and 1831 Census, BPP 1833: xxxvi. p. vi.
[23] The south midlands encompasses the 1,582 rural parishes in the region plus the urban parishes not previously discussed.

Table 12–4.  Employment of adult men in the south midlands, 1676 and 1831

|  | 1676 | 1831 |
|---|---|---|
| Agriculture |  |  |
|   fully employed | 72,801 | 54,976 |
|   rarely employed | 0 | 49,880 |
| Protoindustry | 2,000 | 15,200 |
| Factory industry | 0 | 750 |
| Retail manufacturing |  | 45,137 |
| Labourers |  | 15,849 |
| Building craftsmen | 25.108 | 13,788 |
| Servants and waiters |  | 4,598 |
| Gentlemen |  | 8,507 |
| TOTAL | 99,909 | 208,685 |

*Source:*
1831: The basic source was the 1831 census. The 1,582 rural parishes are included plus urban parishes in that geographical region.

The number of men fully employed in agriculture is from Table 12–1. The number rarely employed equals the total number of men returned in the three categories 'occupiers employing labourers', 'occupiers not employing labourers', and 'labourers employed in agriculture' less the number fully employed.

Proto and factory industry equals the number returned as 'employed in manufacture, or in making manufacturing machinery' plus the 2,100 shoemakers in Northants (otherwise returned in retail and handicraft) that Rickman judged were engaged in production for a national market. (1831 census, BPP 1833: xxxvi. 446–7.) I also added 100 men who made gloves in Woodstock and environs for similar reasons.

Except for the deletions just noted, the census category of those 'employed in retail trade, or in handicraft as masters or workmen' equals the categories of retail manufacturing plus building craftsmen. I divided the category based on the ratio of building craftsmen to total employment in retail and handicrafts for the counties of Beds., Bucks., Oxon., Cambs., Hunts., Northants., Rutland., and Leics.

My three categories of labourers, servants and waiters, and gentlemen correspond, respectively, to the three census categories of 'labourers employed in labour not agricultural', 'male servants 20 years of age [or more]', and 'capitalists, bankers, professional and other educated men.'

Aside from agriculture, 'retail manufacturing' was the biggest employer. This includes bakers, butchers, tailors, blacksmiths, shoemakers, shopkeepers, and publicans as well as carters and other providers of transport services. These trades catered primarily to local markets. The rest of the men fell into four categories—building craftsmen, non-agricultural labourers (who were employed in building and retail production), servants and waiters, and gentlemen including 'capitalists, professionals, and other educated men'.

I made employment estimates for 1676 by working backwards. First, I estimated total employment by reducing the male work-force in proportion to the difference in population between 1676 (the year of the Compton Census) and 1831.[24] Second, I took agricultural employment as equal to the level of full-time employment for the early eighteenth century shown in Table 12–1.[25] Third, I set employment in protoindustry at the generous level of 2,000. Subtracting employment in agriculture and protoindustry from the total work-force leaves only 25,108 men as either irregularly employed farm labourers or as employees in retail production, construction, or one of the remaining categories. Table 12–4 tallies them as non-agricultural.

In 1676, the south midlands were much more agricultural than in 1831. Almost three-quarters of the men were full-time farmers or farm labourers. There were so few men that all must have been fully employed throughout the year. There were not enough men in agriculture to harvest the crops, so many non-agriculturalists and women were employed in the fields in August.

Between 1676 and 1831, male employment increased in all sectors of the economy. Despite the loss of 17,825 full-time jobs in agriculture, the total number of farmers and labourers increased by 32,055. Protoindustry experienced the biggest proportional gain in employment, but the 1676 base was so small that only 13,950 jobs were created—less than the decline in full-time agricultural jobs but also less than the increase in the agricultural work-force. In absolute numbers the biggest increase in employment was in retail production and construction. Recently Wrigley (1986: 296) has argued that these industries were much more important than protoindustry in creating new jobs in the first half of the nineteenth century, and Table 12–4 supports the same conclusion for the eighteenth century as well.

I can also use Table 12–4 to speculate on the history of female employment. There were roughly as many women as men in the south midlands—say, 208,685 in 1831. According to Table 12–1 there were 32,902 women (16 per cent) employed full-time in agriculture. These included the wives of the 19,093 farmers. Indeed, part-time, even harvest work, was largely non-existent for women, so most women—if they were employed at all—were employed in manufacturing and services.

The prospects were grim. Presumably the wives of the gentlemen (8,507) were ladies of leisure. The wives of retail manufacturers (45,137) were probably involved in their husbands' businesses. If the proportion of women

---

[24]  See n. 2.
[25]  Male agricultural employment must have been similar in 1676 and the early 18th cent. since there was little enclosure in that period and since the distribution of farm size was stable.

who were domestic servants was the same in 1831 as it was a decade later, 20,104 were employed as domestics.[26] That leaves 102,035 women (49 per cent)—either single or the wives of labourers, building craftsmen, and waiters—who had to find employment in manufacturing. Outside of the lace district, there was very little.

The wife is no longer able to contribute her share towards the weekly expenses . . . In a kind of despondency she sits down, unable to contribute anything to the general fund of the family, and conscious of rendering no other service to her husband, except that of the mere care of his family. [27]

In 1831 the employment prospects of women in the south midlands were even bleaker than they were for men.

The situation was more favourable at the beginning of the eighteenth century. The number of full-time farming jobs was greater (46,144) and the population was smaller (99,909). Making some allowance for servants under the age of twenty suggests that about 40 per cent of adult women were employed full-time in agriculture. Another 10 per cent or more were probably married to retailers and active in those businesses. Spinning was viable and widespread outside the lace district, so manufacturing work was available throughout the south midlands. The inadequacy of the male agricultural labour force to harvest the corn crop drew many women into farming, if only seasonally. There can be little doubt that the employment opportunities of women were brighter in 1676 than in 1831.

The history of employment between 1676 and 1831 has unexpected implications for labour release. First, Fortrey and Marx were right that technical change in agriculture created the potential for labour release since farm employment of all sorts declined over the eighteenth century. Second, that potential, however, was not realized, at least for men, since the number of farmers and labourers increased over the eighteenth century. Third, while this growth in numbers is consistent with the views of Young and Chambers, they were mistaken as to the cause. It was not a greater demand for labour due to 'improved husbandry'. Instead, agriculture was acquiring the chief characteristic of traditional farming in Lewis' schema; in particular, the marginal product of male agricultural labour was zero for most of the year. Fourth, the employment of women in agriculture declined,

---

[26] In the counties of Beds., Bucks., Cambs., Hunts., Oxon., Rutland, Leics., and Northants, there were 25,343 female domestic servants, 20 years or over, in 1841. The total male population 20 years or over in the same counties was 278,864. Applying that proportion to the total male population of the places totalled in Table 12–4 implies that there were 20,104 women domestic servants in 1831.

[27] *Reports of the Society for Bettering the Condition of the Poor*, iii. 91. Quoted by Pinchbeck (1930: 59).

but they were not redeployed into manufacturing. Instead, they became 'housewives'—another pool of surplus labour identified by Lewis. Fifth, retail production may also have harboured surplus labour. Little is known of what these people were doing, but for every substantial publican, there may have been a half-time baker or carter—'petty retail traders' in Lewis's parlance—and his third category of surplus labour.

The most salient characteristic of the south midlands in the eighteenth century was, thus, not the shift of labour from agriculture to industry, nor was it manufacturing growth consequent upon population growth. Instead, the most striking feature of the eighteenth-century labour market was the growth of unemployed and surplus labour.[28]

## *Protoindustry at the Local Level*

The regional evidence shows that the creation of surplus labour is not *sufficient* for the growth of manufacturing or industrial employment. This result contradicts the usual Agrarian Fundamentalist claim that proto-industry is the result of the release of labour by agriculture. One can further explore the relationship between surplus labour and manufacturing employment with local evidence about population change, industrial history, and wages. Indeed, Chambers (1957: 13) used this sort of evidence to argue for the importance of surplus population as a cause of the growth of framework knitting in Nottinghamshire: 'The industry seems to have spread first among the villages with ample waste . . . where, as the Hearth Tax of 1674 shows, the population was already large and labour plentiful.' The history of the south midlands tells a different story. Its local history establishes that surplus labour was not a *necessary* condition for the development of protoindustry.

Tables 12–5 to 12–8 show the relationship between protoindustry, population density, and enclosure history for framework knitting in Leicestershire and rural manufacturing in Northamptonshire. Tables 12–5 and 12–7 break down the number of male manufacturing workers by the period of enclosure.[29] In both counties, villages enclosed before the middle of the sixteenth century had hardly any protoindustry, villages enclosed between the mid-sixteenth century and 1676 had some but not a lot, while

---

[28] See Redford (1926), Hobsbawm and Rudé (1968), Pollard (1978), Collins (1987, 1989) for discussions of the origins and operation of the surplus labour economy.

[29] For Leics. the number of manufacturing workers is the number returned in the 1831 census. The same source cannot be used for Northants since by then the worsted industry had been eliminated, so the number of manufacturers shown on the 1777 militia lists is used instead.

*Table 12–5. Enclosure and manufacturing employment in Leicestershire, 1831*

| Enclosure period | Number of men per parish in manufacturing | Number of parishes |
|---|---|---|
| Before 1563 | 0.6 | 28 |
| 1563–1676 | 11.1 | 77 |
| 1676–1801 | 32.5 | 113 |
| After 1801 | 33.8 | 11 |

the bulk of the protoindustrial workers were in villages enclosed after that date. Thus, when the framework knitting industry located in Leicestershire, it usually settled in open field villages, as did weavers, frameknitters, and shoemakers in Northamptonshire. This pattern supports the Fortrey/Marx theory of manufacturing development if it is amended to encompass protoindustrialization—and if open field villages become the 'towns' to which the surplus population of enclosed villages migrated.

And migration, not an initial surplus population, was the source of their work-force. Table 12–6 and 12–8 contrast the population histories of agricultural and industrial villages. In 1676, there was hardly any difference in population density between the two types of villages in Leicestershire. In Northamptonshire, industrial villages had a slight excess that may reflect the fact that protoindustrialization had already begun. In both counties, over the eighteenth and early nineteenth centuries the populations of the industrial villages grew much more rapidly. Unless rates of natural increase differed between the villages, there must have been net migration from the agricultural to the industrial villages. There is direct evidence of these movements in two cases. For the borough of Leicester, a quarter to a half of the people listed in the Freeman's Register for 1697–1702 and the Register of Apprentices for 1678–1682 and 1718–1770 were migrants, mainly from other parts of the county, often from the enclosed eastern parishes (*VCH*, Leics. iv. 193). Likewise, the settlement certificates of Kettering show that

*Table 12–6. Population and frameknitting in Leicestershire, 1676–1831*

| Type of village | Average population density | | | Number of villages |
|---|---|---|---|---|
| | 1676 | 1801 | 1831 | |
| Agricultural | 0.1432 | 0.1825 | 0.2368 | 193 |
| Frameknitting | 0.1588 | 0.3158 | 0.4443 | 29 |
| Overall | 0.1452 | 0.1999 | 0.2639 | 222 |

*Note*: Frameknitting villages are those for which there was evidence of knitting in the period 1700–51, as listed in *VCH*, Leics., iii. 20–3. A list for the period 1751–1800 and Blackner's list for 1812 are also given. Very similar results derive from using either of these lists.

*Table 12–7. Enclosure and manufacturing employment in Northamptonshire, 1777*

| Enclosure period | Number of men per parish in manufacturing | Number of parishes |
|---|---|---|
| Before 1563 | 1.4 | 8 |
| 1563–1676 | 2.3 | 40 |
| 1676–1801 | 6.5 | 100 |
| After 1801 | 7.7 | 60 |

Note: 'The number of men in manufacturing' is the number of manufacturers returned on the 1777 Northants militia list. Exempt men are excluded since they were irregularly reported. Manufacturers include all weavers, combers, wool staplers, sorters, spinners, winders, framework knitters, and shoemakers in excess of 5 in the village. I assumed that shoemakers were protoindustrialists if there were more than 5 in a village; otherwise, I assumed they were local cobblers.

Source: Manufacturing employees as given in Hatley (1973).

many of the apprentices came from elsewhere in Northamptonshire (Randall 1971/2: 350). After 1676 migrants from enclosed villages went to open, manufacturing villages or to the county towns, also centres of protoindustry.

This pattern is consistent with the explanation for population density in 1801 developed in Chapter 3. There I showed that the number of owners in 1801 was the decisive factor, and that villages with many owners were more receptive to immigration than were villages with only a few. The industrial villages in Leicestershire and Northamptonshire had many more owners than the agricultural villages, and so accommodated the growing demand for manufacturing labour.

Indeed, the history of real wages indicates that the migration to the industrial villages was in response to a growth in the demand for manufacturing labour, not to a push from agriculture. I showed earlier that both of the supply-side theories of manufacturing imply downward pressure on wages. Figure 12–1 shows the history of real wages in the framework knitting industry and the winter wage for men in agriculture in Leicestershire.[30] Frameknitting wages showed no upward or downward trend over the eighteenth century and were usually above the real winter wage of farm

*Table 12–8. Population and manufacturing in Northamptonshire, 1676–1831*

| Type of village | Average population density | | | Number of villages |
|---|---|---|---|---|
| | 1676 | 1801 | 1831 | |
| Agricultural | 0.1287 | 0.1564 | 0.2044 | 142 |
| Manufacturing | 0.1688 | 0.2762 | 0.3509 | 53 |
| Overall | 0.1396 | 0.1890 | 0.2442 | 195 |

[30] The construction of the real wages in Fig. 12–1 is explained in Appendix III.

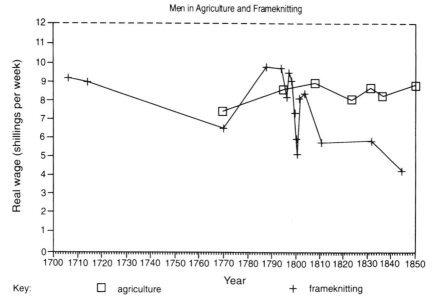

FIG. 12–1 Real Wages in Leicestershire, 1700–1850: Agriculture and Frameknitting

workers. The redeployment of labour from agriculture to framework knitting was not simply the result of labour being 'pushed' by enclosure and population growth: the demand for knitters was growing as rapidly as the supply.

The wages and the population histories are also suggestive as to when surplus labour became a serious feature of the southern labour market. Outside of Northamptonshire and Leicestershire, unemployed male labour probably accumulated soon after the population began to grow in the middle of the eighteenth century, but in Northampton and Leicester, the growing demand for manufacturing labour absorbed the population until 1800. Even in Northampton and Leicester, however, the populations of agricultural villages grew rapidly from 1801 to 1831. Redundant agricultural labour was emerging. In the same period, the real wage of frameknitters collapsed. By 1844, it was half the agricultural wage. In the hosiery industry, labour supply was growing more rapidly than labour demand. The male labour market became saturated with surplus labour by the end of the French Wars, if not earlier, in the less industrialized counties.[31]

[31] Huzel (1989: 761–72) has shown that per capita poor expenditures began rising after 1750 and that by the 1770s regular references to relieving able-bodied adults began appearing in parish records. This evidence suggests that surplus labour was appearing in some areas by 1750 and certainly by the 1770s.

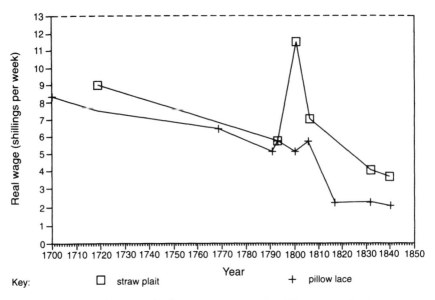

FIG. 12–2  Straw Plaiters and Pillow Lace Knitters: Real Wages, 1700–1850

A lack of employment for women became a serious problem at about the same time. In Leicestershire, Northamptonshire, and Oxfordshire the demand for women in spinning probably grew in the eighteenth century as the knitting and weaving industries expanded. The situation reversed abruptly, however, with the advent of factory spinning in the early years of the nineteenth century. There were then no jobs for women. In the lace district, labour demand grew somewhat less rapidly than supply throughout the eighteenth century. Figure 12–2 shows that the real wage of lace knitters and straw plaiters declined continuously from 1700 to 1800.[32] The straw plait boom during the French Wars offered a temporary respite. All the commentators at the time recounted how the industry bid labour away from agriculture and lace making, lifting the wage in each. Priest (1813: 346) observed in Buckinghamshire:

Women and children here make great earnings by making lace and platting straw, unfortunately to the disadvantage of agriculture; for whilst they can earn by such work from 7s. to 30s. per week, . . . it can scarce be expected they would undertake work in the field at such a rate as the farmer could afford to pay.

Priest quotes no day wage rates for women, which may indicate that farmers did not find it profitable to pay enough to retain women as day labourers.

---

[32]  The construction of the real wages in Fig. 12–2 is explained in Appendix III.

However, farmers bid up the wage of servants to keep their dairymaids. 'The wages of dairy-maids have risen much within a few years, on account of the lace and straw manufactories and it is with difficulty they are procured' (Priest 1813: 335). Their cash earnings were frequently £8 to £10 per year in Buckinghamshire, while their counterparts earned only £3 to £6 in Northamptonshire, Leicestershire, Cambridgeshire, and Huntingdon. (Priest 1813: 334–5; Pitt 1809*a*: 305; Pitt 1809*b*: 293; Gooch 1813: 285; Parkinson 1811: 267). Young's (1804*c*: 222) description of the Hertford-shire labour market was similar to Priest's: 'The farmers complain of it [the chance to earn high wages plaiting straw], as doing mischief, for it makes the poor saucy, and no servants can be procured, or any field-work done, where this manufacture establishes itself.' He added generously, 'There may be some inconvenience of this sort, but good earnings are a most happy circumstance, which I wish to see universal.' The happy circumstance did not last. At the end of the French Wars real wages in both lace and straw plait collapsed. In the 1830s and 1840s, the women of the lace district worked long hours for a miserable income.

## *The Failure to Protoindustrialize*

The regional and the local evidence together shows that reductions in farm employment and the growth of surplus labour were neither necessary nor sufficient for the growth in manufacturing. While this result may be a surprise to Agrarian Fundamentalists, it is common for cheap labour to fail to find employment in industry. In backward countries this is called 'underdevelopment'; in advanced countries, it is a 'regional problem'. Canada's maritime provinces, southern Italy, the southern American states (until the 1950s) all suffered from high unemployment and low wages and yet failed to industrialize. Industry remained located in high wage areas. Likewise, many regions in early modern Europe had no protoindustry although wages were low and there was redundant labour.

A variety of factors that determine industry output have been advanced, in one case or another, as explanations for the failure of a low wage region to industrialize. These factors include:

1. Demand for the product: a region is too isolated to sell any manufac-tures. Improvements in transport, then, might lead to industrial growth.
2. Economies of large-scale production: if costs fall as the scale of production rises, then it may be cheaper to locate an industry in a region of high population where demand is large. Even if wages are high in the metropolis, the economies of scale may allow the large producer to

overcome the advantage which low wages and the protection of transport costs confer on the would-be small-scale producer in the periphery.

3. Raw materials: it has often been claimed that the manufacturing development of France, for instance, was curtailed by the lack of coal deposits and the consequent high price of fuel.

4. Capital: regional backwardness may be due to the shortage or high cost of capital. Lewis's model of economic growth emphasizes the necessity for a rise in the savings rate to increase the capital stock, output, and the demand for labour. Many historians, including Gerschenkron (1962), have emphasized the establishment of banks as the factor that set off the burst of regional growth.

5. Technology: technology may have to be imported. If it was designed for a high wage economy, it may economize on labour and be prodigal with capital. This orientation might neutralize the competitive advantage of an economy with low wages and high capital costs.

6. Entrepreneurship: the lack of entrepreneurs in the backward region may retard its development. Landes (1949) has advocated this possibility for nineteenth-century France.

7. External economies: even if large-scale production confers no economies, the presence of many small firms in a region may confer an advantage on all of them if specialized suppliers appear. Such advantages are called 'external economies of scale'. Jacobs (1984) has recently argued that the creation of an urban system with external economies is crucial to self-generating growth.

8. Labour: low wages may not translate into low labour costs if the labour is unskilled or lacks the habits of mind to perform manufacturing work.

Historians of protoindustry tend to regard their subject as *sui generis*, but this list is a good place to look for an explanation of why protoindustry did not spread throughout the south midlands.[33] Arthur Young again provides a trenchant point of departure. While travelling from Cambridge to Bedford in 1791, he remarked:

The local position of manufactures is not easily accounted for. The wools of Northamptonshire, Bedfordshire, &c. go to Bury to be spun, and to Norwich to be woven, yet St. Neot's and Bedford are populous places, the former without any manufactures, and the latter little, except a scattering of lace. Why should not these

[33] See Pollard (1981: 3–12) for a similar discussion. Hudson (1989) also discusses these issues. The collection of papers published in *Annales: économies, sociétés, civilisations* provides an important perspective in which the discussion revolves around labour availability and intensity. See, in particular, from 1984, Deyon, Ho, Dewerpe, Vandenbroeke, Chao, and Mendels.

wools be spun and woven here, upon a considerable navigation, and much nearer to London, one great market for the goods of Norwich? (Young 1791: 483, punctuation modernized)

This observation highlights a distinctive feature of protoindustrialization —its regional concentration. Each protoindustrial region concentrated on a single product or a narrow range of products. Regions had clear boundaries, and there were large gaps between regions—including Bedford and St Neots. Neither agriculture, population densities, nor wage levels were notably different in Huntingdonshire and Bedfordshire from the adjoining protoindustrial regions. Why didn't protoindustry spread to fill in the gap?

Many of the possible explanations can be discarded straight away. Thus, there was no shortage of demand—indeed, Young suggests that Bedford was better placed to serve London than was Norwich. There were not increasing returns to scale, at least in production. This is an important difference between protoindustry and modern industry. There is no reason to believe that Huntingdonshire suffered more from a shortage of capital or entrepreneurs than did Leicestershire or Norfolk, and protoindustrial technology was as suitable for use in the non-industrial counties as in the industrial. Likewise, raw materials were not a constraint—indeed, Young notes that St Neots was closer to the raw wool than Bury. The constraint on the spread of protoindustry lay elsewhere.

The regional pattern of protoindustrialization was due to the remaining two items on the list: external economies of scale and labour quality.

The external economies of scale arose since a social infrastructure of marketing and distribution was necessary for the worldwide sale of protoindustrial products. Recent research has exposed this network for the woollen and worsted industry of the West Riding, one of the most successful protoindustries of the eighteenth century. As early as 1741, Cowper (1741: 39) observed that 'Yorkshire hath rivall'd them [Kettering weavers] . . . and very much decreased their Trade, as also lower'd their Prices'. Worsted production grew rapidly in Yorkshire only after the mid-eighteenth century when it seriously pressed East Anglia (Hudson 1986: 29) and drove the small Northampton industry out of business.

A distinctive feature of the West Riding industry (as of any successful protoindustry) was the large number of firms involved. In 1800 Edward Law observed that 'from the wool grower to the consumer a piece of broadcloth passes through an hundred different hands' (Hudson 1986: 107). Some manufacturers bought wool direct from growers, but most wool was handled by wholesalers and exchanged at fairs and markets. Midland wool was carted to East Anglian ports (when it passed Bedford and St Neots) and then shipped to Hull where it was dispatched to the West

Riding. Much of the wool was handled by staplers who transported it, stored it, and sold it in small quantities to the clothiers who purchased it, as they needed it, often on credit. The wool was then carded, spun, woven, fulled (in the case of wool but not worsted), dyed, and dressed. These activities typically involved different firms. The finished cloth was sold in large markets in Leeds, Bradford, etc. By the end of the eighteenth century, two-thirds of the production was exported and the growth of the industry had depended on the cultivation of European markets. Factors purchased the cloth at the fairs. It was stored in warehouses owned by yet other firms. Merchants shipped the cloth overseas.

The West Riding wool industry is an example of Adam Smith's (1776: 17) maxim that 'the division of labour is limited by the extent of the market'. The large volume of business allowed a multiplicity of specialized firms to emerge at each stage. Competition at every level provided service at lowest cost. The cumulation of these economies meant that the West Riding could produce cloth at a lower price than other districts, even those like Northamptonshire that had an advantage due to proximity to the raw wool. Conversely, in the West Riding, wool production was more lucrative than any other protoindustry since the external economies gave every woollen business a cost advantage that was not available to a non-woollen business. External economies thus produced regional specialization in protoindustry.

The effect of external economies was reinforced by the way skills were transmitted. Many protoindustrial jobs took years to learn. Indeed, like farming, one had to be brought up to the trade to do it well. Thus, skills were passed on within protoindustrial families and through apprenticeship. In a few cases (e.g. lace knitting) the skills were taught in schools, but then the teachers were knitters. Thus the principal way to become a skilled protoindustrialist was to grow up in a protoindustrial region. Protoindustry could not take root in a new district without difficulty since qualified workers were not available. The wage rate of labour might have been low, but the cost of the labour to a protoindustrialist was high since it was untrained and inefficient.

The protoindustrial system also created adverse incentives that limited its spread. Developing an international business required establishing a reputation for quality. While it was in the interest of every producer that this reputation exist, it was also in the interest of each to cheat and pass off shoddy goods if they could not be detected. Several strategies were available to deter this behaviour. In some cases, the trade as a whole sought to police its members through the machinery of guilds. In other cases, buyers could be relied on to do the job, but in that case they had to inspect the goods. That was the course adopted in the West Riding woollen industry. Huge cloth halls were built in Leeds, Wakefield, Halifax, and so on, where cloth was displayed and sold. These markets brought the scattered producers and the factors together and provided a means for the factors to inspect the goods they bought. The need to transport goods to these markets, however, meant that production could not extend far from the markets. Clothiers bore the cost of carting their goods to the cloth halls. Since transport costs increased with the distance hauled, there was a distance beyond which it was not profitable to produce and ship cloth. Thus, the atomistic nature of the protoindustrial system set a geographical limit to its extension across the countryside.

These considerations suggest that there was a dynamic of successful protoindustrialization. Initially, industrial location may have been random. Of the Staffordshire potteries, Young (1771a: iii. 253) remarked 'There is no conjecture formed of the original reason of fixing the manufacture in this spot, except for the conveniency of plenty of coals'—hardly a unique feature. Once, however, production was under way, the competitive advantage of a region increased. Each increment in size lowered costs as external economies increased. Further, a work-force was brought up to the trade. The region's costs declined and it was able to expand at the expense of competing regions; this expansion further lowered costs.

To return to Arthur Young's query: the reason worsted was not made in Bedfordshire was because no one knew how to do it. There was no way to learn the skill in the county because there was no one to teach it. Moreover, even if someone had learned how to weave, he would have had an incentive to go to Norwich or Yorkshire where his skill would have been more valuable.

## Conclusion

At the end of the seventeenth century, the labour market in the south midlands had been reasonably tight. Three-fourths of the men were fully employed in agriculture. Most women, indeed, worked full-time on the farm. A very large share of the rural population was needed to bring in the harvest. Manufacturing, building, and shopkeeping probably provided jobs for the rest.

After Waterloo, the situation was entirely different. Half the men were still employed in agriculture, but many were unemployed except for the harvest. For most of the year, the marginal product of labour in agriculture was zero. The sector had turned into a source of 'unlimited labour' in the Lewis sense. The decline of female employment had been at least as extensive, and the retreat of women into the home had created another reservoir of surplus labour. Retailing had almost certainly become overstocked.

The emergence of a surplus labour economy was not a 'natural' event—it was the result of social forces. These included the reduction in the agricultural demand for labour, the rise in the fertility rate, the decline in the rate of emigration, and the slow growth in the demand for non-agricultural labour in the south midlands. The first three were probably related to the shift from a society of owner-occupying family farmers to a society based on the great estate, the large farm, and the landless labourer. First, as I have argued, these changes reduced the demand for labour. Second, they also lowered the incomes of the bulk of the agricultural population and deprived them of the ownership of property. (See Chapters 5 and 14.) The fall in income and wealth probably reduced the ability of rural people to migrate even as the gains to migration were increasing. Third, there is also a case that the rise in the fertility rate was related to a fall in the age of marriage due to proletarianization. More research is necessary to investigate these relationships, but to the degree that they prove substantial, the emergence of the surplus labour economy can be linked to the rise of the great estate.

During the eighteenth century agriculture in the south midlands was modernized, so that labour on the farm would be used efficiently. This reorganization did not accelerate the rate of economic growth, however. All it did was to turn the midlands into an even more perfect example of an underdeveloped country. The problem was not overcome, the surplus labour not sopped up, until the end of the nineteenth century. The basic problem was that the agricultural revolution came too soon. It preceded the region's industrial revolution by at least a century. Instead of contributing to the growth of manufacturing, the premature release of labour from agriculture caused nothing but poverty.

# 13

# Agrarian Change and Economic Growth

All things must be done, that may effectually increase the value of rents and price of land, which will add true strength to the nation.

Charles Davenant[1]

WHEN Professor Postan addressed the Second International Conference of Economic History in 1962, he offered a historical lesson to the less developed countries of the twentieth century: 'For these countries, as for England in the 18th century, development of agriculture is, and must for some time remain, the source of industrial development—its *sine qua non*' (Postan 1965: 16). Whatever its wisdom for the poor countries of today, there was little connection between agricultural development and England's industrial revolution. While the agrarian changes before 1700 probably did raise the national income, enclosures and the movement to bigger farms in the eighteenth century did little to accelerate industrial growth. The reason is clear from earlier chapters—the biggest effect of eighteenth-century agrarian change was to release labour rather than increase output. Since that labour was not successfully re-employed, manufacturing output did not rise. Hence, agrarian reorganization made little contribution to the growth of national income during the Industrial Revolution.

Discussions of agriculture's contribution to economic development usually begin with a broad range of possibilities, including the following:

1. an increase in agricultural output that helped feed the expanding urban population
2. the release of labour to other sectors
3. the release of capital to other sectors
4. the expansion of the home market for manufactures.

---

[1] As quoted by Arthur Young (1815: 171). Young indicates that his source is Davenant's *Works* (1771), ed. Whitworth, i. 71.

In this chapter, I will show that English agriculture performed only the first two roles and that those contributions were made mainly in the seventeenth century.

## *What English Agriculture Did Not Do*

To clear the decks, I begin with what English agriculture did *not* do. It did not release capital to other sectors, and it did not provide a home market for manufactures.

There are two ways in which agriculture might have released capital. The first was by reducing its *demand* for capital. In fact, the reverse occurred in England. While the shift to large farms economized on draught animals and implements (thus reducing the demand for capital), the conversion of arable to pasture raised herd sizes, thereby increasing the demand. The resource costs of enclosures boosted the demand for fixed capital. The combined effect of all of these changes was to raise agricultural capital in the south midlands by 21 per cent over the seventeenth and eighteenth centuries. (Table 11–10.)

The second way agriculture might have released capital was by increasing savings, the *supply* of capital. Much of the rental income of England accrued to a small group of rich landowners who were consequently in a position to save. Development theorists entertain the hope that such income will be directed to capital formation. In surplus labour models (like Lewis's) industrial employment and output depend on the industrial capital stock. Since labourers are too poor to save and since, at the outset, the economy is predominantly agricultural, industrial investment depends on the savings of landlords. Therefore, Fei and Ranis, who have developed the Lewis model in great detail, assert, 'the landlord's performance in terms of his *ex ante* willingness to save plays a major role in determining the rate of capital formation' (Fei and Ranis 1964: 167). Japanese landlords are cited as paragons of virtue.

English landlords, however, were never so abstinent. Throughout the early modern period, they were regarded as profligate rather than frugal. That was certainly the view of the classical economists, who dismissed the aristocracy and gentry as sources of savings for industry, and few later historians have argued otherwise. Crouzet (1972: 56) concluded his summary of this literature by endorsing the view 'that "surprisingly little" of the wealth of rural England "found its way into the new industrial enterprises"' Crafts' (1985: 122–5) more recent calculations of the sources of British savings support this position, at least until the second quarter of the nineteenth century when agriculture may have become a net lender.

Until then, however, agrarian change made little, if any, contribution to non-agricultural capital formation. Fei and Ranis (1964: 164) remarked that if 'landlords choose to consume a considerable portion of their income or to indulge in nonproductive investment (e.g. luxury housing), to that extent a potentially valuable investment fund originating in the agricultural surplus is dissipated'. The English countryside is decorated with the 'luxury housing' built with the agricultural surplus.

Likewise, English agriculture never provided a buoyant market for manufactures, especially mass produced goods. Probate inventories are the main source for studying rural consumption in the seventeenth century. Hoskins (1957: 310) concludes that 'farming had been revolutionised since the early 1600s, but the farmer's household probably enjoyed little more in the way of comforts and amenities in 1820 than their ancestors had in 1620'. Except for the much greater use of glass in windows and a few more carpets, furnishings were not more lavish.

Agrarian demand for manufactures remained quiescent in the eighteenth century. Certainly, much of the production of the new, factory industries was exported, in particular, three-fifths of the output of the cotton industry (Crafts 1985: 143.) O'Brien (1985: 780) has estimated that the rise in agrarian incomes in the eighteenth century increased manufacturing sales by 29–44 per cent.[2] Over the same period, the output of manufactured goods increased 470 per cent. By this reckoning, the contribution of the agrarian 'home market' to industrial growth was negligible. Using different data and methods, Crafts (1985: 133–4) imputed a larger—but still not decisive—share of manufactured goods sales to the agrarian market in the eighteenth century, and, again, a negligible share in the nineteenth. He estimated

that for the eighteenth century . . . 28.6 per cent of incremental industrial output was sold to agriculturalists . . . from 1801 onward [sales to] agriculturalists represent a rather small percentage of extra industrial output sales, probably less than 10 per cent of the total from 1801–51.

Exports, non-agricultural investment, and consumption by the bourgeoisie were the principal markets for industrial output. Agriculture did not provide a 'home market' for the industrial revolution.

These results are not surprising in view of the distributional character-istics of the agricultural revolution. In the next chapter I will show that real wages and the returns to farm capital were static from the seventeenth to the nineteenth century. The number of farmers and labourers declined. These were the groups that might have bought the output of manufacturers, but

---

[2] This increase includes sales to displaced agricultural labourers and farmers as well as sales to people receiving agricultural income.

their spending power was diminished. Instead, the gains from agricultural progress accrued to landowners as higher rents. Further, the ownership of property was becoming more concentrated. The gentry and aristocracy spent lavishly, indeed, but the object was servants, stately homes, and custom handicrafts—not mass produced commodities or industrial plant and equipment.

## What English Agriculture Did Do

The contributions that agrarian change made to economic development lay in increasing output and releasing labour to other sectors, although the importance of these developments varied over time and among regions. Increasing output was probably the most important contribution in the seventeenth century throughout the kingdom. Labour release was more important in the eighteenth, although its contribution to growth varied across the country depending on the effectiveness with which the freed labour was redeployed outside agriculture.

In the seventeenth century, crop yields and real output per acre increased substantially in the south midlands, as I have shown earlier. Overton's (1979) investigation of East Anglian probate inventories points to the same conclusion for that region.[3] Jackson's (1985: 346) reworking of the aggregate evidence suggests that between 1660 and 1740 agricultural output increased at 4.5–10 per cent per decade in England as a whole. The seventeenth century was therefore probably a time of rising yields and output in most parts of England.

On the other hand, labour release was not an important aspect of seventeenth-century agrarian change. My estimates show that full-time farm employment in the south midlands fell only 8–14 per cent. Furthermore, since enclosure in the midlands was much more likely to have led to migration to London—the booming non-agricultural centre—the reallocation of labour to manufacturing was probably not important in England as a whole.

The character of agrarian change reversed in the eighteenth century. In earlier chapters I showed that yield growth slackened in the south midlands, and with it the growth in total agricultural output. Between 1740 and 1790, Jackson's calculations show agricultural output growing at −2.8 to 2.0 per

---

[3] Overton's (1979) estimates show without a doubt a substantial increase in yields over the 17th cent. It is likely, however, that his estimates understate yields at all dates. Since his figures understate yields at the beginning of the 18th cent., comparison of those estimates with late 18th-cent. figures conveys the misleading impression that yields continued to rise over the 18th cent. See Allen (1988b).

cent per decade in England as a whole. Since it is often claimed that parliamentary enclosure in this period was increasing the cropped acreage, Jackson's calculation suggests that corn yields were growing very little. My estimates for the south midlands and Jackson's calculations with price, wage, and population data are consistent with Fussell's (1929) conjecture that English yields grew little over the eighteenth century.

In contrast, the increase in average farm size and enclosure with conversion to pasture were the principal developments in the south midlands in the eighteenth century, and both resulted mainly in the release of labour. My estimates show that farm employment in the south midlands declined much more rapidly in the eighteenth century than it had in the seventeenth. While enclosure was not as important outside of the midlands in the eighteenth century, the investigations of Mingay (1962) and Wordie (1974) have shown that farms were being amalgamated in other parts of the country, and those changes probably also reduced farm employment. If agrarian change raised GDP, the rise was realized mainly by re-employing this freed labour in manufacturing or services. In the south midlands, most of this labour remained redundant although in the industrializing North the surplus labour may have been productively re-employed.

After the Napoleonic Wars, the character of agrarian change altered again. Labour shedding had been completed. Wrigley's estimates of the adult male agricultural work-force show an 11 per cent increase between 1811 and 1851 (Wrigley 1986: 332). At the same time agricultural production was rising again—at 20 per cent per decade between 1801 and 1831 according to Crafts's (1985: 42) estimates. The installation of hollow drains in the heavy arable district is an example of the sort of change that raised yields, but progress was much more general. In most districts yields in 1850 were considerably higher than in 1800 (Allen and O Grada 1988). By this time improvements in livestock breeding were probably widespread enough to be raising the productivity of grasslands.

## Agrarian Change and the Growth in Gross Domestic Product

Increasing output and releasing labour both contributed to economic growth by increasing England's gross domestic product—the country's total production of goods and services. My analysis of the contribution of agrarian change turns on the distinction between its *direct* and *indirect* effects. In the first case, if productivity growth raised agricultural output, then GDP increased by the value of that amount.[4] That is a *direct* effect.

---

[4] More properly, if technical change in agriculture increased *value added* in agriculture, then GDP increased by that amount. However, the increase in value added and in output amounted

Alternatively, if productivity growth reduced capital and labour employed in farming, and if those freed resources were successfully re-employed elsewhere in the economy, then the output (value added) of other sectors expanded and increased national income. In that case, the effect of agricultural productivity growth was *indirect* in that it was manifest by a growth in non-agricultural GDP. The indirect effects were more problematic than the direct since they were more complex—the indirect effect required the re-employment of labour rather than just its release. Clearly, agricultural productivity growth could have had both direct and indirect effects, in which case the total contribution to growth is the sum of the two.

The matter can be put differently with interesting results. In Parts II and III, I was concerned with whether enclosure and bigger farms were efficient from the point of view of farming *per se*. I measured efficiency by changes in cropping, yields, and labour productivity, but the best summary measure was the rise in rent since that indicated the change in total factor productivity or overall economic efficiency. The question I am asking now is whether enclosure and the increase in farm size raised GDP. Juxtaposing the definitions immediately prompts the following questions: Is there a relationship between the rise in rent (the measure of farm efficiency) and the rise in GDP (the output of the whole economy)? Is it necessary to make any 'corrections' to the rise in rent in order to use it as a measure of the contribution of agrarian change to the rise in GDP? Were landlords, in pursuing their private interest by reorganizing their estates to raise rents, also causing economic growth by raising GDP?

I propose the following answers to these questions. In the sixteenth and seventeenth centuries the rise in rent did, indeed, indicate the rise in GDP due to agrarian change. The increase in farm output from the rise in yields directly raised GDP, while the labour released by enclosures was successfully re-employed. Hence, enclosures made an indirect contribution to the growth in GDP. In the eighteenth century, however, the effect of agrarian change was far less beneficial. Such increases in output as occurred still directly raised GDP, but they were minor. The biggest contribution of

to the same thing in early modern England. GDP is the sum of value added in all sectors in the economy. Value added is the value of output in a sector less purchases from other sectors. Value added thus equals returns to land, labour, and capital employed in the sector. Increased agricultural output will equal the increase in value added in agriculture and thus the increase in GDP only if purchases by agriculture do not also increase. That is a defensible assumption for early modern farming where implements were the main purchased input. Their use was small in any event and probably increased little as output rose. The situation today is very different where increased farm output is often the result of additional purchases of machinery, fertilizer, pesticides, etc. The greater use of these inputs will expand value added and GDP less than output.

agrarian change was the release of labour, but it was not successfully re-employed—these people became the legions of paupers who so distressed upper-class opinion during the Industrial Revolution. Since both the direct and indirect effects of agrarian change were small, it did not raise GDP much. The rise in agricultural rent, therefore, does not indicate an increase in the national income. Instead, it represents a transfer from the rest of the community to the landowners. Indeed, I will show later, this transfer was far greater than the rise in poor expenses, which may be thought of as a small recompense.

## Social Savings Calculations

Several techniques are available to assess the contribution of changes such as enclosure and farm amalgamation to the growth in GDP. These include social savings calculations and the use of shadow prices to compute the social profitability of the change. Both approaches evaluate the direct and indirect effects of the change. I will begin with social savings calculations.

My point of departure is McCloskey's (1972, 1975) application of social savings theory to eighteenth-century enclosures. McCloskey's principal theoretical contribution was the suggestion that increases in rent were proxies for increases in output following the enclosure. 'The increase in rent . . . can be used as an estimate (although biased downwards by not including the value of the increased employment of the other, mobile factors of production) of the increase in the value of output resulting from the enclosure' (McCloskey 1972: 33). This insight is an advance, but it is not quite on the mark: more precisely—and more pertinently—the rise in rent equals the rise in GDP (including any changes in manufacturing output) due to the enclosure, given McCloskey's assumptions.

McCloskey cast his analysis in the mould of Tory fundamentalism; like Young and Chambers he wrote as though enclosures increased employment, capital, and output.

Since labor and capital were mobile, the increase in their productivity would reveal itself in an increase in their employment, not an increase in their prices: if they were paid more, more would flow into the village, and continue to flow until the previous wages of labor and returns to capital were reestablished. (McCloskey 1972: 32)

Few enclosures in the south midlands, in fact, conformed to that pattern. In itself, this is not a difficulty, for the link between rent changes and GDP changes depends on other assumptions—in particular, profit maximization by all concerned and 'perfect' factor markets. The labour market, for instance, was perfect if (1) employment losses (or gains) in agriculture were matched by gains (or losses) in manufacturing, and (2) the wage was the same in two sectors. Only if these assumptions were true did rent changes equal changes in GDP.

To see why these assumptions are critical and what happens when they are violated, it is necessary to develop the theory step by step. For clarity, I write out the key steps in equations. As developed here, the model abstracts from many features of the eighteenth century economy. For instance, I make no allowance for the sales of intermediate goods (e.g. raw wool) between agriculture and manufacturing, nor for intersectoral capital flows. These transactions could be accomodated but only at the cost of obscuring the critical labour market assumptions.

The question is: how did enclosure and farm amalgamation affect GDP. To answer it, GDP must be defined:

$$Y = P_A Q_A + P_M Q_M \qquad \text{(Equation 13–1)}$$

where $Y$ is GDP, $P_A$ and $Q_A$ are the price and output of agricultural goods, and $P_M$ and $Q_M$ are the price and output of manufactured goods.[5] The agricultural revolution had direct effects on GDP by raising $Q_A$ and indirect effects by releasing labour that shifted to manufacturing and raised $Q_M$. Using $d$ to indicate the change in the quantity of a variable, I can write the change in GDP due to agrarian change as:

$$dY = P_A dQ_A + P_M dQ_M \qquad \text{(Equation 13–2)}$$

The change in GDP ($dY$) equals the price of agricultural goods ($P_A$) multiplied by the change in agricultural production ($dQ_A$) plus the price of manufactured goods ($P_M$) multiplied by the change in manufacturing production ($dQ_M$).

In this formulation the prices of agricultural and manufactured goods are assumed to remain unchanged despite enclosure or the increase in farm size. This assumption may be inappropriate since enclosures often changed the structure of agricultural output. In the pasture district, for instance, enclosure reduced the production of corn and increased the production of meat, butter, cheese, and wool. Consequently, the enclosure movement *as a whole* might have changed the prices of these goods. The simplest solution to this dilemma—and the one McCloskey adopts—is to analyse the enclosure of a single village. In that case, the output changes were negligible compared to the whole English market, so price changes were also negligible. In the parlance of cost-benefit analysis, McCloskey treats the

---

[5] Strictly speaking, GDP is the sum of value added in the two sectors. Value added is the value of output of the sector minus purchases from other sectors; equation 13–1 thus presumes that there are no intersectoral transactions.

enclosure as a 'small project'. But the problem that remains is how to generalize from the results of a single enclosure to the results of the enclosure movement as a whole. The 'small project' assumption renders suspect social savings calculations for England as a whole.

Introducing the concept of the marginal product of labour allows us to make explicit the link between labour absorption in manufacturing and the growth in manufacturing output. The marginal product of labour ($mpp_L$) is the increase in manufacturing output ($dQ_M$) caused by employing another worker in manufacturing. If the increase in manufacturing employment is 'small', then the marginal product does not vary with the number of workers hired, and the increase in manufacturing output is given by the following equation:

$$dQ_M = mpp_L dL_M \qquad \text{(Equation 13–3)}$$

where $dL_M$ is the increase in manufacturing employment. Hence, the growth in GDP equals

$$dY = P_A dQ_A + P_M mpp_L dL_M \qquad \text{(Equation 13–4)}$$

To establish the connection between rent changes and changes in GDP, I next define the rent of land as its Ricardian surplus, that is farm revenue minus the cost of labour and capital:

$$sT = P_A Q_A - w_A L_A - i_A K_A \qquad \text{(Equation 13–5)}$$

Here $s$ equals rent (Ricardian surplus per acre), $T$ equals the acreage farmed, $w_A$ is the wage in agriculture, $L_A$ is agricultural employment, and $i_A$ and $K_A$ are the price and quantity of agricultural capital. Enclosure might change capital, labour—hence output—and rent. To focus on the problematic labour market assumptions underlying McCloskey's procedure, I will assume that enclosure did not change agricultural capital. Hence,

$$dsT = P_A dQ_A - w_A dL_A \qquad \text{(Equation 13–6)}$$

The change in rent per acre ($ds$) multiplied by the area enclosed ($T$) equals the price of agricultural products ($P_A$) multiplied by the change in agricultural output ($dQ_A$) minus the agricultural wage rate ($w_A$) multiplied by the labour released from agricultural employment ($dL_A$).

I can now state and prove McCloskey's social savings theorem. If

$$P_M mpp_L = w_M \qquad \text{(Equation 13–7)}$$

$$w_M = w_A \qquad \text{(Equation 13–8)}$$

$$dL_M = - dL_A \qquad \text{(Equation 13–9)}$$

then the rise in GDP due to the agrarian revolution equals the rise in rent:

$$dY = sT \qquad \text{(Equation 13–10)}$$

The algebra of the proof is straightforward: equations 13–7, 13–8, and 13–9 are substituted sequentially into equation 13–4 until its right side is identical to the right side of equation 13–6. Equation 13–10 follows as a result.

How confident can one be that the rise in rent equals the rise in GDP due to an agrarian revolution? The answer depends on the realism of the assumptions. Equation 13–7 is the familiar condition for profit maximization—firms expand their use of labour (whose productivity is assumed to diminish as employment increases) until the value of the marginal product of labour equals the wage. Therefore, the truth of equation 13–7 requires that manufacturing firms be profit maximizers, and sell their products and hire their workers in competitive markets. Equation 13–7 is probably a reasonable assumption, given the competitive organization of most proto- and factory industries.

The real problems arise with the other two equations, which describe the labour market. Equation 13–8 states that the manufacturing and agricultural wages are identical, and equation 13–9 states that the labour released from agriculture is re-employed in industry. Together equations 13–8 and 13–9 define a 'perfect' labour market. Indeed, male wages in agriculture, protoindustry, and construction were of the same order in the early eighteenth century (Figures 12–1, 14–5, 14–6, 14–7), so equation 13–8 is a reasonable approximation for that period. Likewise, equation 13–9 was also reasonably accurate early in the eighteenth century before the emergence of the surplus labour economy. By the end of the eighteenth century, however, labour released from farming remained unemployed, so equation 13–9 is manifestly unrealistic.

What is the relationship between rent increases and increases in GDP, when surplus labour is present? The answer depends on how agrarian change affected output and employment. Consider two polar cases. In the first case, suppose that agrarian change increased agricultural output without affecting employment. In that case $dQ_A$ is positive and $dL_A$ and $dL_M$ are zero. According to equation 13–6, the rent rise equals the rise in the value of agricultural output, as does the rise in GDP (equation 13–4). Hence, the rent increase continues to measure the rise in national income due to agrarian change.

In the second case, suppose that agrarian change decreases farm employment without raising output. Now $dQ_A$ is zero, $dL_A$ is negative, but $dL_M$ is also zero since labour release increases the army of the unemployed rather than the manufacturing work-force. According to equation 13–4, GDP does not increase, while equation 13–6 indicates that rent rises by an amount equal to the reduction in the agricultural wage bill. The important conclusion is that an agricultural revolution that simply sheds labour when there is already unemployment does not increase GDP. Further, the rise in rent is a transfer of income from labourers (including, at this level of abstraction, farmers) to landlords, not an indicator of a higher national income. In the presence of surplus labour, landlords who pursued their own advantage were no longer led 'as if by an invisible hand' to maximize social output.

## The Social Savings of Enclosure in the South Midlands

Restrictive as the assumptions underlying the social savings model are, it is still worth applying it to the English agricultural revolution. I will discuss two calculations—the first to assess the contribution of agrarian change in the sixteenth and the seventeenth centuries; the second for the eighteenth and early nineteenth.

The assumptions of the social savings model are reasonably well satisfied for the south midlands before the eighteenth century. Surplus labour was not yet an important feature of English rural life, so the labour market assumptions of the model are plausible. In any event, output increases were a more important consequence of productivity growth than were employment reductions, so the difficulties involved in shifting labour from agriculture to industry did not loom large.

I compute the rise in GDP as the rise in rents. At the end of the middle ages, 4 per cent of the south midlands was enclosed; in 1700, 32 per cent was enclosed.[6] In the early eighteenth century, enclosed land let for about 16s. per acre and open field land at about 10s. In the seventeenth century, however, before the big rise in yields, open field rents had been 5s. per acre.[7] The 'prerevolutionary' rent of land was, therefore, £0.272 per acre,[8] while the 'postrevolutionary' rent was £0.602[9]—a gain of £0.330 per acre. In the south midlands, there were about 2.6 million acres of farm land, so the rent gain for the region was £858,000. The most recent revision of Gregory King's national income estimate puts it at £54 million in 1688 (Lindert and Williamson 1982: 393). Hence, the agricultural revolution in the south midlands raised GDP by 1.6 per cent. This in itself is not a large number, but

---

[6] These percentages apply to the land whose enclosure history I can date.

[7] The increase was not due to a rise in agricultural prices.

[8] $0.272 = (0.04 \times 16 + 0.96 \times 5) \div 20$.

[9] $0.602 = (0.34 \times 16 + 0.66 \times 10) \div 20$.

the south midlands was not a large region. It contained just under 10 per cent of England's farm land.[10] If agrarian change raised rents by the same order elsewhere, then the increase in the national income due to the pre–1700 agricultural revolution was 15 per cent.[11] This is a big gain. While it is only the roughest of estimates, it provides a perspective for the development in the south midlands and shows that the accomplishments of the region were of a high order.

National calculations of social savings are difficult due to regional variation in the rate and bias of technical change, and because changes in the volume of production in the kingdom as a whole may have been substantial enough to vary prices—an effect that is not allowed for in the conventional social savings model. Recognizing the very approximate nature of the calculations, it is nevertheless interesting to probe the national estimates more deeply. For instance, one can decompose the national social saving due to agrarian change into the direct effect (the increase in the volume of output) and the indirect effect (the increase in GDP due to the reallocation of released labour). Thus, the most recent estimate of the share of income originating in agriculture c.1700 is about 0.4 (Crafts 1983: 189), so value added in agriculture, which also equals the value of agricultural output, was £21.6 million at the end of the seventeenth century. Table 11–9 shows that agricultural output per acre in the south midlands c.1700 was 37 per cent above medieval levels.[12] Hence the 'pre-revolutionary' value of output was £15.8 million. The increase in output, £5.8 million, was less than the increase in rent (£7.9 million). The increase in output was mainly due to the increase in yields achieved by yeomen farmers. The difference between the increase in rent and the growth of output (£2.1 million) represents the increase in the value of other goods and services produced by the labour released from agriculture. While the calculations are fragile, they indicate that agrarian change before 1700 made a significant contribution to the growth of the English economy, and agriculture made that contribution primarily by increasing output, not by reducing employment.

What of agrarian change during the classical agricultural revolution of the eighteenth and nineteenth centuries? McCloskey (1972) has offered a rough calculation of the rise in GDP from enclosure in this period. He reasoned that rent rose from 10s. to 20s. when land was enclosed—those

---

[10] Gregory King (1936: 35) estimated that there were 24 million acres of arable, meadow, pasture, common, and parks in England and Wales. This includes some forest, but excludes woods, coppices, heath, moors, waste, etc. Whatever happened on this land is excluded from the calculation.

[11] The rent gain is £7.92 million = £0.858 × 24 ÷ 2.6. The percentage increase in GDP is therefore $0.15 = 7.92 \div 54$.

[12] $1.37 = £3.51 \div £2.57$.

were conventional values as discussed in Chapter 9—and speculated that about 14 million acres were affected. Total rent, therefore, increased by £7 million, which was 5 per cent of Arthur Young's estimate of GDP *c*.1770.[13]

McCloskey's calculation is very approximate. Uncertainties attach to both the rent increase and the area to which it applies. The rent increase is about the right magnitude for the rise in real rent in the midlands during the eighteenth and nineteenth centuries, as I will show. However, this rent gain includes the effects of the increase in farm size as well as enclosure, so a calculation based on it overstates the importance of the enclosure movement *per se*. On the other hand, the rent change due to enclosure in other districts is not well established and may have been larger. In particular, some enclosures may have resulted in land reclamation and in the conversion of rough pasture and waste into highly productive corn land on light upland soils. The rise in land values from enclosing these lands may have been considerably higher than the rent increase in areas like the south midlands that were already highly productive as open fields (Jones 1965). The 14 million acres by which McCloskey multiplied the rent gain includes 6 million acres of parliamentary enclosure and 8 million acres enclosed privately after 1700. The recent work of Wordie has shown that the total enclosed in the eighteenth and nineteenth centuries was only about 7.6 million acres.[14] To compute the rise in England's rental from enclosure and other features of the agricultural revolution properly, it is necessary to proceed district by district and to match rent changes carefully with the acreages to which they apply. The sum of these improvements would yield the social saving for England as a whole.

I can offer a calculation for the rise in the rental in the south midlands due both to enclosure and to the growth in farm size. A large sample of estate surveys and contemporary accounts indicate that the average rent in the south midlands rose from 11.08*s*. in 1700–24 to 31.16*s*. in 1825–49.[15] This rent gain encompasses the effects of both enclosure and the increase in farm size; it is also the result of inflation. If I express the rents in prices prevailing during 1750–74, they become 12.06*s*. and 19.33*s*. for a rent gain of 7.27*s*.[16] If I multiply this gain by the acreage of farm land in the south

---

[13] McCloskey reduced the gain further by subtracting an allowance for the resource cost of enclosing.

[14] Wordie (1983: 502) indicates that 24.4 per cent of England was enclosed between 1700 and 1900. Multiplying this by 31 million acres of farm land gives 7.6 million acres enclosed in those centuries.

[15] These rents are weighted averages reflecting the shares of land in each natural district and the proportions of land enclosed in each district in each period. See Table 9–1 for the rents of open and enclosed land in the three natural districts in the south midlands.

[16] To express rents in the prices prevailing in 1750–74, I used the same rent deflator described in Ch. 11 to infer total factor productivity changes from rent changes.

midlands (2.6 million acres), as I did for the pre–1700 calculation, the total rent increase is £945,000—not much more than the rise in rents over the seventeenth century. Dividing the rent gain by Arthur Young's estimate of GDP in *c.*1770 (£130 million)[17] indicates that the rise in GDP due to enclosures and farm amalgamations was in the order of 0.7 per cent of GDP. This is less than half the magnitude of the sixteenth- and seventeenth-century agricultural revolutions. This social savings calculation shows that agrarian change in the eighteenth and nineteenth centuries made a smaller contribution to the growth in GDP than did agrarian change before 1700.

Moreover, 0.7 per cent overstates the contribution of agrarian change due to the emergence of surplus labour. The main effect of enclosure and the growth in farm size was to reduce agricultural employment. Such an agricultural revolution raises GDP only if the freed labour is re-employed in manufacturing. In 1700 labour markets were tight, but as the century progressed, non-agricultural employment opportunities either disappeared or became far less remunerative than farm work. Since agriculture was releasing labour that was not re-employed, the rise in rent overstates the contribution of agrarian change to the growth in English GDP.

## Shadow Pricing Labour

One cannot evaluate the effect of the eighteenth- and nineteenth-century agricultural revolution without recognizing the growth of surplus labour. The use of shadow prices to calculate rents is a standard approach for evaluating projects in economies characterized by surplus labour. I will use this method to re-evaluate the contribution to growth of enclosure and the increase in farm size during the late eighteenth and nineteenth centuries.

The logic of shadow prices is straightforward. If an input such as labour is in surplus, then more of that input fails to increase output. Under that circumstance, additional labour has no value when GDP is used as the measure of economic performance. If a project frees labour GDP does not rise; if a project uses labour the workers are drawn from the pool of unemployed instead of reducing output elsewhere in the economy. There is no cost in reduced GDP from using an unemployed resource and no gain in producing more of it. In the jargon of cost-benefit analysis, the unemployed resource has no social opportunity cost. Hence, in deciding whether projects increase the national income, labour should be costed at a wage of zero, if labour is surplus. This conclusion is true even though businesses pay a positive wage to the workers they actually employ.[18]

---

[17]  As reworked by Deane and Cole (1969: 156).

[18]  See Berry and Cline (1979: 54–8, 127–34) for an application of shadow prices within an index number framework.

To determine whether farm amalgamations and enclosure increased GDP in the conditions of the early nineteenth century, I repeat the statistical exercises of Parts II and III to see whether the rise in Ricardian surplus that accompanied the increase in farm size and some enclosures remains when labour is valued at its social opportunity cost or shadow price. I know that landowners usually found these changes to be privately profitable. These calculations indicate whether they were also socially profitable.

These calculations require shadow wages. In the last chapter, I showed that half the male agricultural work-force was idle outside the harvest. In the slack period its opportunity cost was, therefore, zero. In Buckinghamshire, Bedfordshire, and Northamptonshire, women could make lace if they did not work on a farm, but they earned a very low wage. Outside of those counties there was little for them to do. The lack of employment was quite general for children. In view of these considerations, I adopt the following shadow wages. Outside of the harvest, male agricultural labour is valued at zero; during the harvest, I value it at wages paid. (I use average values from Young's *Tours* for consistency with the other wages and prices.) I value the labour of women at £3.72 per year, which was the equivalent in 1770 prices of the typical earnings of a poor woman making lace, according to the 1834 *Poor Law Report.*[19] I valued the labour of boys at zero. I experimented with other valuation procedures as well, but they did not alter the results in any material way.

I begin with the increase in average farm size. In Chapter 11 I used the data set derived from Arthur Young's *Tours* to compare the Ricardian surplus of large and small farms. In these calculations labour was valued at prevailing wages. To measure the social efficiency of farm amalgamation, I first computed the 'shadow rent' of each of Young's sample farms using these shadow wages. Then I correlated shadow rent with size. For comparative purposes, Table 13–1, Part A, repeats the regression results that showed that large farms were privately profitable. When labour is valued at its market wage, Ricardian surplus per acre increased with size but at a diminishing rate. Part B reports the corresponding regressions when labour is valued at its social opportunity cost. Notice that the coefficients of acres and acres-squared are driven towards zero and become statistically insignificant. Large farms are no longer more efficient than small farms. At the end of the eighteenth century, large farms were privately profitable but not socially efficient.

---

[19] Following the Rural Queries for Bucks., Beds., and Northants, I assume that the women in 1834 could earn £5 per year, i.e. 2s. per week for 50 weeks. Deflating £5 by the relative change in the male winter agricultural wage in Oxon. between Young's *Tours* and the 1834 *Poor Law Report* yields £3.72.

*Table 13–1. Scale economies in farming (t-ratios in parentheses)*

| | Constant | Acres | Acres squared | Commons | $R^2$ |
|---|---|---|---|---|---|
| *Part A: Labour valued at market prices* | | | | | |
| 1. Arable | 0.77309 | 0.00592 | $-0.57053 \times 10^{-5}$ | | 0.09 |
| | (2.587) | (2.319) | (−1.504) | | |
| 2. Arable | 0.51296 | 0.00621 | $-0.67250 \times 10^{-5}$ | 0.51788 | 0.18 |
| | (1.756) | (2.560) | (−1.860) | (3.795) | |
| 3. Pasture | 0.45569 | 0.00292 | $-0.41591 \times 10^{-5}$ | | 0.04 |
| | (2.903) | (1.929) | (−1.586) | | |
| 4. Pasture | 0.09571 | 0.00485 | $-0.64839 \times 10^{-5}$ | 0.23579 | 0.28 |
| | (0.646) | (3.601) | (−2.817) | (6.138) | |
| *Part B: Labour valued at shadow prices* | | | | | |
| 5. Arable | 1.91306 | 0.00223 | $-0.16242 \times 10^{-5}$ | | 0.03 |
| | (6.635) | (0.914) | (−0.448) | | |
| 6. Arable | 1.56952 | 0.00269 | $-0.29531 \times 10^{-5}$ | 0.62323 | 0.20 |
| | (5.778) | (1.212) | (−0.890) | (5.002) | |
| 7. Pasture | 1.62995 | −0.00212 | $0.21244 \times 10^{-5}$ | | 0.04 |
| | (10.472) | (−1.399) | (0.813) | | |
| 8. Pasture | 1.22371 | 0.00018 | $-0.06923 \times 10^{-5}$ | 0.26018 | 0.35 |
| | (8.725) | (0.141) | (−0.316) | (7.173) | |

*Note*: The dependent variable in the regressions is Ricardian surplus per acre.

The social profitability of enclosure can also be measured by valuing labour at its social opportunity cost. This recalculation does not change my previous evaluation of enclosure in the arable districts since enclosure had no significant impact on employment in those regions. The agricultural revolution on the heavy clays, for instance, raised output by raising yields without significantly affecting farm labour. GDP rose because the value of output rose; changing the valuation of labour does not alter that. In the light arable district, the efficiency gains were small anyway, and they do not get much larger by changing the value of labour. Enclosure for the purpose of growing turnips and clover made a negligible contribution to efficiency, either private or social.

In the pasture district, the gains from enclosure are reduced when labour is valued at its shadow price since reductions in employment were an important cause of the rise in Ricardian surplus. Only the early enclosures in Rutland remain socially efficient. Their superior efficiency was primarily due to the big rise in output per acre. As with enclosed farms in the heavy arable district, this advantage persists when labour is valued at its social opportunity cost. Other enclosed farms in the pasture district were not as productive as these—the returns to enclosure depended on the structure of agricultural prices and were negligible when corn fetched a high price during the French Wars. Since these enclosures mainly caused a reduction in employment, they look an even worse bet when the boost to private profits from displacing labour is reduced by valuing the labour at its opportunity cost. The rent increases following these enclosures were transfers from farm labourers as a class to landlords.


## Conclusion

Agrarian Fundamentalists argue that enclosure and the transition to large farms not only increased agricultural productivity but also contributed to industrialization and overall economic development. Four links have been proposed: agrarian reorganization expanded food production, released labour, increased savings, and provided a home market for manufactures. Agrarian change in England failed to perform the latter two functions. In the seventeenth century, especially, it did lead to an increase in production, and in the eighteenth to a release of labour. These changes created the potential to raise GDP. That potential was realized in the seventeenth century but not in the eighteenth.

There are two reasons that agrarian change made a much more substantial contribution to English economic growth before 1700 than after. First, technical change raised farm output considerably in the seventeenth century. Aside from the draining of heavy clays, the eighteenth and early nineteenth centuries brought no comparable advances. Second, the English population was small enough before the late eighteenth century, so that the people who lost jobs in agriculture could be productively re-employed elsewhere in the economy, particularly in London. The labour market was glutted by the end of the eighteenth century, so the release of agricultural labour resulted in unemployment for many and falling wages for the few who could find work.

This conclusion means that the agrarian changes of the eighteenth century were premature. Bigger farms and enclosure pushed people out of agriculture before there were productive, alternative jobs for them. In continental European countries the agricultural work-force was not run down until the twentieth century, indeed, until after the Second World War. By then, the industrial sector was large enough and wages high enough that the reallocation of labour from farm to factory raised incomes generally. Thus, Denison (1967: 201–24, 335–7) concluded that the transfer of labour from agriculture to industry made a significant contribution to raising growth rates on the continent between 1950 and 1962. In England, agriculture shed labour early, before it had productive alternatives. As a result, GDP did not rise and the standard of living of the rural population fell. In France, agriculture shed labour late, when productive alternatives were widespread; GDP rose sharply, as did the standard of living of the farmers and displaced workers.[20]

[20] See O'Brien and Keyder (1978) for a discussion of this issue.

# PART V

# The Distribution of the Benefits of Technical Progress

# 14

# Who Gained from the Agricultural Revolution?

> But in like manner as, through a proportionate excess of land, rent may
> be almost unknown; and of capital, profits may be annihilated; so in the
> case of labourers, excess of numbers has a *tendency* to reduce their
> wages to the borders of famine; the other classes in either case gaining
> by the loss of one, so long as industry is uncorrupted by hopelessness,
> and security maintained without excessive cost.
>
> J. Fletcher, Secretary to the Commission on Handloom
> Weavers, BPP 1840: xxiv. 195

NOT everyone gained from the agricultural revolution. Indeed, most of the
benefits of agricultural productivity growth accrued to landlords as rising
rents—consumers, farmers, and labourers gained barely, if at all. Marxist
Agrarian Fundamentalists explain this in terms of the impact of enclosure
and farm amalgamation on output and employment. This analysis is correct
as far as it goes, but it is incomplete. Tory Fundamentalists explain the
inequality trends with Ricardo's theory of income distribution, Young's
theory of capital-intensive agriculture, and Malthus's principle of population. The first is problematic over periods as long as a generation, and the
other two are erroneous. The main reasons that the gains of agrarian change
were channelled to landlords were, first, that the output increases of the
agricultural revolution were too small to meet the growth of demand so real
prices rose in the long run; and, second, real wages fell from 1450 to 1600
and remained roughly constant thereafter. Rent—the difference between
farm revenue and cost—was the only free variable, and it expanded to
absorb all the income gains of technical progress and rising product prices.

## Trends in Rural Inequality, 1450–1850

I analyse changes in inequality in terms of people's relationship to
agriculture. There were four classes of potential beneficiaries: the consumers
of farm products and landowners, labourers, and the owners of capital.

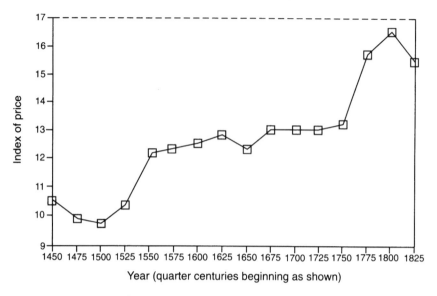

FIG. 14–1  Real Price of Farm Products, 1450–1825

Consumers would have benefited had the price of farm products *decreased*, while the other three groups would have benefited had the rent of land, the wage of labour, or the return to capital, respectively, *increased*. Figures 14–1 to 14–4 show the histories of these prices from 1450 to 1850.[1] Clearly, only landowners benefited from the technical progress.

Consumers did not benefit since real product prices rose over the long term, although at an irregular rate (Figure 14–1). While real prices dropped in the late fifteenth century, they rebounded in the early sixteenth to post a 15 per cent rise between 1450 and 1550. For the next two centuries, which include the yeomen's agricultural revolution, real prices rose very slowly, if at all. After 1750, when the landlords' revolution was in full swing, they grew rapidly, reaching a peak during the Napoleonic Wars. Prices stayed high after Waterloo, but at a slightly lower level than during the Wars.

The rising real price of food burdened the poor especially since they spent a higher share of their income on food than did the rich. As a result, the inequality of purchasing power increased across English society as a whole. This is the reverse of the pattern characterizing agricultural development today. Thus, Timmer, Falcon, and Pearson (1983: 286) report that in poor countries 'improved technology in food production during the past thirty years has benefited the poor significantly, primarily through cheaper food'.

---

[1]  See Appendix IV for the sources and methods by which these indices were constructed.

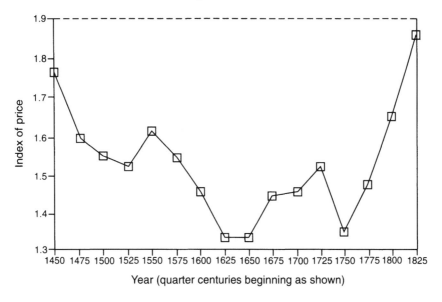

F<small>IG</small>. 14–2   Real Rental Price of Animals and Tools, 1450–1825

This favourable pattern of development was not anticipated by the English agricultural revolution, especially the landlords' revolution.

The owners of labour and capital also failed to reap gains. Thus, farmers (as owners of implements and livestock) made no sustained advance (Figure 14–2) since the real return to farm capital fluctuated within narrow limits over the four centuries. Furthermore, labourers and farmers (to the extent that their income was a return to their labour) did not benefit since the real wage fell sharply and remained low (Figure 14–3).

Landowners are the only remaining category of potential beneficiaries, and indeed, they were the *only* gainers of the agricultural revolution: real rents increased *sevenfold*[2] from the middle of the fifteenth century to the middle of the nineteenth (Figure 14–4). In earlier chapters I showed that enclosure and farm amalgamation often raised rents given an unchanging structure of farm prices. From 1450 to 1850 real product prices rose, however, which gave a further boost to agricultural incomes, so rents rose even more than my earlier calculations suggested. Before the eighteenth century, these gains were probably divided between manorial lords, on the one hand, and small-scale copyholders and beneficial leasees, on the other,

---

[2] In this calculation, I reduced the rents of enclosed land in the 19th cent. by 10 per cent to adjust for the changes in taxation and lease terms that occurred in the preceding 5 centuries. If that adjustment is not made, then real rents rose eightfold. See Allen (1988*a*: 44–5).

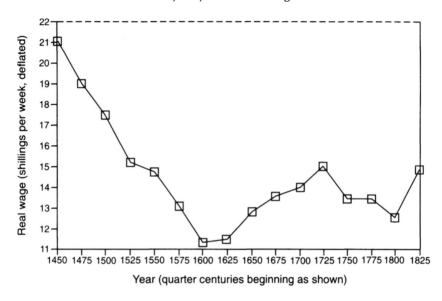

FIG. 14–3 Real Wage Rate, 1450–1825

since it is unlikely that copyholders and beneficial leasees paid fines equal to the full value of their property. As copyholds and long leases were run out in the eighteenth century, however, the gentry and aristocracy became the

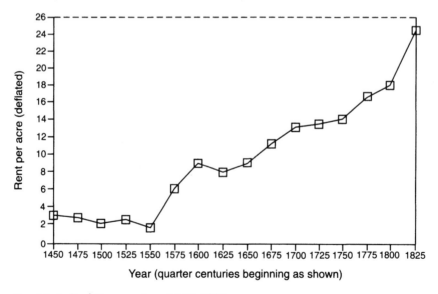

FIG. 14–4 Real Rent per Acre, 1450–1825

exclusive beneficiaries of the rise in land values. All of the gains of the agricultural revolution were thereby channelled into the pockets of this privileged group.

## The Labourer's Standard of Living

Figures 14–1 to 14–4 indicate the main changes in inequality, but they do not reveal all of the details. In particular, the well-being of agricultural labourers declined much more than Figure 14–3 suggests. There were both economic and non-economic aspects, including the following.

### UNEMPLOYMENT

Over the eighteenth and early nineteenth centuries, agricultural employment fell and surplus labour accumulated. It was impossible for all labourers to find work at the prevailing wage. Hence, the welfare of labourers as a group deteriorated even though their wage did not fall.

### ELIMINATION OF COMMONS

Commons provided grazing and firewood to many households in open field villages. Enclosure eliminated commons, and awarded land in compensation. If a family had access to the common by virtue of the ownership of a farm or a cottage, it received an allotment in consequence. Historians have debated whether these awards really equalled the value of the lost commons rights and whether they were swamped by the cost of fencing the allotment. If a family owned no property, it received no allotment. Humphries (1990) has shown that the loss of income was often substantial.[3]

### FAMILY CHANGES

The economic developments of the eighteenth century changed the economic structure of families. The husband, the wife, and the children were all productively employed on the farms of seventeenth-century yeomen. Eighteenth-century farm amalgamations reduced the employment of women and children much more than that of men. The development of surplus male labour further reduced the demand for female and child labour in agriculture since men were given priority over women in hiring. In the nineteenth century, the demand for women's labour in manufacturing also declined, reducing both employment and earnings. Together these developments rendered many rural women and children unemployed, and left

[3] Hammond and Hammond (1932) is the classic discussion of these issues. See also Neeson (1984).

families dependent principally on the earnings of the husband and public charity. These developments reduced welfare in two ways: first, family income declined even though male wages remained constant, and, second, women lost economic independence.

SEASONALITY AND SERVANTS

Agricultural employment became less permanent over the eighteenth century. The decline of service is the most obvious manifestation of this change. In the late seventeenth century, a high share of farm labourers were hired on one-year contracts as servants. They were typically unmarried and received room and board from the farmer. In the first half of the eighteenth century, servants were probably as numerous as day labourers in the south midlands. By 1851, however, servants had almost disappeared. Their place was taken by day labourers (Kussmaul 1981: 11–27).

The shift from servants to casual labourers also denoted an increase in the seasonality of farm employment. Seasonality was partly natural—the harvest in particular implied a sharp rise in the demand for labour—but was also partly artificial. When labour was scarce at the end of the seventeenth century, farmers had to offer steady employment over the year to secure their work-force for the peak periods. The farmers responded to annual hiring by rescheduling work to keep their servants employed evenly over the year. When surplus labour appeared—and with the Poor Law supporting redundant workers—the farmers could schedule work exactly as they chose and without the constraint of providing year-round employment. Hence, irregularity, casualness, and the seasonality of employment increased.[4]

TWO-TIERED WAGE STRUCTURE

The 1834 Poor Law Report argued that the Poor Law, by providing a guaranteed income, eliminated the labourer's incentive to work. The result was low productivity and 'demoralization'. This analysis, however, is simplistic. The farmers could motivate their workers by creating permanent jobs that paid a premium over the Poor Law. Since the poor were so very poor, a small wage gain was enough to create a very desirable job. Indeed, the difference between the winter wage and the level of Poor Law support was often very small. To keep the job, however, the worker had to work hard and be self-supervising. Labour contracts of this form, and for this reason, are very common today in poor agricultural countries (Eswaran and Kotwal 1985).

___

[4] See Hobsbawm and Rudé (1968: 38–48) for a good discussion of this issue. See Boyer (1985) on the Poor Law and seasonality.

To some degree, the possibility of obtaining permanent jobs influenced the behaviour of workers who failed to land them. Perhaps if they did work hard when employed, they would be kept on and thereby reap a small income gain and escape the supervision of the Poor Law officials. But for many, such behaviour was pointless. The result was a class of poorly motivated, unreliable, casual labourers. The Poor Law was only one element—and not the main element—in the emergence of this class and this pattern of behaviour. Since Poor Law policy itself was a response to surplus labour (Blaug 1963, 1964; Digby 1975; Boyer 1986a), the principal causes of 'demoralization' were enclosure, farm amalgamation, and depriving the labouring people of ownership of land.

SKILLS

Reducing the number of farms and reducing their occupiers to landless labourers amounted to the deskilling of a considerable fraction of the rural population. The yeoman had had to manage his farm, and that stimulus to mental exertion was lost when he was reduced to a labourer. As I discussed in Chapter 11, the yeoman farmer had to be able to perform all tasks on his farm. In the nineteenth century, most labourers were ploughmen with more limited skills. Even these skills were difficult to acquire since they could no longer be assimilated by children helping their parents. The deskilling of the female population was even more extensive. Yeomen's wives had run dairies and performed many tasks on the farm; by the nineteenth century, they had no trade. Depriving the yeomen of their land and lowering them to the status of ploughmen (or, as with women, making them redundant) reduced the intellectual demands in their lives leaving those abilities as merely latent talents.

DECISION MAKING

Enclosure also contributed to a narrowing of the mental life of the rural population. The open field village was run as a co-operative. The fields were managed by the farmers of the village; they deliberated the rules within the framework of the manorial court and provided the necessary public officials to manage the system. For that reason, Brailsford (1961: 423) pointed out that Leveller demands for national democracy were an extension of village democracy rather than a whole new departure. As the open fields were enclosed, this political forum vanished. Furthermore, as the yeomen lost their copyholds and beneficial leases, they ceased to be ratepayers and so lost the public responsibilities that followed from that obligation.

Together these developments meant that the condition of the agricultural labouring population deteriorated over the eighteenth and nineteenth centuries even though the real winter wage of men, which is graphed in Figure 14–3, remained constant. Part of the decline was a loss of income, but an important part was a change in mental outlook. Management responsibility, skills, and participation in public life all declined. The incentive that the owner-occupying farmer had to work hard was radically diminished. This reduction in mental demands, narrowing of horizons, and reduction in motivation were important causes of the malaise that upper-class observers described as the 'demoralization' of the agricultural labourer. The symptoms can, of course, be debated, but the standard diagnosis, which blamed the Poor Law for the condition, is surely wrong. The Poor Law was not the cause of the problem—it was caused by the transformation of agriculture.

## The Split Between Consumers and Producers

I will analyse why the benefits of the agricultural revolution were channelled to landlords, in two steps. The first step is to explain why product prices rose over the long term, so that agricultural producers gained at the expense of consumers. The second step is to explain why landlords, rather than farmers or labourers, received the augmented income.

There are two possible explanations for the rise in product prices that reflect two contradictory positions in the literature about the influence of international trade on the English market. At one extreme are writers like Mokyr and Savin (1976: 202), and Williamson (1985: 111) who argue that England was a small part of a large 'world' market. English producers could sell all they chose at the world price. Consequently, the English price equalled the world price and did not depend on English supply and demand. Changes in English supply and demand resulted in changes in the balance of trade in agricultural products rather than changes in English prices. If this model were true, the long-term history of English prices was independent of the English agricultural revolution. The failure of consumers to benefit from that revolution is explained by the assimilation of the English market to the international market and by the rise of prices in that market.

The other extreme assumption is that the English market was independent of foreign influences, so that English prices depended solely on English supply and demand. This view of the English market is implicit in the analyses of English prices undertaken by Crafts (1976) and Jackson (1985) and discussed in the last chapter. Their models treat the demand facing English producers as depending only on English population and per capita income. Foreign income, population, and prices do not influence the English. These analyses explain the constancy of the real price of English agricultural products between 1660 and 1740 in terms of a slow growth in demand that was met by an increase in output. Prices rose in the middle of the eighteenth century as output growth slackened and demand accelerated. Thus, these models imply that the yeomen's agricultural revolution of the seventeenth and early eighteenth centuries benefited consumers by increasing production and offsetting the tendency for prices to rise as demand grew. No such benefits, however, flowed from the landlords' agricultural revolution of the eighteenth century since it resulted in labour release rather than output expansion. The rapid growth in demand that accompanied the demographic and industrial revolutions of the late eighteenth century resulted in the real price rises shown in Figure 14–1.

Price history provides the strongest evidence for the integration of markets, but this support is far from conclusive. Abel (1980: 304–5) has shown that wheat prices in England, France, Belgium, and Italy rose in a broadly similar pattern from the middle ages to the nineteenth century. There were, however, often wide discrepancies between the countries, which are hard to explain if markets were integrated. John (1976: 59) has shown that the movements of Danzig and London wheat prices were very similar in the first half of the eighteenth century. While suggesting market integration, these correspondences could also result from similar weather patterns, monetary developments, and so on.

The strongest evidence against market integration is the small volume of trade. Only in Holland did trade loom large relative to production. In the first half of the eighteenth century, England provided an appreciable fraction of this grain, but, compared to the English harvest, the quantity was small. In the peak years, England may have exported 10 per cent of its wheat production, but usually net trade was only a few percent of output (John 1976). A few percent is such a small quantity that it is difficult to credit changes in the direction of trade with cancelling out national discrepancies in supply and demand—thus equating prices.

Clearly, more research is needed to determine the effectiveness of international trade in arbitraging European grain markets. The small volume of trade, however, casts doubt on its importance and suggests that English prices reflected mainly English supply and demand. The rise in corn yields can therefore probably take the credit for the constancy of real product prices over the seventeenth and early eighteenth centuries. While the yeomen's agricultural revolution did not lower the real cost of food, it certainly checked its growth, and to that degree prevented inequality from rising as much as it would have otherwise. To this degree, the English agricultural revolution conformed to the pattern of more recent output-augmenting agricultural revolutions. On the other hand, enclosures and farm amalgamations in the eighteenth century failed to raise output much, so food prices—and inequality—rose.

## *The Ricardian–Malthusian Model of Income Distribution*

With productivity increasing and with real agricultural product prices constant or rising, it was possible to increase the returns to farm labour, capital, and land. Yet only the rent of land rose. Both Marxists and Tories have offered explanations, but a satisfactory solution must go beyond them.

The important insight of Tory analysis is Ricardo's treatment of rent as surplus, the difference between revenue and the cost of the inputs applied to the land. Rent rose since revenue increased more rapidly than cost. Malthus's theory of population is the usual Tory explanation for the stagnation of costs, which are treated ultimately as wages. According to Malthus, the population expanded whenever the wage rose above subsistence and continued to increase until the wage was forced back to the subsistence level. Hence, the rise in labour demand that followed the spread of capital-intensive agriculture in the eighteenth century resulted in population growth rather than rising wages. The inevitable result was constant costs and rising revenue—hence rising rent.

The Marxists are, of course, right to point out that the theory of capital-intensive agriculture was often false, but the more fundamental reason for rejecting the theory is the inadequacy of Malthus's demography. His views are not consistent with the population history of developed countries for the nineteenth and twentieth centuries since fertility and population growth have fallen as incomes have risen. The question is how well his theory fits the pre-industrial period. The study of this problem has been transformed by the demographic estimates of Wrigley and Schofield's *Population History of England 1541–1871*. While they concluded that Malthus was right before 1800 (Wrigley and Schofield 1981: 412), other investigators have disputed that result. Several econometric investigations of the Wrigley–Schofield data have failed to show that population adjusted to maintain a constant wage as labour demand changed (Lindert 1985; Lee 1985).

There is an elementary feature of the data that needs emphasis in assessing this debate, and that is the time period at issue. It begins in 1541 (when parish registers—the essential primary source—become available), and ends about 1800. In the perspective of *la longue durée* this was a period in which the real wage was remarkably constant: the collapse in the fifteenth century is excluded as is the rise that began about the middle of the nineteenth century. One reason that the statistical analysis of the Wrigley–Schofield data has become highly technical is that it involves correlating vital rates with a real wage series that exhibits little variation. Far more powerful tests of Malthus's theory of population and wage determination would use the later or earlier data. Of course, Malthus's theory is emphatically rejected for post-1850 England. What of the late medieval period?

Malthus's theory also performs badly before 1550. Both the posited link from high wages to rapid population growth and the link from a growing population to falling wages are extremely problematic. So far as the first is concerned, the outstanding feature of the fourteenth and early fifteenth centuries was the coexistence of exceptionally high wages and a non-buoyant population. As I argued in Chapter 3, the population of the south midlands was lower in the mid-sixteenth century than at the time of the poll tax in the late fourteenth century, and B. Campbell (1981) has argued that this decline characterized England as a whole. Hatcher (1977: 44) agrees that the population fell from 1377 to the mid-fifteenth century but adds that it then began to grow again. However, 'it would be wrong to assume that the population increased rapidly and continuously thereafter, for there are many signs that this early vitality was temporarily undermined some time before the end of the century' (Hatcher 1977: 63). One can try to salvage Malthus by claiming that fertility was at a record high but that the population declined because of the frequent reappearance of plague—a piece of bad luck unrelated to economic conditions (Hatcher 1977: 55–7). But other historians believe that fertility restraint was common (Helleiner 1967: 69–71; Duby 1968: 309–10). There is no evidence on vital rates to settle the dispute, but Malthus's opinions are open to serious question.

Furthermore, the real wage decline that began after 1450 is hard to explain in Malthusian terms since the population was growing so little, if at all. Hatcher (1977: 64–6) warns us that 'we must be careful to distinguish any early tentative upward movements in numbers [in the late fifteenth century] from the well-documented population pressure experienced under the later Tudors and early Stuarts'. Further, even by the mid-sixteenth century, the population was so low that there was 'a great deal of slack' in the economy and England was not 'forced against the final limits of subsistence'. If the population was not growing from 1450 to 1550, or growing only a little, how can rising labour supply account for the huge fall in the real wage? The proximate cause of the decline was a rise in the price level. Lindert (1985) has urged us to consider monetary developments—rather than diminishing returns in agriculture—as an explanation for the fall in real wages. A wider search of explanations is certainly needed.[5]

[5] Lee (1973) has estimated an econometric model that implies that the real wage fell between 1450 and 1550 due to diminishing returns to labour. The data with which the model was estimated indicate that the English population rose from 2.4 million in 1450 to 3.6 million in 1550 (Lee 1973: 606). Such an increase is not supported by the more recent research discussed here.

## Wages in the Surplus Labour Economy

The period of the Industrial Revolution also poses a conundrum for Malthus's and Ricardo's theories of the agricultural wage. In their usual formulations, they assume that wages were determined by the demand and supply for labour; in other words, the wage rose or fell as necessary to clear the labour market. Such a theory is patently inconsistent with the persistent unemployment and constant real wage that characterized the surplus labour economy.

How can the steadiness of wages in the face of rising unemployment be explained? When Sir Arthur Lewis described disguised unemployment in agriculture and the huge number of stalls in petty retailing—each doing a tiny volume of business—he maintained that these activities would generate a subsistence income for all concerned. He added:

Twenty years ago one could not write these sentences without having to stop and explain why in these circumstances, the casual labourers do not bid their earnings down to zero, or why the farmers' product is not similarly all eaten up in rent, but these propositions present no terrors to contemporary economists. (Lewis 1954: 141)

The allusion to Robinson (1933) and Chamberlin (1935), unfortunately, does not explain the phenomenon. Product differentiation may increase the number of firms (and thus jobs), but it cannot prevent the pressure of a growing labour supply from depressing wages. The question remains: What would have kept real wages from falling to zero?

Posing the question like this is unorthodox in view of most recent literature on the history of real wages during the Industrial Revolution, for the consensus has shifted towards the view that real wages were rising. At the outset of the modern debate, Hobsbawm (1957) advanced the 'pessimist' case that 'it is not improbable' that the standard of living of the British working class declined from the late eighteenth century to the 1840s. The attempts by Flinn (1974) and Lindert and Williamson (1983) to bring together wage and price data have suggested that real wages did not change much over the period of the Napoleonic Wars and began to rise by the 1820s. While the magnitude of the increase has been disputed, this evidence appears to exclude the possibility that living standards were deteriorating. However, these studies rely on a relatively narrow set of wages. The wage index of Lindert and Williamson, for instance, includes mainly men employed in agriculture and the modern sector. The experiences of protoindustrialists, retail manufacturers, and women in general are excluded. The sorry history of cotton handloom weavers shows that some

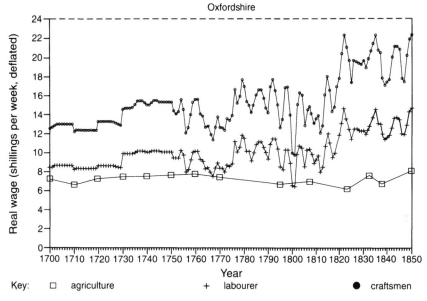

FIG. 14–5  Male Agricultural and Construction Wages, 1700–1850: Oxfordshire

people failed to share in the economic advance. The question is how general that experience was. Recently, Mokyr (1988) has used consumption data to argue that real wages in aggregate were probably not rising until the 1840s.

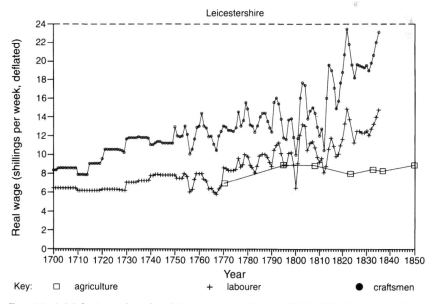

FIG. 14–6  Male Agricultural and Construction Wages, 1700–1850: Leicestershire

Here I will argue that the history of the south midlands contains little evidence to support an optimistic assessment of the Industrial Revolution and much to support an assessment like Hobsbawm's original conjecture of deterioration.

To give the optimists their due, several groups experienced constant— even rising—real wages. Agriculture was perhaps the most important. Figures 14–5 and 14–6 show that the real winter wages of men in agriculture were remarkably stable in Leicestershire and Oxfordshire from 1700 to 1850.[6] The figures also show that real construction wages were constant in Oxford and Leicester during the eighteenth century and then increased (especially in Leicester) in the first half of the nineteenth. Figure 14–7 shows that weavers' wages in Witney fell between 1767 and 1807, but were constant thereafter.[7] The living standards of these trades did not deteriorate, although they did not exhibit that sustained increase in average real earnings that Lindert and Williamson measured for their sample of workers.

The real wages of many domestic workers in the south midlands did fall, however. Framework knitting, the largest male protoindustry, is a prime

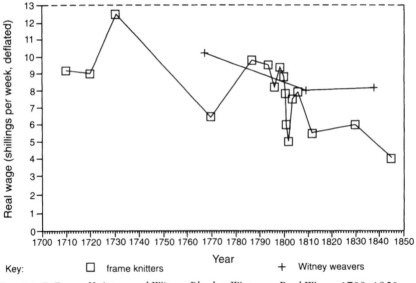

Key:          □  frame knitters                    +  Witney weavers

FIG. 14–7  Frame Knitters and Witney Blanket Weavers: Real Wages 1700–1850

[6] The construction of the real wages in Figs 14–5 to 14–7 is explained in Appendix III.

[7] The fall in the 18th cent. may have been spurious—Young, who recorded the 1767 figure, was impressed that the wage was exceptionally high for weaving. It may represent the response to a temporary surge in the demand for blankets. In that case, the real wage for weavers, like that of farm workers and builders, did not fall in spite of the surplus labour in the region.

example. (See Figure 14–7.) Real wages were broadly constant over the eighteenth century and usually above the real winter wage of farm workers in Leicestershire. However, during and after the French Wars, wages collapsed and were only half of farm labourers' wages in the 1830s and 1840s.

The women's trades, where there was a great growth in surplus labour, also experienced substantial falls in real wages. (See Figure 12–2.) The wages of pillow lace knitters declined by two-thirds over the period. Straw plaiting wages also declined, but the decline was interrupted by the wage surge during the French Wars when high demand drew women, children, and even some men into the industry. This favourable situation was reversed after Waterloo with the renewal of Italian imports. The demand for English plaiters declined, surplus labour reappeared, and wages fell again.

The divergent wage behaviour is indicative of a segmentation of the south midlands labour market in the nineteenth century. There were the privileged sectors (agriculture, weaving, building) where wages were constant and, in the case of agriculture, surplus labour was hoarded, and there were the deprived sectors (framework knitting, pillow lace knitting, and straw plaiting) where wages declined. These sectors were highly competitive, with prices and wages upwardly and downwardly flexible, in response to changes in supply and demand. Unlike agriculture, persistent unemployment was not a problem in these industries, for wages fell to clear the labour market— they thus operated in accord with Chambers's theory of protoindustry. People worked long hours at exceptionally low pay. The full consequences of unrestrained competition in an environment of surplus labour are apparent here.

The puzzle is why wages did not fall in the privileged sectors. The immediate answer is that employers did not cut them. It needs to be emphasized that this sort of behaviour is common both in advanced capitalist economies (where it is rare for non-unionized firms to cut wages even when there are unemployed workers eager to work at a lower wage) and in poor agrarian economies (when farmers persist in paying relatively high wages when unemployed people are willing to work for less). There are two popular theories for this behaviour of employers. The first is custom or tradition, and, indeed, traditional ideas of fair prices and wages were widespread in eighteenth-century England (E. P. Thompson 1971). It is not clear, however, that these ideas were powerful enough to determine wages. They were fading away, and certainly their impact was limited, for traditional wages were not maintained in many trades.

The second explanation for constant wages is that they were in the employers' interest. With workers' incomes at the margin of subsistence, reductions in wages reduced their food consumption and rendered them incapable of performing heavy work. It was more profitable to pay a high wage and have a well-nourished work-force than to pay a low wage to ill-fed and unproductive workers. This theory is widely used to explain wages in developing countries; it is called the 'efficiency wage' theory.

There is considerable evidence in favour of this theory. Thus, pillow lace knitters trying to support themselves on 2s. 6d. or frame knitters trying to support a family on 5s. a week were wretched and not capable of strenuous labour. George Shaw, a Leicester doctor, made this point to the Commission on Framework Knitters in 1845:

How do you regard the frame-work knitters as to their physical powers?—In physical powers I consider the frame-work knitters much below the average of even the manufacturing districts in the north. They make worse soldiers, in fact, which is generally applied to as a proof.

Do you consider that the physical debility of which you have spoken is occasioned by long hours of work, in many instances, more than by irregularity of diet and living?—I think more from the deprivation of the necessaries of life.[8]

People this poor were useless for ploughing fields, moving bricks, or even weaving heavy fabrics. Hence, employers in agriculture, building, and some branches of weaving had an incentive to pay their employees a higher wage than frame knitters earned.

These employers can be thought of as investing in the capacities of their employees. Such an investment was worthwhile to the employer, however, only if he could reap the return. That required preventing the productive employee from moving to another employer. A variety of arrangements were developed to tie workers to their employers.

Consider the problem in agriculture where roughly half the male work-force was employed continuously and the other half was employed only at the harvest. The regularly employed men were paid a high enough wage to maintain their strength. Conceivably a farmer could have paid a harvester enough to live a whole year, but he might leave the village before the next harvest and the farmer's investment would have been lost. The danger was averted by supporting the labourer under the Poor Law. The settlement provisions gave him (and the regularly employed worker) an incentive to remain in the village. Furthermore, collectively supporting all the harvesters meant that individual farmers did not need to maintain continuous relations with individual labourers, who could shift among the farmers without threatening their investments. Contrary to the 1834 Poor Law Report, the

---

[8] BPP 1845: xv, appendix to report, p. 175, questions 3057–8.

old Poor Law probably increased labour productivity by ensuring a minimal standard of nutrition.[9]

Some manufacturing workers also received high wages when strength was necessary for the performance of their duties. J. C. Symons, an Assistant Commissioner on Handloom Weavers, distinguished four categories of weaving. In terms of increasing wages, they were: (1) those requiring neither strength nor skill, (2) those requiring skill but no strength, (3) those requiring strength but no skill, and (4) those requiring both skill and strength. The first category received about 5s. per week in the late 1830s— about half the winter wage in agriculture and about what framework knitters earned. Weaving that required skill received more, but Symons believed that weaving that required strength commanded a higher wage. 'Men cannot work hard without good food, or have good food on low wages.'[10] Coarse broadcloths and blankets required strength, and, indeed, the wage of a Witney blanket weaver was a bit higher than that of an agricultural worker.

If the Witney blanket manufacturers were paying a high wage, they were investing in the productivity of their workers. To realize a return, they had to keep them from leaving. They paid constant wages—'The same wages have been paid upwards of twenty years'—and rarely laid off their employees. Edward Early, a manufacturer, described the arrangement:

About February trade was dull, yet the manufacturers did not dismiss the men, but continued making in order to employ labour. This dullness continued to the end of July. The men were aware of the depression in trade, and feeling a common interest with their masters, took what work could be given thankfully and quietly. They saw the warehouses loaded with goods, they evinced a good sense, joined with good feelings, towards the masters.

The workers valued these practices. Thus Richard Stiff, a weaver, testified:

Work is more regular here, and he prefers this district. Some time ago when trade was slack here he had an offer from Mr Derrett, of Wootton-under-Edge, in Gloucestershire, to work for him; he offered to employ him and his family, and it was calculated that their earnings would be 30s. a-week, besides paying all the expenses of moving; but witness considered the cloth trade to be so unsettled, and the masters to be so very different here to what they are in Gloucestershire with regard to kindness, that he preferred remaining in Witney, though at lower earnings, rather than be subject to a Gloucester master. (*Reports from Assistant Hand-Loom Weavers' Commissioners*, Part v, report from W. A. Miles, BPP 1840: xxiv. 548, 549)

[9] Boyer (1985) develops arguments along these lines.

[10] J. C. Symons, 'Note to be Appended to Mr Symons's Report', *Reports from Assistant Hand-Loom Weavers' Commissioners*, pt. v, BPP 1840: xxiv. 616.

Providing a steady income was the cheapest way of securing a well-fed work-force.

Once surplus labour appeared in the south midlands, supply and demand are not sufficient to explain the level of wages. In the privileged occupations, that is those that required physical strength to do the job, employers paid a wage above the market clearing level. They did this to secure a work-force that was sufficiently nourished to do a hard day's work. A variety of institutions emerged to tie workers to their employers in order to protect the employers' investments. In the disadvantaged trades, where strength was not a consideration, wages fell to very low levels. As neither the employer nor the employee was making an investment in the other, there was no incentive to create long-term attachments and a casual labour market operated. In the casual labour market, supply and demand set the wage. In the privileged labour markets, employers maintained higher wages that did not clear the market. This higher wage was a subsistence wage. Nutritional requirements, not population pressure as in Malthus's model, kept the wage at that level.

## Altering the Income Distribution: the Poor Law

Productivity growth creates the possibility to raise the income of everyone in society. Who actually benefits depends on the particular institutions of the society. In England, the market system and the law of property were the principal institutions that determined the division. Reliance on the market meant that most of the benefits accrued to landowners, and the law of property ensured that there were not many of them.

One strategy to spread the benefits of economic growth more widely would have been to tax the increase in land values and distribute the proceeds equitably over the population. Such a proposal is often favoured by economists on the grounds that it would not distort resource allocation. Indeed, such a proposal was advanced in 1797 by Tom Paine in his pamphlet 'Agrarian Justice'. He argued that all people had a natural right to the value of land, although not to improvements. He proposed taxing that income to provide a pension of £10 per year to everyone over 50 and a payment of £15 to everyone reaching the age of 21 to help them begin adulthood independently. Even the rich would receive these payments 'to prevent invidious distinctions' and because they were a matter of right. Paine anticipated that the plan would cost about £5.5 million per year (1797: 332, 336).

Paine's proposal was not simply fanciful, for it bears some resemblance to the English Poor Law. That law, originally proclaimed in Elizabeth's reign and subsequently much amended, required each parish to support its poor through a local property tax. Indeed, at the time Paine made his proposal about £4 million per year was disbursed in poor relief (Mitchell and Deane 1971: 410). By the late eighteenth century, rates were rising rapidly, and the law was radically amended in 1834 to check this growth. The rich were concerned that high rates would devour the full value of their land. If that were true, then the Poor Law might have redistributed enough income to reverse the inequitable trends in factor returns. Did that happen?

Let us begin with landowners since it was their fears that prompted the question in the first place. Table 14–1 shows poor expense per acre in 1783–5, 1803, 1813, 1821, and 1831 for villages in the south midlands surveyed in the 1834 Poor Law Report. This period encompasses the main rise in expenditures. They rose on average from £0.08 to £0.28 per acre. There was little increase in the single village enclosed before the mid-sixteenth century. (Other evidence suggests this constancy was not spurious.) The rise was most pronounced in villages that were open in 1676. There was not much difference between villages enclosed between 1676 and 1800 and those enclosed thereafter. (Almost all the villages shown in Table 14–1 were enclosed by 1831.)

While rates per acre did triple between 1783/5 and 1831, their magnitude was small compared to the commercial value of the land. In the period 1825–49, land in the south midlands had an average value in excess of £1.5 per acre, so poor rates were about one sixth of the value of the land inclusive of rates. In the early eighteenth century, before the parliamentary enclosures and widespread farm amalgamation, the average rent in the south midlands had been about £0.5 per acre. The increase in rates of £0.2 from 1783/5 to 1831 was only 20 per cent of the rise in land values associated with the agrarian changes of the period. Despite the hysteria about rising rates, they went only a very small way in reducing the income gains of the landed classes.

Table 14–1. *Poor expense per acre, 1783/5–1831 (£ per acre)*

| Enclosure period | 1783/5 | 1803 | 1813 | 1821 | 1831 | Number of parishes |
|---|---|---|---|---|---|---|
| Enclosed before *c*.1550 | 0.07 | 0.07 | 0.12 | 0.10 | 0.11 | 1 |
| Enclosed *c*.1550–1676 | 0.04 | 0.13 | 0.20 | 0.15 | 0.15 | 11 |
| Enclosed 1676–1801 | 0.07 | 0.18 | 0.28 | 0.28 | 0.27 | 40 |
| Open in 1801 | 0.08 | 0.17 | 0.29 | 0.27 | 0.32 | 33 |
| ALL | 0.08 | 0.17 | 0.28 | 0.26 | 0.28 | 120 |

*Note*: A village was reckoned as enclosed if more than 75% of the land was enclosed and as open if less than 25% was enclosed. 'All' villages include 35 additional villages that went through a partly enclosed phase.

If we examine the matter from the point of view of the agricultural labourers, we reach the same conclusion. The simplest test is to compare poor relief expenditures to agricultural wage payments. In Chapter 8 I found that labour costs ranged from about £0.9 per acre in pastoral districts to about £1.4 per acre in regions specializing in corn in the early nineteenth century. Certainly in the corn districts total relief expenditures were small compared to agricultural wage payments. This conclusion would be strengthened significantly if payments to agricultural labourers and their families could be distinguished from payments to the aged, the non-able-bodied, and manufacturing workers since payments to the latter groups were substantial. In the pastoral district, the average poor expenditure of £0.3 per acre is one third of agricultural labour costs, but it must be remembered that the areas most devoted to livestock husbandry were the old enclosures where Poor Law disbursements were only £0.15 per acre or less.

The Old Poor Law did redistribute income from landowners to labourers and thus served to offset the inequitable distribution of the benefits of the agricultural revolution. However, the magnitude of the redistribution was small. It came nowhere close to reversing the increase in inequality.

# 15

# The Yeoman Alternative

The English are still imbued with that doctrine, which is at least
debatable, that great properties are necessary for the improvement of
agriculture, and they seem still convinced that extreme inequality of
wealth is the natural order of things.

Alexis de Tocqueville, *Journey to England*, 1833: 72

AGRARIAN fundamentalists regard inequality as the inevitable price of
growth. In the last chapter, I showed that the landlords' agricultural
revolution—the enclosure movement, the increase in farm size, and the
concentration of landownership—did, indeed, increase inequality. Would it
have been possible to avoid or mitigate those adverse effects while still
realizing the growth in efficiency? This question is worth pursuing for three
reasons. First, the landlords' revolution was unpopular, and many radicals
favoured preserving the yeoman system. Would its preservation really have
improved the lives of most people? Second, even if preserving the yeomanry
is dismissed as ahistorical, contrasting the likely results of that development
path with the actual course of English history highlights the consequences of
the landlords' agricultural revolution. Third, the inevitability of a trade-off
between growth and equity is a widely learned lesson of English history. If it
is false, the record needs to be set straight.

## Radical Proposals

Not everyone believed that enclosures and large farms were a good design
for England's development. In the Tudor period, objections were founded
on Christian morality—enclosures were abuses by the rich who gained at
the expense of the poor. As Sir Thomas More (1516: 48) remarked 'Thus, a
few greedy people have converted one of England's greatest natural
advantages into a national disaster.'

The labouring population of England was usually hostile to the
landlords' agricultural revolution. This disaffection surfaced many times

during the Civil War of 1640 and 1660. Three demands were reiterated in popular protests: the first was the prevention or reversal of enclosures; the second was the enfranchisement of copyholds; the third was an 'agrarian law' to set an upper limit to the property that any individual could own. The Commonwealth rejected these demands and set the basis for the landlords' revolution in the eighteenth century.

Enclosures had provoked popular unrest for two centuries and had prompted royal commissions, statutory regulation, and judicial intervention. The outbreak of the Civil War resulted in riots to break down enclosures throughout the country (Brailsford 1961: 427; Hill 1972: 94). Richard Overton urged that open fields which had been enclosed should be 'laid open again to the free and common use and benefit of the poor' (Brailsford 1961: 432; Hill 1972: 96). This agitation came to naught. After 1649 'the Rump of the Long Parliament did nothing to encourage agrarian reform' (Hill 1972: 44). In 1656 parliament rejected a bill to check enclosures (Brailsford 1961: 426. Cf. Hill 1972: 45; Hill 1961; 149).

The security of tenure of copyholds had also been a source of friction and litigation in the fifteenth and sixteenth centuries. The formula that the courts finally settled on—to enforce manorial custom 'subject to reason'—was only a partial victory for the tenants. Copyholds for lives, for instance, remained ultimately precarious since they were subject to arbitrary fines at renewal and even to non-renewal if the lord chose to repossess the land. During the Civil War there were numerous petitions to secure the tenant's interest by converting copyholds to freeholds. Clause 16 of *A New Engagement or Manifesto* in 1648 proposed

that all the ancient and almost antiquated badge of slavery, viz. all base tenures by copies, oaths of fealty, homage, fines at the will of the lord, etc. (being the Conqueror's marks upon the people) may be taken away; and to that end that a certain valuable rate be set, at which all possessors of lands so holden may purchase themselves freeholders, and in case any shall not be willing or able, that there be a prefixed period of time after which all services, fines, customs, etc. shall be changed into and become a certain rent . . . (Quoted by Brailsford 1961: 440)

John Lilburne added a more radical statement to the second version of the Levellers' *Agreement of the People* in 1648: 'That the next representative be earnestly desired to abolish all base tenures' (Brailsford 1961: 449). The sentiment was repeated in the fourth version of the *Agreement* in 1653, which urged 'all servile tenures of lands, as by copyholds and the like, to be abolished and holden for naught' (Brailsford 1961: 449).

These pleas came to nothing. Levellers had hoped that parliament's disposition of confiscated royalist estates would have provided a first occasion to enfranchise copyholds and leaseholds, but nothing was done (Thirsk 1952: 188–9, 199). On the contrary, the demands of copyholders and leaseholders were ignored. In 1646, the Long Parliament passed a resolution that converted the feudal military tenures into 'free and common socage', that is, modern freehold. This was followed by an act of parliament in 1656 to the same effect. That act expressly excluded copyholds, and the same parliament rejected a bill that would have set an upper limit on the entry fines paid by copyholders. The first Restoration parliament confirmed the Act of 1656 (Hill 1961: 148–9). The upshot was to relieve the gentry and aristocracy of burdens such as wardship while denying any redress to the copyholder. It was not until the Law of Property Act, 1922, that copyholds were converted to freeholds. But by then enfranchisement was too late to have any social significance (Megarry and Wade 1975: 32–7).

If copyholds had been converted to freeholds, the property rights of many small proprietors would have been protected. However, the demesnes of the large landowners would have been unaffected and many peasants who held land on beneficial leases would have remained at risk. The rise of the great estate would have been limited but not prevented. Further, the accumulation of property by peasants who bought up copyholds to achieve estates of several hundred acres would have proceeded unimpeded. To preserve and extend a peasant social structure, it was necessary to forestall both of these developments. A law limiting the amount of property that anyone could own was the answer.

It was Harrington who popularized the idea of an agrarian law in *Oceana* in 1656. He was adapting local traditions—in 1646 there were calls within the New Model Army for an agrarian law (Hill 1972: 92)—but also drawing on the republican tradition that went back to Machiavelli's *Discourses* on the Roman Republic (Hill 1972: 93; Pocock 1975). This work analysed the social structure of Rome to determine how it had elicited so much stoic self-sufficiency and public-spiritedness from its citizens. Machiavelli's answer emphasized direct participation in government and war, the leadership of exemplary statesmen, and widespread landowner-ship. Rome had had an 'agrarian law' to limit the land that any one person owned. While Machiavelli had been critical of this particular institution (Pocock 1975: 208–12), Harrington endorsed the proposal.[1] The belief that egalitarian agrarianism fostered the classical virtues spread widely and suggested that large tenant farms worked by landless labourers were not making England a better country. Republicanism provided a language to discuss these issues and a powerful rationale in support of equality.

---

[1] In a peculiarly hollow fashion. See MacPherson (1962: 182–90).

The Roman ideal of an agrarian law remained popular among radicals and some social philosophers in the late seventeenth and eighteenth centuries. Henry Neville was enthusiastic in *Plato Redivivus* (1681: 96–100, 132–3). Walter Moyle, in *An Essay Upon the Constitution of the Roman Government* (1699: 238), claimed that 'the Licinian Law' to limit a person's property to 500 acres 'established the great balance of the commonwealth, and would have rendered it immortal, had the law been effectually put in execution'. Fletcher advocated that landlords be forced to sell land to small farmers willing to cultivate it (Stafford 1987: 112; Robertson 1983: 143). In 1721, Thomas Gordon (under the pseudonym Cato) wrote: 'Liberty can never subsist without equality, nor equality be long preserved without an *Agrarian Law*' (Jacobson 1965: 91–2). Akenside wrote poetry about the ancient polis:

> Have ye not heard of Lacedaemon's fame?
> Of Attic chiefs in Freedom's war divine?
> Of Rome's dread generals? The Valerian name?
> The Fabian sons? The Scipios? matchless line!
> *Your* lot was theirs. The farmer and the swain
> Met his lov'd parton's summons from the plain.
> The legions gathered; the bright eagles flew;
> Barbarian monarches in the triumph mourn'd;
> The conqu'rors to their household gods return'd,
> And fed Calabrian flocks, and steer'd the Sabine plough.[2]

Hutcheson's *System of Moral Philosophy* (1755: i. 327; ii. 247, 259) argued that an agrarian law was often necessary for a democracy. Mrs Macaulay's *History of England* also advocated agrarian laws (Stafford 1987: 112). In 1782, Ogilvie's *An Essay on the Right of Property in Land* restated the case for an agrarian law (Stafford 1987: 107–20).

For many eighteenth-century radicals and republicans, the lesson of the Roman Republic was that a law to equalize the ownership of property was essential for democracy and public-spiritedness. The application to England was subversive: the rise of the great estate and the destruction of the yeomanry produced authoritarianism and subservience.

These ideas were put into practice in America rather than England. Jefferson's party attempted to perpetuate a society of small owner-occupying farmers. His draft constitution for Virginia, written in 1776, provided that public lands be distributed so that 'Every person of full age neither owning nor having owned fifty acres of land, shall be entitled to an appropriation of fifty acres or to so much as shall make up what he owns or

---

[2] *Odes*, XII. ix, as quoted by Stewart (1855: viii. 141).

has owned [to] fifty acres in full and absolute dominion. And no other person shall be capable to taking an appropriation' (Wiltse 1935: 137). Jefferson bought Louisiana from Napoleon to ensure that America remained forever the land of the yeoman: 'The small landholders are the most precious part of a state' (quoted by Wiltse 1935: 138).

English radicals remained deeply concerned about agrarian issues throughout the first half of the nineteenth century, but they no longer advocated yeoman agriculture.[3] The transition to capitalist farming was nearly over, and most of the land was already enclosed. Godwin (1798: 714) dismissed 'agrarian laws . . . as remedies more pernicious than the disease they are intended to cure'. (This conclusion reflected his aversion to any form of coercion, as well as the changing times.) Paine, as discussed in the last chapter, favoured taxing the pure rental value of the land without eliminating the great estate. Thomas Spence, an influential thinker, advocated the collective ownership of land at the village level. Rents would be collected by the parish and distributed among the inhabitants. Further, everyone would be guaranteed work in agriculture (Chase 1988: 28–30).

## An Assessment of the Radical Programme: Material Effects

The radical proposals I am concerned with are those that were current in the seventeenth and early eighteenth centuries—in particular, the prevention of enclosures, the enfranchisement of copyholds, and an agrarian law to set an upper limit to landholding. If enacted, a land reform of this sort would have forestalled the landlords' agrarian revolution and preserved the peasant social structure. How successful would the radical programme have been in improving the lives of farmers and labourers?

In answering this question, I will suppose that the radical programme was accomplished in its most extreme form—let even ancient enclosures be converted back to open fields and planted with corn; preserve not only small farms, but divide the demesnes that were operated as large units since the middle ages. Indeed, I will contrast midlands agriculture as it actually was in 1800 with the way it might have been had the region consisted only of 40-acre farms organized in open fields. Further, I will suppose that landlordism was abolished and the farms were owned by their occupiers. This agrarian reorganization would have had both material and non-material effects. I begin with the former, which would have included changes in employment, family income, and rents.

[3] As late as 1805, Hall discussed an agrarian law as an ideal but impractical solution. See Stafford (1987: 151).

The reorganization of midlands agriculture into 40-acre open field farms would have substantially eliminated the rural unemployment problem in the first half of the nineteenth century. There were 104,856 male farmers and agricultural labourers in the south midlands in 1831. Half of these were infrequently employed. (See Table 12–4.) Had the south midlands been divided into 40-acre open field farms, most of these men would have been employed full-time in agriculture: as Table 15–1 indicates, the full-time employment of men would have expanded from 54,976 to 98,966. The employment opportunities for their wives and children would have improved even more, since their employment had fallen off even more rapidly than that of men as farm size had increased. The family pattern of an irregularly employed husband with a wife and children generally unemployed would have been replaced by one in which most members of agricultural households were employed.

The agricultural reorganization would probably have increased total farm income. The change in farm size would not have affected output. The effects of preserving (and restoring) open fields would have varied with the district. The production of corn would have increased and the production of meat fallen, but the overall effect would have been small and in the direction of increasing total production.

If tenancy were abolished and the ownership of the small, open field farms vested in the farmers, then the incomes of the rural population would have risen very substantially since they would have received the rental value of the land as well as wage income actually paid. Reference to the farm income statements in Chapter 9 (Tables 9–2 to 9–4) indicates that rent per

*Table 15–1. Labour absorption in agriculture*

|  | Number employed in the south midlands |
|---|---|
| *Men* |  |
| 40-acre, open field farms | 98,966 |
| Early 17th-century distribution | 79,135 |
| Early 19th-century distribution | 54,976 |
| *Women* |  |
| 40-acre, open field farms | 74,329 |
| Early 17th-century distribution | 52,148 |
| Early 19th-century distribution | 32,902 |
| *Boys* |  |
| 40-acre, open field farms | 75,185 |
| Early 17th-century distribution | 50,210 |
| Early 19th-century distribution | 27,507 |

*Sources*: For the early 17th century and early 19th century: Table 12–1.
40-acre, open field farms computed with equation 1 in Table 11–5.

acre in open field farms amounted to about £1 per acre, while labour costs were £1 in pastoral farms to £1.5 in arable farms, so the recreation of peasant proprietorship would have increased the total incomes of rural agricultural families by between 67 per cent and 100 per cent.[4] This increase, indeed, equals the rise in income from both the increase in employment and the redistribution of landownership.[5]

The material benefits for the labouring population of restoring peasant proprietorship were substantial. The magnitude of this increase illustrates the inequitable distribution of income that actually obtained in early nineteenth-century England. The radical programme has much to commend it as a device for reducing rural poverty.

## An Assessment of the Radical Programme: Non-material Effects

Radicals also claimed that a society of yeoman farmers had political, social, and psychological advantages. They believed that political equality was effective only if accompanied by economic equality. They also thought that the ownership of property developed the mental abilities of the labourer. These ideas hark back to Machiavelli's concern with eliciting public-spiritedness and virtue. By the nineteenth century, however, the quality of the mental life of the rural population took on a greater urgency. The upper classes were distressed that the Poor Law led to the 'demoralization' of the labouring population, which, in turn, led to more poverty. John Stuart Mill (1848) reformulated the traditional republican concerns in response to this problem. He traced 'demoralization' to the loss of land and argued that it would have been avoided had peasant proprietorship been maintained.

Mill believed that peasant proprietorship would enhance the mental life of England's labourers in four ways. First, their enthusiasm for work would increase. All observers have noticed that peasants work hard, and even Arthur Young attributed this to their ownership of property. Second, the mental abilities of the rural labourers would be developed if they were given land and forced to manage farms. The peasant proprietor 'is no longer a being of a different order from the middle classes; he has pursuits and objects like those which occupy them, and give to their intellects the greatest part of such cultivation as they receive'. In contrast, 'the day-labourer has, in the existing state of society and population, many of the anxieties which have not an invigorating effect on the mind, and none of those which have'.

---

[4] This increase is an understatement since the reconversion of pasture to arable would have raised the wage bill in the pastoral districts.
[5] These calculations assume no change in product prices.

The labourer feared for the loss of his job, but he remained 'a child, which seems to be the approved condition of the labouring classes according to the prevailing philosophy' (Mill 1848: 285–6). Third, the ownership of property would stimulate the peasant's sense of 'prudence, temperance, and self-control' (Mill 1848: 286). Fourth, Mill believed that peasant proprietorship promoted a low birth rate.

Mill's claims for the non-material benefits of peasant proprietorship are not as grand as those of other reformers. Mill claimed only that making the labourers into bourgeois would have inculcated the attitudes that the middle classes found essential to carrying out their roles. This is not an implausible conjecture. If true, then peasant proprietorship might have made some contribution to combating the subservience and alienation of the labourer's lot.

## Conclusion

Enclosures, capitalist farms, and the great estate were the distinctive features of England's agricultural development. Since the days of Arthur Young, they have been proclaimed as the institutional breakthroughs that raised productivity and propelled England into the industrial era. The claims, however, are serious distortions of the facts.

There were two agricultural revolutions in England's history. The big rise in output and much of the increase in labour productivity were accomplished by small-scale, owner-occupying farmers in the yeomen's revolution of the seventeenth and early eighteenth centuries. Imposed upon this revolution was the more famous landlords' revolution, which consisted of enclosures, farm amalgamations, and the concentration of property ownership. This revolution made only a small contribution to the growth in output, but it did release labour. Unfortunately, the redundant workers were not re-employed in industry but remained unemployed. The contribution of the landlords' revolution to the growth in national income was, therefore, negligible.

Despite the fact that the landlords' revolution made little contribution to economic development, it made them rich. The cessation of output growth raised the price of food and, therefore, the value of agricultural output. During the eighteenth century, labour became a surplus factor of production. The wage represented the minimum expenditure to keep the labourers strong enough to work. Only land was in restricted supply, so it was the only input to command a scarcity value. As a result, the income gains from the agricultural revolution accrued to landowners.

The injustice of this outcome raises the question of whether there were other patterns of development with more desirable properties. In this chapter, I have argued that preserving and extending the yeoman system of production would have raised the well-being of the rural poor. Their employment and incomes would have increased substantially. The 'demor-alization' that middle-class reformers complained of would have been alleviated. These improvements could have been realized without sacrificing economic growth. Indeed, agricultural output might have grown faster in the eighteenth century if the yeomen had retained ownership of their land: his greater proprietary interest gave the yeoman a greater incentive to increase production than tenancy at will gave the capitalist farmer.

While re-enforcing the yeoman social system may seem an unusual notion, it is in the mould of many successful twentieth-century rural development programmes in poor countries. Land reforms in countries like Taiwan and South Korea have eliminated landlords and created systems of small, owner-occupied family farms. These countries have achieved high rates of output growth. The benefits of this growth have been much more equally distributed than in countries like the Philippines that have witnessed the concentration of landownership in the hands of the elite and a concomitant rise in tenancy. Indeed, as it has become clear that industry in many poor countries is incapable of providing a large increase in employment, labour absorption (rather than labour shedding) in agriculture has become a policy objective. The yeomen's agricultural revolution in England was consistent with this style of peasant-oriented development, while the landlords' revolution was not.

# Regional Patterns of Peasant Land Tenure

THE yeomen held their lands by copyholds of inheritance, copyholds for lives, and beneficial leases. While there are probably examples of all forms in all counties, there were distinct regional concentrations in the south midlands. Thus, copyholds of inheritance were most common in the eastern part of the region—in Cambridgeshire, Huntingdonshire, and Bedfordshire, for example. Copyholds for lives tended to be the norm in Warwickshire, Oxfordshire, and Berkshire. In Leicestershire and Northamptonshire, there was a mixture of beneficial leases and copyholds of inheritance. In addition, demesnes tended to be held by beneficial lease throughout the region. Kerridge (1969: 32–64) and Clay (1985: 198–214) provide general overviews on the geographical distribution of tenures. In this Appendix, I report evidence bearing on the south midlands in particular.

Copyholds of inheritance tended to be common in the eastern counties. Thus, Parkinson (1811: 31) observed that 'rather more than half the county [of Huntingdon] is of a freehold tenure, the remainder, (with the exception of about one thirtieth part of the whole, which is leasehold) is copyhold'. His parish-by-parish returns indicate a mixture of fixed and arbitrary fines (Ibid. 29–31). Gooch wrote of Cambridgeshire that 'every kind of tenure is in this county', Batchelor wrote of Bedfordshire that 'copyhold estates are numerous in some parishes' (Marshall 1818: iv. 580, 632). Clay (1985: ii. 200) summarized the situation: 'In the eastern part of the country, and in most of the Midlands, copyhold was normally hereditary.'

The importance of heritable copyhold in the eastern part of the south midlands is corroborated by the evidence of many villages. Thus, Spufford (1974: 75–6, 101, 158) found that copyhold of inheritance was the standard customary tenure in the Cambridgeshire parishes of Chippenham, Orwell, and apparently also in Willingham. Copyholds were heritable in the manors of Ramsey Abbey, which were located mainly in Huntingdonshire and Cambridgeshire (Raftis 1964: 198–204). The *Victoria County History* volumes for Cambridgeshire report seven cases of heritable copyhold— Great Abington (vi. 10), Hildersham (vi. 64), Duxford (vi. 210), Ickleton (vi. 239), Sawston (vi. 254), Bassingbourn (viii. 19, 20), and Foxton (viii. 171). The *VCH* also indicates that Great Abington had copyholds for life in the sixteenth century, as did Croxton (v. 40/1). Copyholds for lives were not otherwise noted in Cambridgeshire.

In contrast, copyhold for lives was common in Warwickshire, Oxfordshire, and in the Vale of White Horse, as Clay (1981: 83; 1985: v, pt II, p. 200) noticed. Young (1813: 472) wrote of Oxfordshire: 'Freehold and copyhold leases for lives remain here: and church and college leases, both for lives and years, abound greatly.' Murray (1813: 26) and Mavor (1813: 52) also noted the existence of copyhold in Warwickshire and Berkshire without describing its form. Copyhold for lives extended into Buckinghamshire as Priest (1813: 24) reported. Middle Claydon, Buckinghamshire, was an example discussed by Broad (1972).

The *Victoria County History* of Oxfordshire provides many examples of copyholds for lives, in particular: Chesterton (vi. 97), Hampton Poyle (vi. 164), Lower Heyford (vi. 188), Islip (vi. 213), Kirtlington—1 life (vi. 226), Wendlebury (vi. 343), Weston-on-the Green—lifehold (vi. 350), Clifton Hampden (vii. 22), Dorchester (vii. 47), Burcot (vii. 68), Drayton St Leonard (vii. 77)—also life leasehold, South Stoke (vii. 101), Great Milton (vii. 131, 135), Adwell (viii. 5), Pyrton (viii. 164), Shirburn (viii. 188, 190), Stoke Talmage (viii. 204–5), Watlington (viii. 228), Hanwell (ix. 117)—copy and lifeholders, Horley and Hornton (ix. 129), Shenington (ix. 145), Tadmarton (ix. 154), Wroxton (ix. 181), and Neithrop (x. 54). Upper Heyford (iv. 201), Burcot (vii. 68), Thame (vii. 191–2), and Duns Tew (xi. 217) were also probably copyholds for lives, but the text is not clear.

Some copyholds of inheritance were also found in Oxfordshire. Examples include Ambrosden (v. 21), Launton (vi. 236), Lewknor (viii. 108), Bloxham (ix. 68–9), Drayton (ix. 107), Horley and Hornton (ix. 129), Charlbury (x. 141), South Newington (xi. 152), and Wootton (xi. 276). In addition to the villages with leases for lives noted above, this tenure also was found in Horton (v. 69), Holton (v. 174), Hampton Poyle (vi. 164), Wendlebury (vi. 343), Dorchester (vii. 47), Shirburn (viii. 190), Stoke Talmage (viii. 205), and Great Tew (xi. 237).

Finally, leases for long terms of years were reported in Holton—70 years (v. 174), Piddington—2,000 years (v. 255), Fringford—2,000 years (vi. 129), Hampton Gay—99 years (vi. 157), Mixbury—21, 40 years (vi. 256), Drayton St Leonard—40 years (vii, p. xx), Waterstock—21 years (vii. 224), Bloxham—40 years (ix. 66), Prescote—21 years (x. 209), North Aston—61 years (xi. 13), Glympton—30 years (xi. 127), Duns Tew—21 years (xi. 217).

As the lists indicate, some villages included more than one tenure. The variety reflects differences in policy for demesnes and customary lands and in changes from one form to another.

Beneficial leases were common in three circumstances. First, demesnes were often let on beneficial leases. Second, copyholds for lives were sometimes replaced by beneficial leases in the western part of the south midlands; Clay (1985: 203) says this was done after the mid-seventeenth century but the persistence of copyhold of inheritance into the eighteenth century shows that replacement proceeded slowly. Third, beneficial leases were a common way in which customary land was held in the upland pasture district—in Northamptonshire, Leicestershire, and the adjoining portions of Rutland, although copyhold was also important. It was usually of inheritance. Pitt (1809*b*: 23) wrote of Northamptonshire that 'The tenures of land, in this country, are, I am informed, mostly freehold; but, there are some copyhold and some leasehold.' He wrote virtually the same thing about Leicestershire (Marshall 1818: iv. 213). Parkinson (1808: 23–4) reports both freehold and copyhold land in parishes in the upland part of Rutland. In the seventeenth century, most of the freehold land was probably let on beneficial leases.

Studies of landholding reveal a mixture of copyhold of inheritance and beneficial leases in Northamptonshire, Leicestershire, and adjoining parts of Rutland and Warwickshire in the Tudor/Stuart period. Thus, Lennard's (1916) study of the manors of the Honour of Grafton in the seventeenth century showed that leasehold predominated in eight manors (Grafton, Hartwell, Ashton, Greens Norton, Grimscott, Higham Ferrers (mostly), Holdenby, Stoke Bruerne), copyhold of inheritance in six (Brigstocke, Irchester, Kingscliffe, Raunds, Rushden, Little Welden), and copyhold for lives in two (Moorend and Potters Pury).[1] Two others had very little land or mixed arrangements (Pury and Chalveston cum Caldecott). The leases usually ran for three lives during the reign of Elizabeth and 31 years later (Lennard 1916: 32, 120). Finch's (1956) study of *The Wealth of Five Northamptonshire Families, 1540–1640* gives the impression of a predominance of leasehold land but also notes the existence of copyhold land. The estates discussed by Finch include manors in Rutland and Leicestershire as well as Northamptonshire. The apparent importance of beneficial leases may be due to the five families' being leading enclosers. Hoskins' (1957: 89–115) study of Wigston Magna, Leicestershire, shows that the customary tenants in the sixteenth century were copyholders who then bought the freeholds of their property from their lords in the late sixteenth and early seventeenth centuries. In the case of Oxford Manor, it was sold off to the tenants because they were so successful in litigation against their lord as to render the manor almost valueless. Howell (1983: 63) reports that the

---

[1] There was some leasehold land in both of these manors and the copyhold customs of Potters Pury allowed family members to redeem land after the last life ran out.

sixteenth-century tenants of Merton College in Kibworth Harcourt, Leicestershire, were copyholders who came to hold their land officially on 21-year beneficial leases after 1593 since the manor was owned by an Oxford college. Parker's (1949: 47) study of Cotesbach found that the customary tenants were leaseholders—indeed, leaseholders with expired leases who consequently could not offer a strong defence against enclosure.

APPENDIX II

# The Batchelor–Parkinson Farm Model

THE purpose of this Appendix is to document the method used to produce
the tables comparing open and enclosed villages in terms of land use,
livestock density, labour costs, farm outputs and their values, and the net
income of farm operations in Chapters 6, 7, 8, and 9. As indicated there, the
basic source of data on land use, livestock, seeds, methods, yields, etc., was
the census-like surveys of Rutland and Huntingdonshire executed by
Parkinson (1808, 1811). Some statistics, such as land use and livestock
density reported in Chapter 6, could be computed directly from these data.
The production, labour, and financial statistics reported in the subsequent
chapters required that Parkinson's data be supplemented with information
from other Board of Agriculture reports, in particular Batchelor's *General
View of . . . Bedford* (1808).

The calculation of revenues and costs is not completely straightforward.
The following notes summarize the conventions I have followed.

## Revenue: Crops

I compute the value of production of each crop as the acreage planted,
multiplied by the yield, multiplied by the price. Acreage and yield are from
Parkinson and the price from Batchelor (1808: 107).

## Revenue: Animals and Firewood

Calculating the output of animal products was more difficult than
calculating the output of crops. As with most agricultural censuses,
Parkinson reported the number of each kind of animal, but not the amounts
of meat, wool, cheese, and butter that they produced. I used information
from other Board of Agriculture reports to compute the production per
animal and multiplied those production rates by the corresponding number
of animals to compute farm output. Sheep and cows provided the greatest
difficulties since the flocks and herds were replenished through breeding and
the animals were kept over their full life cycles, so output varied with the age
structure of the flock or herd. I dealt with this problem as follows.

Sheep are the simpler case. For each village Parkinson reported the number of sheep and lambs.[1] Sheep produced meat and wool. Wool production equalled the number of sheep multiplied by the average weight of a fleece. Parkinson reported both values, so the calculation of wool output is simple arithmetic.

I computed mutton output on the assumption that the sheep were managed as self-sustaining breeding flocks.[2] Hence, there were two sources of meat: the female lambs born each year replaced old ewes for breeding, and the culled ewes were sold for mutton. The male lambs were castrated and fattened. These animals (the wethers) were the second source of meat. Evidently, the total production of mutton equalled the number of lambs born each year multiplied by the average quantity of mutton obtained from a wether and a culled ewe. I took these weights from Board of Agriculture reports and livestock manuals. I varied the weights depending on the breed of sheep and the system of management.

The dairy presented more formidable problems. Parkinson recorded the number of calves and cows. These categories included animals of several ages producing several products. Pitt (1809*a*: 227) tells us that in Leicestershire

Dairy farmers generally breed their own cows. Those who wish for a quick return, send their heifers to the bull at two; those who chiefly regard the improvement of the breed, do not send them till they are three years old. Good cows will continue to breed and to give milk till they are 20 years old, but they lose in value after 8 years old, and of course are generally put to fat at about that age.

'Put to fat' meant not breeding the cow but grazing it for a year in preparation for slaughtering it for beef. Priest (1813: 302) rendered a similar account for Buckinghamshire: '*Calves*—Generally are sold from the dairy farms to sucklers [for veal], at an age from four to twelve days old, except a few which are kept to supply the dairy; for the dairymen either breed or buy heifers for the dairy at three years old, and sell them at seven or eight years old.' I interpreted Parkinson's number of 'calves' to be animals one and two years old and 'cows' to be three to seven years old, producing milk and calves for four years, and fattening the last year prior to slaughtering.

[1] Parkinson (1808: 126–8 and 1811: 238–42) reports the numbers of sheep and lambs separately for Rutland but only the combined total for Hunts. I used the average ratio of sheep to lambs in Rutland to divide up the Hunts. totals.

[2] I presume that losses from disease were made good by purchasing new animals rather than by reducing sales. Hence, in my calculations, animal mortality impinges on the cost side of the income statement rather than the revenue side.

Given this interpretation of Parkinson's numbers, the production of dairy products can be computed. Eighty per cent of the cows—those not put to fat —produced butter or cheese and a calf each year. I assigned different production rates of butter and cheese per cow depending on the region and whether the village was open or enclosed. In addition, animals were sold at three ages for meat. The first and oldest group were the fatted, superannuated cows: the number equalled one fifth of the total number of cows. The second group were two-year-old calves in excess of the number required to keep the herd size constant.[3] The third group were one-week-old calves— again the excess over those needed to maintain the stock of one- and two-year-old animals. I took the weights of these animals from Board of Agriculture reports.

I based my estimates on the value and quantity of animal products and the cost of livestock on the following discussions:

Sheep—Parkinson (1811: 45, 61, 70, 76 [note error], 77, 247), Pitt (1809*a*: 280–2), Parkinson (1810: i. 314), Monk (1794: 24, 33), Crutchley (1794: 14), Vancouver (1794: 205), T. Stone (1794: 31–2), Batchelor (1808: 537, 557–9), Culley (1807: 102–17).

Calves—Priest (1813: 302), Young (1807: ii. 278), Pitt (1809*a*: 227), Batchelor (1808: 94).

Sucklers—Pitt (1809*a*: 232), Young (1807: ii. 289, 377), Parkinson (1810: i. 26–39).

Cows—Pitt (1809*a*: 227–33), Young (1813: 270–83), Batchelor (1808: 527–30), Priest (1813: 301–2), Parkinson (1811: 60, 269).

Fatting cattle—Pitt (1809*a*: 234–7).

Store cattle—Parkinson (1811: 60, 70–1).

I estimated the production of firewood from hedges based on the acreage of enclosed fields.

## Seed Cost

In the case of wheat, barley, oats, and beans, I computed the cost of seed as the acreage planted multiplied by the sowing rate multiplied by the price of the crop. For clover, turnips, and tares, I multiplied the acreage by the cost per acre of seed, as given by Batchelor.

---

[3] That is, half the calves less one fifth of the cows. For most villages, the excess was positive and small. There was no significant movement of heifers into or out of the south midlands. This feature of the calculations substantiates the assumptions about the age structure of the dairy herds.

## Labour Cost

I listed the tasks in farming and estimated their cost by multiplying the acreage or number of animals involved by the appropriate labour cost rate. I discuss the procedure in Chapter 8 and give many examples there.

## Capital Cost of Livestock

The capital costs of livestock consisted of two components. The first is interest. I charged the farm 5 per cent interest on the value of all livestock; Young (1770) used this value in calculations of farm profits. The value of each kind of animal was taken from Batchelor; the numbers were taken from Parkinson. The second is the change in value. The value of horses declined as they aged, so I charged the farm depreciation on their value. The other stock appreciated unless it died. Batchelor estimated mortality rates for the various sorts of animals, and I charged the farm an amount equal to the expected loss from mortality. I charged the farm the purchase price of stock that was bought, fattened, then resold.

## Other Livestock Costs

Farmers also incurred various non-capital costs from their livestock. A major one was the value of oats eaten by horses. (This was shown in Tables 7–4 to 7–12.) In addition, there was shoeing, harness repair, patent medicines, and the hire of rams for stud. Colts were charged the decline in the value of the mare's work effort during pregnancy. Batchelor's estimates of these costs per animal were multiplied by the numbers of animals in Parkinson's census to compute the total costs.

## Implements

The farm was charged interest and depreciation for its implements—its ploughs, carts, harrows, barrels, etc. Batchelor listed and valued the implements necessary for 150 acres of arable land; interest and depreciation on that sum per acre of arable was used to estimate arable implement costs. Villages that folded their sheep were charged interest and depreciation on hurdles. Likewise the capital cost of equipment associated with the dairy was estimated as the number of cows multiplied by interest and depreciation on Batchelor's estimate of the value of dairy equipment per cow.

# APPENDIX III

# Sources of Real Wages

THE purpose of this appendix is to describe the sources for the real wage series graphed in Figures 12–1, 12–2, 14–3, 14–5, 14–6, 14–7. In all cases, the real wage equals the nominal wage divided by a consumer price index. I will first list the sources for the nominal wages and then describe the consumer price index.

## Nominal Wages

### MEN IN AGRICULTURE, OXFORDSHIRE AND LEICESTERSHIRE

The wage is the winter wage from Bowley (1898: 704–7) except for 1833, which is abstracted from the 1834 Poor Law Report. I checked Bowley's figures against the original sources to be sure of using winter wage rates inclusive of payments in kind.

### CONSTRUCTION WAGES, OXFORDSHIRE

The wage of building labourers is from Phelps Brown and Hopkins (1955: 205). For the eighteenth century, the wage rate is for Maidstone, which Phelps Brown and Hopkins note was equal to the Oxford wage both before and after. For the nineteenth century, the wage rate is an interpolation based on London wages.

### CONSTRUCTION WAGES, LEICESTERSHIRE

The wage of building labourers as summarized in VCH, Leicestershire, iv. 185.

### FRAMEWORK KNITTERS

The wage is net average weekly earnings. In some cases, I estimated the deductions from gross earnings to compute net earnings. Most figures are for worsted knitters in Leicestershire, but sometimes in Nottinghamshire. James (1968: 211), Henson (1831: 104), Cowper (1741: 8), Chambers (1966: 295), Wells (1935: 79, 87, 91, 93), *Report from the Select Committee [on] the Framework-knitting Trade* (BPP 1812: ii. 59–61).

*Report from the Select Committee [on] the Exportation of Machinery* (BPP 1841: vii, evidence of Felkin, *Report [on] the Condition of the Framework Knitters* (BPP 1845: xv, evidence of Felkin, p. 41). Much other evidence in the parliamentary reports corroborates the nineteenth-century fall in earnings.

BLANKET WEAVERS

Young (1768: 100), Young (1813: 326), evidence of John Earley to the Commission on Handloom Weavers. (*Reports from Assistant Hand-Loom Weavers' Commissioners*, part V, report by W .A. Miles, BPP 1840: xxiv. 546–52.)

STRAW PLAITING

The wage rate is really the earnings (reported daily or weekly) of a 'good' straw plaiter. The main source was Pinchbeck (1930: 216, 221) with additional quotations from Eden (1797) and Batchelor (1808: 594).

PILLOW LACE

As with straw plaiting, the wage rate is really an estimate of the earnings of a 'good' lacemaker. For 1700, I used the earnings for 1699 given by Spenceley (1973: 92). The other earnings figures are from Pinchbeck (1930: 207–8) with the addition of Batchelor (1808: 593).

## Consumer Price Index

The consumer price index (cpi) spans the full period 1450–1850 and is used to deflate the agricultural output and input price series in Figures 14–1 to 14–4 as well as to deflate the wage series listed above. Indeed, three series with slightly different weights were computed for the periods 1450–1658, 1659–1760, 1761–1850 and linked together using average values for a five-year overlap period (1659–64 and 1761–5). The change in weighting reflects the introduction of tea, sugar, and potatoes into the labourer's diet. Within each of the three periods, the weights were unchanged. I will first explain the weighting scheme and then discuss the sources of the price series that were averaged.

## Weights

Following D. Davies (1795) and Eden (1797), I concocted a stylized budget for an agricultural labourer's family in southern England. The budget assumes that the family consists of a man, his wife, and four children. The budget does not reflect actual spending patterns but rather reformers' ideas about the appropriate standard of living for the poor. Thus, Davies (1795: 16) remarked of one budget: 'Very few poor people can afford to lay out this sum in clothes [£6], but they should be enabled to do it.' The consumer price index based on Table III–1, therefore, indexes a poverty line rather than anyone's actual spending pattern.

Appendix Table III–2 lists the various series and the weights implied by Table III–1. I used proxy commodities for a few items: thus, the wheat price series is weighted by the combined expenditure on flour, yeast, and salt; the barley series is weighted by expenditure on malt, hops, beer (at births), and gin. The cloth price series is weighted by expenditures on clothes, thread, and the baby's linen. Finally, I used the wage rate as a proxy for the price of ministers' services, nurses, midwives, medicines, and burials.

Sugar, tea, and potatoes entered working-class diets during the early modern period. To accommodate them, I reallocated the expenditure on

*Table III–1. Labourers' weekly budget*

|  | Weekly consumption and price | Weekly expense | Yearly expense |
|---|---|---|---|
| Flour | 42 lb. = 6 gallons @ 10d. | 5s. 0d. | |
| Yeast | | 2d. | |
| Salt | | 1d. | |
| Bacon | 1 lb. @ 8d. | 8d. | |
| Cheese | ½ lb. @ 6d | 3d. | |
| Butter | ½ lb. @ 8d. | 4d. | |
| Tea | 1 oz. @ 2s. 8d./lb. | 2d. | |
| Sugar | ½ lb. @ 8d. | 4d. | |
| Soap | ⅓ lb. @ 9d. | 3d. | |
| Candles | ⅓ lb. @ 9d. | 3d. | |
| Beer | 1 peck malt @ 1s. 4d./bushel | 4d. | |
| | ¼ lb. hops @ 4d. | 1d. | |
| Potatoes | 1 peck @ 6d. | 6d. | |
| Thread | | 3d. | |
| | | 8s. 8d. | = £22 10s. 8d. |
| Firing | | | 6s. 0d. |
| House rent | | | 1 15s. 0d. |
| Clothes | | | 5 12s. 11d. |
| Medicine and burials | | | 0 10s. 0d. |
| Birth (every two years) | | | |
| children's linen | 3s. 6d. | | 1s. 9d. |
| beer | 10d. | | 5d. |
| gin, 1 bottle | 2s. | | 1s. 0d. |
| midwife | 5s. | | 2s. 6d. |
| nurse | 5s. | | 2s. 6d. |
| minister for christening | 1s. | | 6d. |
| TOTAL expense per year | | | £31 3s. 3d. |

those commodities in the late eighteenth century to other commodities for the years before 1761. These reallocations imply slightly different weights, as indicated in Table III–2.

## Sources of Price Series

The computation of the consumer price index (cpi) requires times series of thirteen variables between 1450 and 1850. The sources are detailed below. In general the series contain many interpolated values and (especially between 1640 and 1749) impute average values over a decade to every year in the decade. The cpi does not, therefore, provide a good measure of price changes from one year to the next, although it does provide an accurate comparison of the price level over longer periods. Since wage rates did not vary from year to year, it is probably better, in any event, to deflate them with a cpi that reflects the level of prices typical of the period of the wage quotation, rather than an index that is sensitive to transitory fluctuations in prices.

WHEAT, PORK, CHEESE, BARLEY, WAGE RATE

These series are the same as those described in Appendix IV.

*Table III–2. Price series and weights*

| | Expense (d./yr.) | Weights | | |
|---|---|---|---|---|
| | | 1761–1850 | 1659–1760 | 1450–1658 |
| Wheat | 3,276 | 0.437 | 0.479 | 0.479 |
| Pork | 416 | 0.056 | 0.056 | 0.056 |
| Cheese | 156 | 0.021 | 0.021 | 0.021 |
| Butter | 208 | 0.028 | 0.028 | 0.028 |
| Tea | 104 | 0.014 | — | — |
| Sugar | 208 | 0.028 | 0.028 | — |
| Candles | 312 | 0.042 | 0.042 | 0.042 |
| Barley | 277 | 0.037 | 0.051 | 0.079 |
| Potatoes | 312 | 0.042 | — | — |
| Cloth | 1,532 | 0.204 | 0.204 | 0.204 |
| Firing | 72 | 0.010 | 0.010 | 0.010 |
| House rent | 420 | 0.056 | 0.056 | 0.056 |
| Wage rate | 186 | 0.025 | 0.025 | 0.025 |
| TOTAL | 7,479 | | | |

*Source*:
Expense: Table III–1, expense per year in pence
Weights: 1761–1850: expense divided by 7,479 (total expense);
        1659–1760, 1450–1658: expenses reallocated.

BUTTER

1450–1649: Bowden (1967: 839–45).

1650–1749: Bowden (1985: 843–6).

1750–1826: Beveridge (1939: 576).

1827–1849: *Report on Wholesale and Retail Prices in the U.K.*, BPP 1903: lxviii, series for Royal Hospital, Greenwich.

TEA

1761–1829: Beveridge (1939: 435), green tea.

1830–1849: extrapolated forward with the series of the price of tea inclusive of duty in the *Report on Wholesale and Retail Prices in the U.K.*, BPP 1903: lxviii. 176–7.

SUGAR

1659–1829: Beveridge (1939: 429–31), 'powder or Lisbon'.

1830–1849: extrapolated forward with the sugar price series in *Report on Wholesale and Retail Prices in the U.K.*, BPP 1903: lxviii.165.

CANDLES

1450–1582: Rogers (1866–1892: iv (1882), pp. 376–80).

1583–1702: Rogers (1866–1892: v (1887), pp. 398–404).

1703–1829: Beveridge (1939: 146–7).

1830–1849: extrapolated forward using McCulloch (1880: 1138–1140).

POTATOES

1760–1829: Beveridge (1939: 427), 'old potatoes'.

1830–1849: extrapolated forward with John's (1989: 984–5) potato series for Preston.[1]

CLOTH

1450–1583: Rogers (1866–1892: iv (1882), pp. 583–8). cloth, second quality, per panni of 24 yards.

1584–1700: Rogers (1866–1892: v (1887), pp. 588–91) Cambridge choristers' cloth.

1701–1729: I extrapolated the series forward using Beveridge (1939: 194) —the price of broadcloth for scholars.

1729–1849: I extrapolated the series forward using Tucker (1936: 78–80). I interpolated the missing values.

[1] I thank Dr Joan Thirsk for kindly making available a copy of John's appendix in advance of publication.

FIRING

My index of the price of firing (i.e. fuel used for heating and cooking) reflects the changing availability of 'free' fuel taken from commons. Young's (1771*a*: iv. 286–7; 1771*b*: iv. 308–10) tours are the points of departure. He collected the cost of firing in many villages he visited, and in many cases it amounted to 30 or 40*s.* per year. However, Young (1771*b*: 309–10) observed that such figures were

> more the expense of the really industrious part of the poor, who expend something in firing, rather than the average of any neighbourhood; for there is scarcely any in the Tour, where great numbers among them, do not depend for this article totally by pilfering, breaking hedges, and cutting trees; and this so general, that if a real average had been gained, I do not apprehend it would amount to 10 s. a year.

In constructing my series of the price of firing, I set its cost at 6*s.* per year in 1780. I extrapolated this figure back to 1450 using the wage rate on the grounds that the wage was the opportunity cost of a labourer's time spent collecting wood. I assumed that by 1849, labourers were totally dependent on purchased fuel. The budget surveys in *Returns Relating to the Poor Law Amendment Act* (BPP 1837–8: xxxviii. 2–3) indicate that agricultural labourers in Leicestershire and Lincolnshire were spending about 1*s.* per week on fuel in the 1830s. I assumed that such expenditures amounted to 52*s.* per year in the 1840s and interpolated values between 6*s.* and 52*s.* for the years between 1780 and 1840.

HOUSE RENT

My rent series is based on estate surveys, tours, and agricultural reports. These provide quotations or assessments of cottage rents for a few years scattered over the four-century span. I assigned typical values from these sources to the years to which they applied and interpolated missing values. I derived cottage rents from the *Grey of Ruthin Valor* of 1467–8 (Jack 1965), a 1552 survey of Cropredy, Oxfordshire (Royce 1879: 9–13), surveys of Bedfordshire manors in the Wade-Grey manuscript of 1623–8 (Beds. RO, CTR 110/34), surveys of manors in Buckinghamshire, Oxfordshire, and Berkshire in 1728 (Bodleian Library MS Top Oxon, c.381), Young's tours (1771*a*: iv. 286–7; 1771*b*: iv. 308–10), Davies's returns (1795: 19, 139, 161, 165, 175, 177), Eden's returns (1797: iii, pp. cccxxxix-cccl), Board of Agriculture county reports including Gooch (1813), Young (1813), Mavor (1813), Parkinson (1808), Pitt (1809*b*), Batchelor (1808), and Murray (1813), and Caird's tour of 1851–2 (1967: 113, 437, 468, 472, 474).

Other data corroborate the resulting series. Since a 'cottage' was a house with attached grass, the rental value of a cottage was an average of the rental value of a structure and of grass land. It is reassuring that my series of the rent of cottages inflated more rapidly than the wage rate (which may indicate the rate of inflation of the cost of structures) and less rapidly than the rent of enclosed land in the pasture district, as indicated in Table 9–1.

APPENDIX IV

# Sources of Agricultural Output and Input Price Indices

I CONSTRUCTED time series of agricultural prices, wages, the rental price of capital, and rents and used them for two purposes. The first was to compute 'real rents' as an indicator of total factor productivity growth in Chapter 11. The second was to construct Figures 14–1 to 14–4, which map out changes in income distribution. (Note that Figures 11–1 and 14–4 both plot 'real rent', but that the series are different. The rent series is deflated differently in the two cases since it is intended to answer different questions.)

## The Price of Agricultural Output, 1450–1849

The price index of agricultural output is a weighted average of the price of nine commodities: wheat, barley, oats, beans, wool, beef, mutton, pork, and cheese. Farms produced other products like peas and butter, but I assumed that their prices changed like those of beans and cheese. I interpolated many missing values, especially for early years. I explain the weighting scheme later; here I give the sources of the price series.

WHEAT, BARLEY, OATS, BEANS

1450–1639: I used Bowden's (1967: 815–21) national indices of the prices of wheat, barley, oats, and beans for 1450–1649 to extrapolate the corresponding average south midlands prices for 1640–9 back to 1450.
1640–1749: Bowden (1985: 864, 865, 867, 868). I applied the decennial averages to every year in the decade. I used the values for the south midlands.
1750–1770: I extrapolated the average values for 1740–9 forward using price indices kindly made available to me by Dr P. K. O'Brien.
1771–1849: I used the London *Gazette* prices as given in the manuscript version of John (1989: 974–5). I filled in the missing prices from summaries of the London *Gazette* prices in the *Annals of Agriculture* and *Gentleman's Magazine*.

WOOL

1450–1639: I assumed that the average value for wool in 1450–99 was 8s. per todd of 28 lb. in view of the quotations for southern England in Lloyd

(1973: 43–4). 1450–99 is the base period for Bowden's (1967: 839–45) index of the price of wool, and I used that index to extrapolate the price forward.

1640–1749: I used Bowden's (1985: 855) decennial average indices of the price of wool to continue the earlier series forward. I linked them at the overlap in 1640–9.

1750–1820: I interpolated the average price for 1740–9 forward using O'Brien's index of the price of wool.

1821–49: John's (1989: 990–3) first price of southdown wool expressed in terms of a todd of 28 lb.

BEEF

1450–1540: I took the price to be ½d. per pound (0.58s. per stone of 14 lb.) based on Rogers (1866–1892: iv (1882), p. 332).

1541–59: 1d. per pound (Rogers 1866–1892: iv (1882), p. 332).

1562–82: Rogers (1866–1892: iv (1882), p. 333).

1583–1699: Rogers (1866–1892: v (1887), pp. 347–53).

1700–1749: I assumed the average price was 3.5s. per stone of 14lb. for the period 1640–1749 and interpolated values for 1700–49 using the indices of decennial averages given in Bowden (1985: 855). Prices were quite stable over the period, and this procedure provides a good link with the earlier values.

1750–1820: I extrapolated the average price for 1740–9 forward using O'Brien's index of the price of beef.

1821–49: average price paid by St Thomas's Hospital as reported in *Accounts Relating to Wheat*, BPP 1850: lii. 313.

MUTTON

1450–1639: I treated the price of beef as the price of mutton. Their prices tend to be similar later.

1640–1749: I assumed that the average price was 3.5s. per stone of 14lb. for the period 1640–1749 and interpolated values for that period using the indices of decennial averages given in Bowden (1985: 855). Prices were quite stable over the period, and this procedure provides a good link with the earlier values.

1750–1820: I extrapolated the average price for 1740–9 forward using O'Brien's index of the price of mutton.

1821–49: average price paid by St Thomas's Hospital as reported in *Accounts Relating to Wheat*, BPP 1850: lii. 313.

PORK

1450–1639: I treated the price of beef as the price of pork. Their prices tend to be similar later.

1640–1749: I assumed that the average price was 3.5s. per stone of 14 lb. for the period 1640–1749 and interpolated values for that period using the indices of decennial averages given in Bowden (1985: 855). Prices were quite stable over the period, and this procedure provides a good link with the earlier values.

1750–1820: I extrapolated the average price for 1740–9 forward using O'Brien's index of the price of pork.

1821–30: Beveridge (1939: 426).

1831–9: interpolated.

1840–9: price of 'large hogs', *Accounts Relating to Wheat*, BPP 1850: lii. 310–12.

CHEESE

For 1450–1639, I used Bowden's (1967: 839–45) index of the price of cheese. I extended the series to 1750 with Bowden's (1985: 855) national decennial averages. I applied each average to every year in the decade. I extended the series to 1830 with the series for Cheshire cheese prices given in Beveridge (1939: 428–9). For 1831–49, I extrapolated the series forward using the price of Cheshire cheese from John (1989: 1002–3).

## The Rental Price of Farm Capital, 1450–1849

I distinguished six kinds of farm capital for which I computed rental prices: implements, horses, cows, beef cattle, sheep, and pigs. The overall rental price of farm capital was a weighted average of these six prices. The weights are described later. I computed the rental prices of implements, horses, and cows by multiplying an index of the price of each item by an interest rate plus 10 per cent for depreciation. I computed the rental price of beef cattle, sheep, and pigs as an index of the price of each multiplied by the interest rate. I did not assess depreciation for these inputs since they were usually sold after a year and, indeed, appreciated in value over the period the farmer held them. In all of these calculations, the interest rate used was the legal maximum lending rate as summarized in Homer (1977: 117, 126, 157). The sources for the purchase prices of the inputs were as follows.

IMPLEMENTS, 1450–1849

I assumed that the price of implements moved in proportion to the wage rate, the sources for which are given below.

SHEEP, BEEF, CATTLE, COWS, 1450–1749

For these animals, I used the decennial average prices for wethers, bullocks, and cows reported by Bowden (1985: 871, 873, 874) for the south midlands. I assigned the decennial averages to every year in the decade and interpolated missing values based on trends in adjoining regions. I extended these series back to 1450 using the national index numbers for cattle and sheep computed by Bowden (1967: 822–8).

HORSES, PIGS, 1450–1749

I used Bowden's (1985: 854) decennial average national indices (which I applied to every year in the decade) and extended them backwards using Bowden's (1967: 834–8) indices of the prices of horses and pigs. I assumed that the average price of a pig was 20s. in 1740–9 to convert the indices to actual values that could be linked to values after 1749. For horses, I used Bowden's (1967: 869) base price for 1450–99, which provided a good fit with post–1750 values.

ALL ANIMALS, 1750–1849

I extended all of the series forward from 1749 to 1849 by using values reported for livestock in Young (1771*a*, 1771*b*), Batchelor (1808), Pitt (1809*a*, 1809*b*), Parkinson (1808, 1811), Low (1838), Tomson (1847: 36), Wratislaw (1861: 169, 170, 177), and Morton (1855: i. 379–88).

## The Wage Rate, 1450–1849

1450–1639: The average computed by Bowden (1967: 864).

1640–1749: I took the agricultural wage rate to have been 12*d*. per day throughout. This seemed appropriate in view of the scattered information reported in Bowden (1985: 877).

1750–1849: I interpolated the wage rate based on the sources for the winter wage of men employed in agriculture in Oxfordshire described in Appendix III.

## The Rent of Land, 1450–1849

I computed the average rent of land in the south midlands as the average rent of open and enclosed land in each of the three natural districts. Rents per acre at quarter-century intervals from Table 9–1. I interpolated a few missing values. I estimated the shares of open and enclosed land in each natural district from the data set of village enclosure histories underlying Figure 2–1.

## Weights for the Price Indices of Farm Products and Capital

I derived weights from the farm accounts developed by Young in his Eastern Tour (1771*b*: iv. 455–62). Young also developed accounts in the Northern tour which imply slightly different weights. It was necessary to reorganize Young's accounts to convert capital costs into flows of capital services and to eliminate intermediate products like turnips. The result is shown in Table IV–1.

*Table IV–1. Young's national farm accounts*

|  | Amount (£) |
| --- | --- |
| *Revenue* | |
| Wheat | 16,770,048 |
| Barley | 10,495,540 |
| Oats | 5,635,645 |
| Peas/beans | 4,049,888 |
| Dairy products | 4,078,426 |
| Sheep—meat | 8,000,000 |
| —wool | 4,943,551 |
| Fatting and young cattle | 3,479,501 |
| Swine | 1,283,400 |
| TOTAL | 58,745,999 |
| *Costs* | |
| Labour | 15,989,191 |
| Seed: | |
| wheat | 1,746,880 |
| barley | 1,147,949 |
| oats | 630,301 |
| peas/beans | 519,072 |
| clover | 800,356 |
| turnips | 85,561 |
| Oats for horses | 6,297,317 |
| Capital costs: | |
| implements | 4,704,000 |
| horses | 1,026,737 |
| cows | 778,609 |
| fatting and young cattle | 490,553 |
| sheep | 832,086 |
| swine | 51,336 |
| TOTAL | 35,099,948 |
| *Surplus to land* | 23,646,051 |

*Notes:*
Surplus to land = revenues minus costs.
The capital cost of implements, horses, and cows equals the value of each kind of stock multiplied by the interest rate (5%) plus an assumed depreciation rate (10%). The other capital costs equal the value of the stock multiplied by the interest rate.

*Source:* Young (1771*b*: iv. 455–62).

I computed weights for the agricultural product price index as the value of output of each commodity in Table IV–1 divided by the total value of farm output, i.e. the share of wheat equals 16,770,048 ÷ 58,745,999. I used the share of fatting and young cattle as the weight for beef and the share of swine as the weight for pork. Likewise, I computed weights for the index of the price of capital as the ratio of the cost of each item to the total cost of capital, e.g. the share of implements was 4,704,000 ÷ 7,883,321.

## Computing Real Rents in Figure 11–1

The real rent series plotted in Figure 11–1 equals the nominal rent of land divided by an index of the price of farm outputs and inputs. This deflator was a weighted average of the farm product and farm capital indices just described and the wage rate. The weights were the corresponding farm revenues and costs divided by the surplus to land in Table IV–1.

## Computing Figures 14–1 to 14–4

The indices of agricultural product prices, wages, the rental price of capital, and rents were deflated with the consumer price index described in Appendix III.

# BIBLIOGRAPHY

*Manuscripts*

Bedfordshire County Record Office
Bedford; de Grey; Honour of Ampthill; Ossory; Wade-Grey; surveys and maps; land tax assessments; glebe terriers; parish history abstracts.

Birmingham Central Library, Local Studies Department
Surveys and maps.

Bodleian Library, Oxford
Bertie; Harcourt; MS Top Oxon.

British Museum
Harleian Manuscripts.

Buckinghamshire County Record Office
J. G. Jenkins papers; surveys and maps; Land tax assessments; glebe terriers.

Cambridgeshire County Record Office
Land tax assessments; glebe terriers.

Cheshire Record Office
Leases and maps.

Hertfordshire Record Office
Enclosure awards; tithe awards; maps.

Humberside Record Office, Beverley
Enclosure awards; surveys and maps.

Huntingdon Record Office
Manchester; Connington; Torkington; Earl of Sandwich; enclosure awards; land tax assessments.

Kendall Record Office
Maps.

Lancashire Record Office
Glebe terriers; leases and maps.

Leicester County Record Office
Surveys and Maps; land tax assessments.

Lincolnshire Archives Office
Enclosure awards.

Northamptonshire Record Office
Andrew of Addington; Cartwright of Aynho; Fitzwilliam of Milton; Dryden of Canons Ashby; Wake of Courtenhalle; Montague of Weekly; Brudenell of Deene;

Tyron of Bulwick; Honour of Grafton; Wake Collection; surveys and maps; enclosure awards; land tax assessments; glebe terriers.

Northumberland Record Office
Enclosure awards; surveys, leases, and maps; Bell.

North Yorkshire County Record Office
Enclosure awards; surveys and maps.

Nottingham Record Office
Enclosure awards; maps.

Oxford County Record Office
Dillon; probate inventories; land tax assessments; parish register abstracts.

Public Record Office, London
Tithe commutation files; 1607 depopulation returns; 1795 inquiry concerning yields.

Sheffield Public Library
Enclosure awards; Fairbank collection.

Staffordshire Record Office
Surveys and maps.

University Library, Cambridge
Glebe terriers.

Wallington House, Northumberland
Estate maps.

Warwickshire County Record Office
Land tax assessments.

William Salt Library, Stafford
Compton Census.

Worcestershire Record Office, St Helen's
Enclosure awards; glebe terriers; surveys and maps.

*Published Maps*

Berkshire: John Rocque, *A Topographical Map of the County of Berks*, 1762.

Cumberland: T. Donald, *The County of Cumberland*, 1774.

Dorset: Isaac Taylor, *The County of Dorset, Reduced*, 1796.

Durham: *The County Palatine of Durham Surveyed by Capt. Armstrong and Engraved by Thomas Jeffreys*, 1769.

Essex: John Chapman and Peter André, *A Map of the County of Essex*, 1774.

Hampshire: T. Kitchin, *A New Improved Map of Hampshire*, 1753; W. Faden, *Hampshire*, 1796.

Hertfordshire: A. Drury and J. Andrews, *A Topographical Map of Hartford-Shire*, c.1766.

Kent: J. Andrews, A. Drury, and W. Herbert, *A Topographical Map of the County of Kent*, 1769.

Northumberland: Armstong & Son, *A Map of the County of Northumberland*, 1769.

Oxford: Richard Davis, *A New Map of the County of Oxford*, 1797.

Somerset: W. Day and T. Masters, *The County of Somerset, Surveyed by Day and Masters*, 1783.

Sussex: T. Yeakell and W. Gardner, *An Actual Topographic Survey of the County of Sussex*, 1778–1783.

Westmorland: *The County of Westmoreland, Surveyed Anno MDCCLXVIII and Engraved by Thomas Jeffreys*, 1770.

*British Government Papers*

JOURNALS OF THE HOUSE OF COMMONS

'An Act for the Better Cultivation, Improvement, and Regulation of the Common Arable Fields, Wastes, and Commons of Pasture, in this Kingdom', 13 Geo. III, c. 81, in Danby Pickering (ed.), *The Statutes at Large* (Cambridge, John Archdeacon, 1773), pp. 253–61.

*Abstract of Answers and Returns Pursuant to Act 41 Geo. III, for taking an Account of the Population of Great Britain in 1801*, BPP 1801–2, vi, vii.

*Abstract of Answers and Returns under Act 43 Geo. III, relative to the Expense and Maintenance of the Poor in England*, BPP 1803–4, xiii.

*Report from the Select Committee [on] the Framework-knitting Trade*, BPP 1812, ii.

*Abstract of the Population Returns of Great Britain, 1831*, BPP 1833, xxxvi–xxxviii.

*Report from His Majesty's Commissioners for Inquiring into the Administration and Practical Operation of the Poor Laws*, BPP 1834, xxvii–xxxix.

*Report from the Select Committee [on] Handloom Weavers*, BPP 1834, x, 1835, xiii.

*Returns Relating to the Poor Law Amendment Act*, BPP 1837–8, xxxviii.

*Reports from Assistant Hand-Loom Weavers' Commissioners*, BPP 1839, xlii, 1840, xxiii, xxiv.

*Report from the Select Committee [on] the Exportation of Machinery*, BPP 1841, vii.

*Report of the Commissioners on Handloom Weavers*, BPP 1841, x.

*Report . . . into the Condition of the Framework Knitters*, BPP 1845, xv.

*Accounts Relating to Wheat, Barley, Oats, Flour, Butcher's Meat, Pork, Bacon, Cattle, Sheep, Swine, Butter, Cheese, Foreign Wool, Gold, and British Shipping*, BPP 1850, lii.

*Census of Great Britain, 1851*, BPP 1852–3, lxxxv–lxxxix.

*Report on Wholesale and Retail Prices in the U.K.*, BPP 1903, lxviii.

Board of Agriculture and Fisheries, *Report on the Decline in the Agricultural Population of Great Britain, 1881–1906*, Cd. 3273, BPP 1906, xcvi.


*Other Government Papers*

*Census of the Canadas*, 1851–2.

Institut Internationale d'Agriculture, *Annuaire internationale de statistique agricole*.


*Books and Articles*

ABEL, Wilhelm (1980). *Agricultural Fluctuations in Europe from the Thirteenth to the Twentieth Centuries*, trans. Olive Ordish (London, Methuen).

ADDINGTON, S. (1772), *An Enquiry into the Reasons For and Against Inclosing Open Fields* (Coventry, J. W. Piercy).

ALLEN, Robert C. (1982), 'The Efficiency and Distributional Consequences of Eighteenth Century Enclosures', *Economic Journal*, 92: 937–53.

——(1983), 'Collective Invention', *Journal of Economic Behavior and Organization*, 4: 1–24.

——(1986), 'Enclosure, Depopulation, and Inequality in the South Midlands, 1377–1801', University of British Columbia Department of Economics Discussion Paper No. 86–36.

——(1988a), 'The Price of Freehold Land and the Interest Rate in the Seventeenth and Eighteenth Centuries', *Economic History Review*, 2nd series, 41: 33–50.

——(1988b). 'Inferring Yields from Probate Inventories', *Journal of Economic History*, 38: 117–25.

——(1988c), 'The Growth of Labour Productivity in Early Modern English Agriculture', *Explorations in Economic History*, 25: 117–46.

——(1989a), 'Enclosure, Farming Methods, and the Growth of Productivity in the South Midlands', in George Grantham and Carol S. Leonard (eds.), *Agrarian Organization in the Century of Industrialization: Europe, Russia, and North America*, (*Research in Economic History*, Supplement 5), pp. 69–88.

——(1989b), 'Le due rivoluzioni agrarie inglesi, 1450–1850', *Rivista di storia economica*, NS, 6: 257–82.

——(1990), 'Agrarian Fundamentalism and English Agricultural Development', *Rivista di storia economica*, NS, 7: 153–61.

——(1991), 'The Two English Agricultural Revolutions, 1450–1850', in B. Campbell and M. Overton (eds.), *Productivity Change and Agricultural Development* (Manchester, Manchester University Press), pp. 236–54.

——and O GRADA, Cormac (1988). 'On the Road Again with Arthur Young: English, Irish, and French Agriculture during the Industrial Revolution', *Journal of Economic History*, 38: 93–116.

ALLISON, K. J., BERESFORD, M. W., and HURST, J. G. (1965), *The Deserted Villages of Oxfordshire* (Leicester University Department of English Local History, Occasional Paper 17).

——(1966) *The Deserted Villages of Northamptonshire* (Leicester University Department of English Local History, Occasional Paper 18).

*Annals of Agriculture and Other Useful Arts* (1784–1815), vols. 1–46 (London).

ANSCOMB, J. W. (n.d.), 'Notes on the Parliamentary Acts and Awards for Northamptonshire', deposited in Northamptonshire Record Office.

ARBUTHNOT, John (1773), *An Inquiry into the Connection Between the Present Price of Provisions, and the Size of Farms* (London, T. Cadell).

ARMSTRONG, W. A. (1981), 'The Influence of Demographic Factors on the Position of the Agricultural Labourer in England and Wales, c.1750–1914', *Agricultural History Review*, 29: 71–82.

AUSTEN, Jane (1813), *Pride and Prejudice* (New York, The New American Library, 1980).

BACON, Richard Noverre (1844), *The Report on the Agriculture of Norfolk* (London, Ridgways).

BAIROCH, P. (1965), 'Niveau de développement économique de 1810 à 1910', *Annales: économies, sociétés, civilisations*, 20: 1091–1117.

BAKER, A. R. H. and BUTLIN, R. A., (eds.) (1973), *Studies of Field Systems in the British Isles* (Cambridge, Cambridge University Press).

BAKER, George (1822–41), *The Histories and Antiquities of the County of Northampton* (London, J. B. Nichols and Son).

BAKER, J. R., (ed.) (1978), *The Reports of Sir John Spelman*, 2 vols., 1976–8, vol. ii (London, Selden Society).

BALLARD, A. (1908), 'Notes on the Open Fields of Oxfordshire', Oxfordshire Archaeological Society, Publications No. 54, *Reports*, pp. 22–31.

BARDHAN, Pranab K. (1984), *Land, Labor, and Rural Poverty* (New York, Columbia University Press).

BARRATT. D. M. (1955), *Ecclesiastical Terriers of Warwickshire Parishes* (Oxford, Dugdale Society Publications).

BARTON, John (1971), 'Review of Kerridge, *Agrarian Problems in the Sixteenth Century and After*', *English Historical Review*, 86: 618–19.

BATCHELOR, T. (1808), *General View of the Agriculture of the County of Bedford* (R. Phillips).

BATEMAN, John (1883), *The Great Landowners of Great Britain and Ireland* (New York, Augustus M. Kelley, 1970).

BATES, Robert H. (1988), 'Lessons from History, or the Perfidy of English Exceptionalism and the Significance of Historical France', *World Politics*, 40: 499–516.

BAUGH, D. A. (1975), 'The Cost of Poor Relief in South-East England, 1790–1834', *Economic History Review*, 2nd series, 28: 50–68.

BEAN, J. M. W. (1971), 'Review of Kerridge, *Agrarian Problems in the Sixteenth Century and After*', *Renaissance Quarterly*, 24: 389–91.

BECKER, Lawrence C. (1977), *Property Rights: Philosophic Foundations* (London, Routledge and Kegan Paul).

BECKETT, J. V. (1984), 'The Pattern of Landownership in England and Wales, 1660–1880', *Economic History Review*, 2nd series, 37: 1–22.

——(1986), *The Aristocracy in England, 1660–1914* (Oxford, Basil Blackwell).

BECKINSALE, R. P. (1963), 'The Plush Industry of Oxfordshire', *Oxoniensia*, 28: 53–67.

BENNETT, M. K. (1935), 'British Wheat Yield Per Acre for Seven Centuries', *Economic History*, 3: 12–29.

BERESFORD, M. (1949), 'Glebe Terriers and Open Field Leicestershire', in W. G. Hoskins (ed.), *Studies in Leicestershire Agrarian History* (Leicester, Leicestershire Archaeological Society), pp. 77–126.

——(1950), 'The Deserted Villages of Warwickshire', *Transactions* (Birmingham Archaeological Society) 66: 49–106.

——(1954), *The Lost Villages of England* (London, Lutterworth Press).

——(1961), 'Habitation versus Improvement: The Debate on Enclosure and Agreement', in F. J. Fisher (ed.), *Essays in the Economic and Social History of Tudor and Stuart England* (Cambridge, Cambridge University Press).

——(1977), *Seventeenth Century Chancery Decree Rolls* (unpublished MS, Institute for Historical Research).

——and HURST, J. G. (eds.), (1971), *Deserted Medieval Villages: Studies* (London, Lutterworth Press).

BERG, Maxine (1985), *The Age of Manufactures, 1700–1820* (London, Fontana Press).

BERRY, R. A., and CLINE, W. R. (1979), *Agrarian Structure and Productivity in Developing Countries* (Baltimore, Johns Hopkins University Press).

BEVERIDGE, W. (1937), 'Wages in the Winchester Manors', *Economic History Review*, 1st series, 7: 22–43.

——(1939), *Prices and Wages in England from the Twelfth to the Nineteenth Century* (London, Longmans, Green, and Co.).

BHARADWAJ, K. (1974), *Production Conditions in Indian Agriculture* (Cambridge, Cambridge University Press).

BLAUG, Mark (1963), 'The Myth of the Old Poor Law and the Making of the New', *Journal of Economic History*, 23: 151–84.

——(1964), 'The Poor Law Report Reexamined', *Journal of Economic History*, 24: 229–45.

BLISS, C. and STERN, N. (1978a), 'Productivity, Wages and Nutrition; Part I: The Theory', *Journal of Development Economics*, 5: 331–62.

——(1978b), 'Productivity, Wages and Nutrition; Part II: Some Observations', *Journal of Development Economics*, 5: 363–98.

——(1982), *Palanpur: The Economy of an Indian Village* (Oxford, Clarendon Press).

BLITH, W. (1652), *The English Improver Improved* (London, John Wright).

BLOCH, Marc (1931), *Les Charactères originaux de l'histoire rurale française* (Paris, Armand Colin, 1988).

BLYN, George (1966), *Agricultural Trends in India, 1891–1947: Output, Availability, and Productivity* (Philadelphia, University of Pennsylvania Press).

BONFIELD, Lloyd (1983), *Marriage Settlements, 1601–1740* (Cambridge, Cambridge University Press).

BOOTH, A. and SUNDRUM, R. M. (1985), *Labour Absorption in Agriculture* (Oxford, Oxford University Press).

BOWDEN, Peter (1952), 'Movements in Wool Prices, 1490–1610', *Yorkshire Bulletin of Economic and Social Research*, 4.

——(1957), 'Movements in Wool Prices—a Reply', *Yorkshire Bulletin of Economic and Social Research*, 9.

——(1962), *The Wool Trade in Tudor and Stuart England* (London, Macmillan).

——(1967), 'Agricultural Prices, Farm Profits, and Rents', in Joan Thirsk (ed.), *The Agrarian History of England and Wales*, iv, *1500–1640* (Cambridge, Cambridge University Press), pp. 593–695.

——(1985), 'Agricultural Prices, Wages, Farm Profits, and Rents', in Joan Thirsk (ed.), *The Agrarian History of England and Wales*, v, Pt II, *1640–1750: Agrarian Change* (Cambridge, Cambridge University Press), pp. 1–118, 827–902.

BOWEN, H. C. (1961), *Ancient Fields* (London, British Association for the Advancement of Science, republished by S R Publishers Ltd, 1970).

BOWLEY, A. L. (1898), 'The Statistics of Wages in the United Kingdom During the Last Hundred Years. (Part I.) Agricultural Wages', *Journal of the Royal Statistical Society*, 61: 702–22.

BOYER, George R. (1985), 'An Economic Model of the English Poor Law circa 1780–1834', *Explorations in Economic History*, 22: 129–67.

——(1986*a*), 'The Old Poor Law and the Agricultural Labour Market in Southern England: An Empirical Analysis', *Journal of Economic History*, 46: 113–35.

——(1986*b*), 'The Poor Law, Migration, and Economic Growth', *Journal of Economic History*, 46: 419–30.

BRAILSFORD, H. N. (1961), *The Levellers and the English Revolution*, ed. C. Hill (London, The Cresset Press).

BRANDON, P. F. (1972), 'Cereal Yields on the Sussex Estates of Battle Abbey during the Later Middle Ages', *Economic History Review*, 2nd series, 25: 403–20.

BRENNER, R. (1976), 'Agrarian Class Structure and Economic Development in Pre-Industrial Europe', in T. H. Aston and C. H. E. Philpin (eds.), *The Brenner Debate* (Cambridge, Cambridge University Press, 1985), pp. 10–63.

BROAD, J. F. P. (1972), 'Sir Ralph Verney and his Estates', Oxford University, D.Phil. thesis.

——(1980), 'Alternate Husbandry and Permanent Pasture in the Midlands, 1650–1800', *Agricultural History Review*, 28: 77–89.

BURGESS, J. H. (1978), '*The Social Structure of Bedfordshire and Northamptonshire, 1524–1674*', York University, D.Phil. thesis.

BURN, R. (1763), *Ecclesiastical Law* (London, A. Millar).

CAIRD, James (1851), *English Agriculture in 1850–51*, ed. G. E. Mingay (New York, A. M. Kelley, 2nd edn., 1967).

CAMPBELL, Bruce M. S. (1981), 'The Population of Early Tudor England: A Re-evaluation of the 1522 Muster Returns and 1524 and 1525 Lay Subsidies', *Journal of Historical Geography*, 7/2: 145–54.

——(1983), 'Agricultural Progress in Medieval England: Some Evidence from Eastern Norfolk', *Economic History Review*, 2nd series, 36: 26–46.

——(1987), 'Arable Productivity in Medieval English Agriculture', paper presented at the conference Pre-Industrial Developments in Peasant Economies: The Transition to Economic Growth, The Huntingdon Library, Los Angeles, 22–24 May 1987.

CAMPBELL, Mildred (1942), *The English Yeoman Under Elizabeth and the Early Stuarts*, Yale Historical Publications, Studies, vol. 14 (New Haven, Yale University Press).

CANNADINE, David (1977), 'Aristocratic Indebtedness in the Nineteenth Century: The Case Reopened', *Economic History Review*, 2nd series, 30: 624–50.

CHAMBERLIN, Edward (1935), *The Theory of Monopolistic Competition*, Harvard Economics Studies, vol. 38 (Cambridge, Mass., Harvard University Press).

CHAMBERS, J. D. (1953), 'Enclosure and Labour Supply in the Industrial Revolution', *Economic History Review*, 2nd series, 5: 319–43.

——(1957), *The Vale of Trent* (*Economic History Review*, Supplement No. 3).

——(1965), 'The Rural Domestic Industries during the Period of Transition to the Factory System, With Special Reference to the Midland Counties of England', *Deuxième conférence internationale d'histoire économique, Aix-en-Provence, 1962* (Paris, Mouton), pp. 429–55.

——(1966), *Nottinghamshire in the Eighteenth Century* (London, F. Cass, 2nd edn.).

——and MINGAY, G. E. (1966), *The Agricultural Revolution 1750–1880* (London, B. T. Batsford).

CHAO, Kang (1984), 'La Production textile dans la Chine traditionnelle', *Annales: économies, sociétés, civilisations*, 39: 957–76.

CHAPMAN, S. D. (1965), 'The Pioneers of Worsted Spinning by Power', *Business History*, 7: 97–116.

——(1972), 'The Genesis of the British Hosiery Industry, 1600–1750', *Textile History*, 7–50.

——(1974), 'Enterprise and Innovation in the British Hosiery Industry', *Textile History*, 14–37.

CHARTRES, J. A. (1985), 'The Marketing of Agricultural Progress', in Joan Thirsk (ed.), *The Agrarian History of England and Wales*, v, pt II, *1640–1750: Agrarian Change* (Cambridge, Cambridge University Press), pp. 406–502.

CHASE, Malcolm (1988), *The People's Farm: English Radical Agrarianism, 1775–1840* (Oxford, Clarendon Press).

CHATWIN, C. P. (1961), *British Regional Geology: East Anglia and Adjoining Areas* (London, HMSO, fourth edn.).

CHEN, Nai-Ruenn and GALENSON, Walter (1969). *The Chinese Economy Under Communism* (Chicago, Aldine).

CHIBNALL, A. C. and WOODMAN, A. V. (1950), *Subsidy Roll for the County of Buckingham, Anno 1524* (Bedford, Buckinghamshire Record Society Publication, vol. 8).

CHORLEY, P. (1981), 'The Agricultural Revolution in Northern Europe, 1750–1880: Nitrogen, Legumes, and Crop Productivity', *Economic History Review*, 2nd series, 34: 71–93.

CLARK, Gregory (1987), 'Productivity Growth without Technical Change in European Agriculture before 1850', *Journal of Economic History*, 42: 419–32.

——(1989), 'Productivity Growth in English Agriculture, 1300–1850', paper delivered at the conference Land and Labour Productivity in the European Past, Rockefeller Conference Centre, Bellagio, Italy.

CLAY, Christopher (1980),' "The Greed of Whig Bishops"?: Church Landlords and their Lessees, 1660–1760', *Past and Present*, 87: 128–57.

——(1981), 'Lifeleasehold in the Western Counties of England, 1650–1750', *Agricultural History Review*, 29: 83–96.

——(1984), *Economic Expansion and Social Change: England, 1500–1700* (Cambridge, Cambridge University Press).

——(1985), 'Landlords and Estate Management in England', in Joan Thirsk (ed.), *The Agrarian History of England and Wales*, v, pt. II, 1640–1750: Agrarian Change (Cambridge, Cambridge University Press), pp. 119–251.

COBBETT, W. (1948), *Rural Rides*, 2 vols. (London, J. M. Dent, Everyman's Library).

COHEN, Jon S. and WEITZMAN, Martin L. (1975), 'A Marxian Model of Enclosures', *Journal of Development Economics*, I: 287–336.

COLEMAN, D. C. (1975), *Industry in Tudor and Stuart England* (London, Macmillan).

COLLINS, E. J. T. (1987), 'The Rationality of "Surplus" Agricultural Labour: Mechanization in English Agriculture in the Nineteenth Century', *Agricultural History Review*, 35: 36–46.

——(1989), 'The "Machinery Question" in English Agriculture in the Nineteenth Century', in George Grantham and Carol S. Leonard (eds.), *Agrarian Organization in the Century of Industrialization: Europe, Russia, and North America (Research in Economic History*, Supplement 5), pp. 203–17.

——and JONES, E. L. (1967), 'Sectoral Advance in English Agriculture, 1850–80', *Agricultural History Review*, 15: 65–81.

COOPER, J. P. (1967), 'The Social Distribution of Land and Men in England, 1436–1700', *Economic History Review*, 2nd series, 20: 419–40.

CORNWALL, J. (1959), 'An Elizabethan Census', *Records of Bucks*, 4: 258–73.

COTTIS, Janie (1985), '*Agrarian Change in the Vale of White Horse, 1660–1760*', University of Reading, Ph.D. thesis.

COWPER, Thomas (1741), *A Short Essay Upon Trade* (London, J. Huggonson).

CRAFTS, N. F. R. (1976), 'English Economic Growth in the Eighteenth Century: A Re-Examination of Deane and Cole's Estimates', *Economic History Review*, 2nd series, 29: 226–35.

——(1983), 'British Economic Growth, 1700–1831: A Review of the Evidence', *Economic History Review*, 2nd series, 36: 177–99.

——(1984), 'Patterns of Development in Nineteenth Century Europe', *Oxford Economic Papers*, 36: 438–58.

——(1985), *British Economic Growth During the Industrial Revolution* (Oxford, Clarendon Press, 1986).

CRAFTS, N. F. R. (1987), 'British Industrialization in an International Context', School of Economic Studies, University of Leeds, Discussion Paper Series A: 87/7.

CROUCHLEY, A. E. (1939), 'A Century of Economic Development, 1837–1937 (A Study in Population and Production in Egypt)', *Egypte contemporaine*, 30: 133–55.

CROUZET, François (1972), 'Editor's Introduction', in François Crouzet (ed.), *Capital Formation in the Industrial Revolution* (London, Methuen), pp. 1–69.

CRUTCHLEY, J. (1794), *General View of the Agriculture in the County of Rutland* (London).

CULLEY, George (1807), *Observations on Livestock* (London, 4th edn).

CUSSANS, J. E. (1870–81), *History of Hertfordshire* (East Ardsley, Wakefield, Yorkshire, EP Publishing, 1972).

DANHOF, Clarence C. (1969), *Change in Agriculture: The Northern United States, 1820–1870* (Cambridge, Mass., Harvard University Press).

DARBY, H. C. (1964), 'The Draining of the English Clay-lands', *Geographische Zeitschrift*, 52: 190–201.

——(1976), *A New Historical Geography of England after 1600* (Cambridge, Cambridge University Press).

DAVENANT, Charles (1771), *The Political and Commercial Works*, ed. Sir Charles Whitworth, 5 vols. (London, R. Horsfield).

DAVENPORT. F. G. (1906), *The Economic Development of a Norfolk Manor, 1086–1565* (Cambridge).

DAVICO, Rosalba (1972), 'Baux, exploitations, techniques agricoles em Piédmont dans la deuxième moitié du XVIIIe siècle', *Etudes rurales*, no. 46: 76–101.

DAVID, Paul A. (1966), 'The Mechanization of Reaping in the Ante-Bellum Midwest', in Henry Rosovsky (ed.), *Industrialization in Two Systems* (New York, John Wiley), pp. 3–39.

——(1971), 'The Landscape and the Machine: Technical Interrelatedness, Land Tenure and the Mechanization of the Corn Harvest in Victorian Britain', in Donald N. McCloskey (ed.), *Essays on a Mature Economy: Britain after 1840* (Princeton, Princeton University Press), pp. 145–205.

DAVIES, C. Stella (1960), *The Agricultural History of Cheshire, 1750–1850* (Manchester, Cheatham Society, 3rd series, x).

DAVIES, D. (1795), *The Case of Labourers in Husbandry Stated and Explained* (Bath and London, C. G. and J. Robinson).

DAVIES, E. (1927), 'The Small Landowner, 1780–1832, in the Light of the Land Tax Assessments', *Economic History Review*, 1st series, 1: 87–113.

DAVIES, William (1941), 'The Grassland Map of England and Wales', *Journal of the Ministry of Agriculture*, 48.

——(1960), *The Grass Crop: Its Development, Use, and Maintenance* (London, E. and F. N. Spon Ltd).

DEANE, Phyllis and COLE, W. A. (1969). *British Economic Growth, 1688–1959* (Cambridge, Cambridge University Press, 2nd edn).

DEFOE, Daniel (1724–6), *A Tour through the whole Island of Great Britain*, ed. Pat Rogers (Harmondsworth, Penguin Books, 1983).

——(1727), *The Complete English Tradesman in Familiar Letters*, 2 vols. (London, Charles Rivington, 2nd edn.).

DENISON, E. F. (1967), *Why Growth Rates Differ: Postwar Experience in Nine Western Countries* (Washington, Brookings Institution).

DE VRIES, Jan (1976), *The Economy of Europe in an Age of Crisis, 1600–1750* (Cambridge, Cambridge University Press).

——(1984), *European Urbanization, 1500–1800* (Cambridge, Mass., Harvard University Press).

DEWERPE, Alain (1984), 'Genèse protoindustrielle d'une région développée: l'Italie septentrionale (1800–1880)', *Annales: économies, sociétés, civilisations*, 39: 896–914.

DEWEY, Clive J. (1974), 'The Rehabilitation of the Peasant Proprietor in Nineteenth-Century Economic Thought', *History of Political Economy*, 6/1: 17–47.

DEWINDT, Edwin B. (1972), *Land and People in Hollywell-cum-Needingworth* (Toronto, Pontifical Institute of Mediaeval Studies).

DEYON, Pierre (1984), 'Fécondité et limites du modèle protoindustriel: premier bilan', *Annales: économies, sociétés, civilisations*, 39: 868–81.

DIGBY, Anne (1975), 'The Labour Market and the Continuity of Social Policy after 1834: The Case of the Eastern Counties', *Economic History Review*, 2nd series, 28: 69–83.

DONALDSON, J. (1794), *General View of the Agriculture of the County of Northamptonshire* (Edinburgh).

DUBY, Georges (1968), *Rural Economy and Country Life in the Medieval West*, trans. Cynthia Postan (London, Edward Arnold).

DUNSDORFS, Edgar (1956), *The Australian Wheat-Growing Industry, 1788–1948* (Melbourne, The University Press).

DYER, Christopher (1980), *Lords and Peasants in a Changing Society* (Cambridge, Cambridge University Press).

——(1981), *Warwickshire Farming, 1349-c.1520: Preparations for Agricultural Revolution* (Oxford, Dugdale Society Occasional Paper No. 27).

——(1989), *Standards of Living in the Later Middle Ages* (Cambridge, Cambridge University Press).

EAGLY, R. V. (1974), *The Structure of Classical Economic Theory* (New York, Oxford University Press).

EDEN, F. M. (1797), *The State of the Poor* (London).

ELTIS, W. A. (1975a), 'Francois Quesnay: A Reinterpretation I. The *Tableau économique*', *Oxford Economic Papers*, 27: 167–200.

——(1975b), 'Francois Quesnay: A Reinterpretation II. The Theory of Economic Growth', *Oxford Economic Papers*, 27: 327–51.

*Encyclopaedia Britannica* (Edinburgh, A. Bell and C. Macfarquhar, 3rd edn, 1797).

ENGLISH, Barbara, and SAVILLE, John (1983), *Strict Settlement: A Guide for Historians* (University of Hull, Occasional Papers in Economic and Social History, No. 10; Hull, University of Hull Press).

ERNLE, Lord (1912), *English Farming: Past and Present* (London, Heinemann Educational Books Ltd and Frank Cass and Company Ltd, 1961).

ESWARAN, M. and KOTWAL, A. (1985), 'A Theory of Two-Tier Labor Markets in Agrarian Economies', *American Economic Review*, 75: 162–77.

EVANS, E. J. (1976), *The Contentious Tithe: The Tithe Problem and English Agriculture, 1750–1850* (London, Routledge and Kegan Paul).

EVANS, L. T. (1981), 'Yield Improvement in Wheat: Empirical or Analytical?' in L. T. Evans and W. J. Peacock (eds.), *Wheat Science—Today and Tomorrow* (Cambridge, Cambridge University Press), pp. 203–22.

FAITH, Rosamond (1984), 'Berkshire: Fourteenth and Fifteenth Centuries', in P. D. A. Harvey (ed.), *The Peasant Land Market in Medieval England* (Oxford, Clarendon Press), pp. 107–77.

FARMER, D.L. (1977), 'Grain Yields on the Winchester Manors in the Later Middle Ages', *Economic History Review*, 2nd series, 30: 555–66.

FEI, John C. and RANIS, Gustav (1964), *Development of the Labour Surplus Economy: Theory and Policy* (Homewood, Illinois, R. D. Irwin).

FEINSTEIN, C. H. (1978), 'Capital Formation in Britain', in Peter Mathias and M. M. Postan (eds.), *The Cambridge Economic History of Europe*, ii, *The Industrial Economies: Capital, Labour, and Enterprise*, pt. I, *Britain, France, Germany, and Scandinavia* (Cambridge, Cambridge University Press), pp. 28–96.

FELKIN, William (1867), *A History of the Machine Wrought Hosiery and Lace Manufactures*, ed. S. D. Chapman (Newton Abbot, David and Charles, 1967).

——(1877), 'Hosiery and Lace', in G. P. Bevan (ed.), *British Manufacturing Industries* (London, Edward Stanford).

FINCH, Mary E. (1956), *The Wealth of Five Northamptonshire Families, 1540–1640* (Oxford, Northamptonshire Record Society, xix).

FINLAY, Roger (1981), *Population and Metropolis: The Demography of London, 1580–1650* (Cambridge, Cambridge University Press).

FISHER, H. E. S. and JURICA, A. R. J. (1977), *Documents in English Economic History: England from 1000 to 1760*, 2 vols., 1976–7, vol. i (London, G. Bell and Sons).

FISHER, Richard Barnard (1794), *A Practical Treatise on Copyhold Tenure* (London, A. Strahan and W. Woodfall).

FITZHERBERT, Anthony (1539), *Surveyenge*, in R. H. Tawney and E. Power (eds.), *Tudor Economic Documents* (London, Longmans, Green and Co., 1924), iii, 22–5.

FLINN, M. W. (1974), 'Trends in Real Wages, 1750–1850', *Economic History Review*, 2nd series, 27: 395–413.

FLOUD, Roderick, and MCCLOSKEY, Donald (1981), *The Economic History of Britain since 1700*, 2 vols., vol. i, *1700–1860* (Cambridge, Cambridge University Press).

FOGEL, Robert W. and ENGERMAN, Stanley L. (1974). *Time on the Cross: The Economics of American Negro Slavery* (Boston, Little, Brown and Company).

FORDYCE, William (1855), *The History and Antiquities of the County Palatine of Durham* (Newcastle-upon-Tyne, Thomas Fordyce).

FORTREY, Samuel (1663) *Englands Interest and Improvement* (Cambridge, John Field).

FOSTER, John (1807), *Observations on the Agriculture in the North of Bedfordshire* (Bedford, J. Webb).

FOX, H. S. A. (1989), 'Conference Report: Winter Conference 1988, "Agriculture and the Village" ', *Agricultural History Review*, 37: 111–12.

FUSSELL, G. E. (1929), 'Population and Wheat Production in the Eighteenth Century', *The History Teachers' Miscellany*, 7: 65–8, 84–8, 108–11, 120–7.

GALLMAN, Robert E. (1972), 'Changes in Total U.S. Agricultural Factor Productivity in the Nineteenth Century', *Agricultural History*, 41: 191–210.

GAZLEY, John G. (1973), *The Life of Arthur Young, 1741–1820* (Philadelphia, American Philosophical Society).

*Gentleman's Magazine* (1731–1907), vols. 1–303 (London).

GERSCHENKRON, Alexander (1962), *Economic Backwardness in Historical Perspective* (Cambridge, Mass., The Belknap Press of Harvard University Press).

GLENNIE, Paul (1988*a*), 'Continuity and Change in Hertfordshire Agriculture, 1550–1700: I—Patterns of Agricultural Production', *Agricultural History Review*, 36: 55–75.

——(1988*b*). 'Continuity and Change in Hertfordshire Agriculture, 1550–1700: II —Trends in Crop Yields and their Determinants', *Agricultural History Review*, 36: 145–61.

——(1989), 'Measuring Gains in Land Productivity in the Agrarian Economies of Southern England, 1550–1700', paper delivered at the conference Land and Labour Productivity in the European Past, Rockefeller Conference Centre, Bellagio, Italy.

GODWIN, William (1798), *Enquiry Concerning Political Justice* (Harmondsworth, Penguin Books 1976).

GOLDSTONE, J. A. (1986), 'The Demographic Revolution in England: A Re-examination', *Population Studies*, 49: 5–33.

GONNER, E. C. K. (1912), *Common Land and Inclosure* (London, Macmillan).

GOOCH, W. (1813), *General View of the Agriculture of the County of Cambridge* (London).

GRANTHAM, George W. (1980), 'The Persistence of Open-Field Farming in Nineteenth-Century France', *Journal of Economic History*, 40: 515–31.

——(1989), 'The Growth of Labour Productivity in the Production of Wheat in the "Cinq Grosses Fermes", 1750–1930', paper delivered at the conference Land and Labour Productivity in the European Past, Rockefeller Conference Centre, Bellagio, Italy.

GRAS, N. S. B. (1915), *The Evolution of the English Corn Market* (Cambridge, Mass., Harvard University Press).

GRAY, C. M. (1963), *Copyhold, Equity, and the Common Law* (Cambridge, Mass., Harvard University Press).

GRAY, H. L. (1910), 'Yeoman Farming in Oxfordshire from the Sixteenth to the Nineteenth Century', *Quarterly Journal of Economics*, 39: 293–326.

——(1915), *English Field Systems* (Cambridge, Mass., Harvard University Press).

GREENALL, R. L. (1973), 'The Rise of Industrial Kettering', *Northamptonshire Past and Present*, 5: 253–66.

GREENALL, R. L. (ed.) (1974), *Naseby: A Parish History* (Leicester, University of Leicester, Department of Adult Education).

GRIGG, D. B. (1963), 'The Land Tax Returns', *Agricultural History Review*, 11: 83–8.

GUY, J. A. (1977), *The Cardinal's Court: The Impact of Thomas Wolsey in Star Chamber* (Hassocks, Sussex, The Harvester Press).

HABAKKUK, H. J. (1940), 'English Landownership, 1680–1740', *Economic History Review*, 1st series, 10: 2–17.

——(1950), 'Marriage Settlements in the Eighteenth Century,' *Transactions of the Royal Historical Society*, 4th series, 32: 15–30.

——(1952–3), 'The Long-Term Rate of Interest and the Price of Land in the Seventeenth Century', *Economic History Review*, 2nd series, 5: 24–65.

——(1960), 'The English Land Market in the Eighteenth Century', in J. S. Bromely and E. H. Kossman (eds.), *Britain and the Netherlands* (London, Chatto and Windus), pp. 154–73.

——(1965), 'La disparition du paysan anglais', *Annales: économies, sociétés, civilisations*, 20: 649–63.

——(1979), 'The Rise and Fall of English Landed Families, 1600–1800: I', *Transactions of the Royal Historical Society*, 5th series, 29: 187–207.

——(1980), 'The Rise and Fall of English Landed Families, 1600–1800: II', *Transactions of the Royal Historical Society*, 5th series, 30: 199–221.

——(1981), 'The Rise and Fall of English Landed Families, 1600–1800: III', *Transactions of the Royal Historical Society*, 5th series, 31: 195–217.

——(1987), 'The Agrarian History of England and Wales: Regional Farming Systems and Agrarian Changes, 1640–1750', *Economic History Review*, 2nd series, 40: 281–96.

HAINS, B. A. and HORTON, A. (1969), *British Regional Geology: Central England* (London, HMSO, 3rd edn).

HALL, Charles (1805), *The Effects of Civilization on the People in European States* (London, privately printed).

HAMMOND, J. L. and HAMMOND, B. (1932), *The Village Labourer, 1760–1832* (London, Longman).

HANSON, Haldore, BORLAUG, Norman, E., and ANDERSON R. Glenn (1982), *Wheat in the Third World* (Boulder, Colorado, Westview Press).

HARLAN, Jack R. (1967), 'A Wild Wheat Harvest in Turkey', *Archaeology*, 20: 197–201.

HARRISON, C. J. (1979), 'Elizabethan Village Surveys: A Comment', *Agricultural History Review*, 27: 82–9.

HART, Keith (1982), *The Political Economy of West African Agriculture* (Cambridge, Cambridge University Press).

HARTLIB, Samuel (1652), *His Legacie* (London, R. and W. Leybourn, 2nd edn).

HARVEY, Barbara (1977), *Westminster Abbey and its Estates in the Middle Ages* (Oxford, Clarendon Press).

HARVEY, P. D. A. (ed.) (1984), *The Peasant Land Market in Medieval England* (Oxford, Clarendon Press).

HASBACH, W. (1908), *A History of the English Agricultural Labourer* (London, P. S. King and Son).

HATCHER, John (1977), *Plague, Population and the English Economy, 1348–1530* (Houndmills, Basingstoke, Hampshire, Macmillan Educational).

HATLEY, Victor A. (1973), *Northamptonshire Militia Lists, 1777*, Publications of the Northamptonshire Record Society, vol. 25.

HAVINDEN, M. A. (1961*a*), 'The Rural Economy of Oxfordshire, 1580–1730', Oxford University, B.Litt. thesis.

——(1961*b*), 'Agricultural Progress in Open Field Oxfordshire', *Agricultural History Review*, 9.

——(1965), *Household and Farm Inventories in Oxfordshire, 1550–1590*, Oxfordshire Record Society, vol. 44.

HAYNES, John (1706), *A View of the Present State of the Clothing Trade in England* (London).

HEAD, Peter (1961–2), 'Putting Out in the Leicester Hosiery Industry in the Middle of the Nineteenth Century', *Transactions of the Leicestershire Archaeological and Historical Society*, pp. 44–59.

HELLEINER, K. (1967), 'The Population of Europe from the Black Death to the Eve of the Vital Revolution', in E. E. Rich and C. H. Wilson (eds.), *Cambridge Economic History of Europe*, iv (Cambridge, Cambridge University Press).

HENSON, G. (1831), *Henson's History of the Framework Knitters*, ed. S. D. Chapman (New York, A. M. Kelley, 1970).

HILL, J. E. C. (1961), *The Century of Revolution: 1603–1714* (New York, W. W. Norton, 1966).

—— (1972), *The World Turned Upside Down* (London, Temple Smith).

HILTON, R. H. (1949), 'Peasant Movements in England Before 1381', *Economic History Review*, 2nd series, 2: 117–36.

——(1957), 'A Study in the Pre-History of English Enclosure in the Fifteenth Century', in R. H. Hilton, *The English Peasantry in the Later Middle Ages* (Oxford, Clarendon Press, 1975), pp. 161–73.

——(1969), *The Decline of Serfdom in Medieval England* (London, Macmillan).

——(1973), *Bond Men Made Free* (London, Maurice Temple Smith).

HO, Samuel P. S. (1984), 'Protoindustrialisation, protofabriques et désindustrialisation: une analyse économique', *Annales: économies, sociétés, civilisations*, 39: 882–95.

HOBSBAWM, Eric (1957), 'The British Standard of Living, 1790–1850', *Economic History Review*, 2nd series, 10: 46–68.

——and RUDÉ, George (1968), *Captain Swing* (New York, W. W. Norton, 1975).

HODGSON, John (1832), *A History of Morpeth* (Newcastle).

HOFFMANN, Richard C. (1975), 'Medieval Origins of the Common Fields', in William N. Parker and Eric L. Jones (eds.), *European Peasants and Their Markets* (Princeton, Princeton University Press).

HOLDERNESS, B. A. (1972), ' "Open" and "Close" Parishes in England in the Eighteenth and Nineteenth Centuries', *Agricultural History Review*, 20: 126–39.

HOLDERNESS, B. A. (1975–6), 'Credit in a Rural Community, 1660–1800', *Midland History*, 3: 94–115.

——(1976), 'Credit in English Rural Society Before the Nineteenth Century, with Special Reference to the Period 1650–1720', *Agricultural History Review*, 24: 91–109.

HOLDSWORTH, Sir William (1937), *A History of English Law*, 2nd edn, vol. vii (London, Methuen).

——(1942), *A History of English Law*, 5th edn, vol. iii (London, Methuen).

HOLLINGSWORTH, T. H. (1969), *Historical Demography* (Ithaca, NY, Cornell University Press).

HOLMES, G. S. (1977), 'Gregory King and the Social Structure of Preindustrial England', *Transactions of the Royal Historical Society*, 5th series, 27: 41–68.

HOMER, Sidney (1977), *A History of Interest Rates* (New Brunswick, NJ, Rutgers University Press, 2nd edn).

HONORÉ, A. M. (1961), 'Ownership', in A. G. Guest (ed.), *Oxford Essays in Jurisprudence* (Oxford, Oxford University Press), pp. 107–47.

HOSKINS, W. G. (1950a), 'The Deserted Villages of Leicestershire', in W. G. Hoskins (ed.), *Essays in Leicestershire History* (Liverpool), pp. 67–107.

——(1950b), 'The Leicestershire Farmer in the Sixteenth Century', in W. G. Hoskins (ed.), *Essays in Leicestershire History* (Liverpool), pp. 123–83.

——(1951), 'The Leicestershire Farmer in the Seventeenth Century', in W. G. Hoskins (ed.), *Provincial England* (London, Macmillan, 1963), pp. 149–69.

——(1957), *The Midland Peasant* (London, Macmillan).

HOWARD, Henry Fraser (1935), *An Account of the Finances of the College of St. John the Evangelist in the University of Cambridge, 1511–1926* (Cambridge, Cambridge University Press).

HOWELL, Cicely (1976), 'Peasant Inheritance Customs in the Midlands, 1280–1700', in J. Goody, J. Thirsk and E. P. Thompson (eds.), *Family and Inheritance* (Cambridge, Cambridge University Press), pp. 112–55.

——(1983), *Land, Family and Inheritance in Transition: Kibworth Harcourt, 1280–1700* (Cambridge, Cambridge University Press).

HUDSON, Pat (1986), *The Genesis of Industrial Capital: A Study of the West Riding Wool Textile Industry c.1750–1850* (Cambridge, Cambridge University Press).

——(1989), 'The Regional Perspective', in Pat Hudson (ed.), *Regions and Protoindustry* (Cambridge, Cambridge University Press).

HUECKEL, Glenn (1976), 'English Farming Profits during the Napoleonic Wars, 1793–1815', *Explorations in Economic History*, 13: 331–45.

HUME, David (1752), 'Of the Populousness of Ancient Nations', in Eugene Rotwein (ed.), *David Hume: Writings on Economics* (Madison, University of Wisconsin Press), pp. 108–84.

HUMPHRIES, E. and HUMPHRIES, M. (1985), *Woodford juxta Thrapston* (Rushden, Northants, Buscott Publications).

HUMPHRIES, Jane (1990), 'Enclosures, Common Rights, and Women: The Proletarianization of Families in the Late Eighteenth and Early Nineteenth Centuries', *Journal of Economic History*, 50: 17–42.

HUNT, H. G. (1959), 'Landownership and Enclosure, 1750–1830', *Economic History Review*, 2nd series, 11: 497–505.

HUTCHESON, Francis (1755), *A System of Moral Philosophy*, 2 vols. (Glasgow, R. and A. Foulis).

HUTCHINSON, William (1794–7), *The History of the County of Cumberland* (East Ardsley, Wakefield, Yorkshire, EP Publishing, 1974).

HUZEL, J. P. (1989), 'The Labourer and the Poor Law, 1750–1850', in G. E. Mingay (ed.), *The Agrarian History of England and Wales*, vi, *1750–1850* (Cambridge, Cambridge University Press), pp. 755–810.

JACK, R. I. (ed.) (1965), *The Grey of Ruthin Valor, 1467–8* (Bedfordshire Historical Record Society Publication No. 46, and Sidney, Sidney University Press).

JACKSON, R. V. (1985), 'Growth and Deceleration in English Agriculture, 1660–1790', *Economic History Review*, 2nd series, 38: 333–51.

JACOBS, Jane (1984), *Cities and the Wealth of Nations* (New York, Random House).

JACOBSON, D. L. (ed.) (1965), *The English Libertarian Heritage* (New York, The Bobbs-Merrill Company).

JAMES, John (1968), *The History of the Worsted Manufacture in England* (London, Frank Cass).

JAMES, William, and MALCOLM, Jacob (1794), *General View of the Agriculture of the County of Buckingham* (London, Colin Macrae).

JOHN, A. H. (1976), 'English Agricultural Improvement and Grain Exports, 1660–1765', in D. C. Coleman and A. H. John (eds.), *Trade, Government and Economy in Pre-Industrial England* (London, Weidenfeld and Nicolson), pp. 45–67.

—— (1989), 'Statistical Appendix', in G. E. Mingay (ed.), *The Agrarian History of England and Wales*, vi, *1750–1850* (Cambridge, Cambridge University Press), pp. 972–1155.

JOHNSON, A. H. (1909), *The Disappearance of the Small Landowner* (Oxford, Clarendon Press).

JOHNSON, D. Gale (1983), 'The World Food Situation: Recent and Prospective Developments', in D. G. Johnson and G. E. Schuh (ed.), *The Role of Markets in the World Food Economy* (Boulder, Colorado, Westview Press), pp. 1–33.

JONES, E. L. (1965), 'Agriculture and Economic Growth in England, 1660–1750: Agricultural Change', *Journal of Economic History*, 25: 1–18.

—— (ed.)(1967), *Agriculture and Economic Growth in England, 1650–1815* (London, Methuen).

—— (1977), 'Environment, Agriculture, and Industrialization in Europe', *Agricultural History*, 51: 491–502.

—— and HEALY, J. R. (1974), 'Wheat Yields in England, 1815–1859', in E. L. Jones (ed.), *Agriculture and the Industrial Revolution* (Oxford, Basil Blackwel)l.

KAIN, Roger J. P. (1986), *An Atlas and Index of the Tithe Files of Mid-Nineteenth-Century England and Wales* (Cambridge, Cambridge University Press).

KAUTSKY, Karl (1900), *La Question agraire*, trans. Edgard Milhaud and Camille Pollack, Bibliothèque Socialiste Internationale (Paris, V. Giard and E. Brière).

KENT, Nathaniel (1794), *General View of the Agriculture of the County of Norfolk* (London, Colin Macrae).

KERRIDGE, E. (1955), 'The Returns of the Inquisitions of Depopulation', *English Historical Review*, 70: 212–28.

——(1967), *The Agricultural Revolution* (London, George Allen and Unwin).

——(1969), *Agrarian Problems in the Sixteenth Century and After* (London, George Allen and Unwin).

KEYFITZ, N. (1983), 'The Evolution of Malthus's Thought: Malthus as Demographer', in J. Dupâquier, A. Fauve-Chamoux, and E. Grebenick (eds.), *Malthus Past and Present* (London, Academic Press).

KING, Gregory (1936), *Two Tracts*, ed. J. H. Hollander (Baltimore, Johns Hopkins University Press).

KINGSTON-MANN, Esther (1981), 'Marxism and Russian Rural Development: Problems of Evidence, Experience, and Culture', *American Historical Review*, 86: 731–52.

KOMLOS, John (1988), 'Agricultural Productivity in America and Eastern Europe: A Comment', *Journal of Economic History*, 48: 655–64.

KOSMINSKI, E. A. (1956), *Studies in the Agrarian History of England in the Thirteenth Century*, trans. Ruth Kisch, ed. R. H. Hilton (Oxford, Basil Blackwell).

KUSSMAUL, A. (1981), *Servants in Husbandry in Early Modern England* (Cambridge, Cambridge University Press).

LANDES, David S. (1949), 'French Entrepreneurship', *Journal of Economic History*, 9: 45–61.

——(1969), *The Unbound Prometheus* (Cambridge, Cambridge University Press).

LANGDON, John (1986), *Horses, Oxen and Technological Innovation* (Cambridge, Cambridge University Press).

LAURENCE, Edward (1727), *The Duty of a Steward to His Lord* (London, John Shuckburgh).

LAURENT, Robert (1976), 'Tradition et progrès: le sector agricole', in Fernand Braudel and Ernest Labrousse (eds.), *Histoire économique et social de la France* (Paris, Presses Universitaires de France), vol. iii/2, pp. 619–738.

LAWRENCE, John (1809), *A General Treatise on Cattle, the Ox, the Sheep, and the Swine* (London, Sherwood, Neely, and Jones, 2nd edn.).

LEADAM, I. S. (ed.) (1897), *The Domesday of Inclosures, 1517–1518* (Port Washington, NY, Kennikat Press, 1971).

——(ed.) (1898), *Select Cases in the Court of Requests* (London, Selden Society, xii).

LEE, Ronald (1973), 'Population in Preindustrial England: An Econometric Analysis', *Quarterly Journal of Economics*, 87: 581–607.

——(1985), 'Population Homeostasis and English Demographic History', *Journal of Interdisciplinary History*, 15: 635–60.

LENIN, V. I. (1899), *The Development of Capitalism in Russia*, English trans. of the 4th Russian edn with corrections, in *Collected Works* (Moscow, Progress Publishers, iii, 1972).

——(1908), *The Agrarian Question and the 'Critics of Marx'*, English trans. of the 5th Russian edn (Moscow, Progress Publishers, 2nd revised edn, 1976).

LENNARD, Reginald (1916), *Rural Northamptonshire Under the Commonwealth*, Oxford Studies in Social and Legal History, vol. 5.

LEVINE, David (1987), *Reproducing Families: The Political Economy of English Population History* (Cambridge, Cambridge University Press).

LEWIS, Naphtali (1983), *Life in Egypt Under Roman Rule* (Oxford, Clarendon Press).

LEWIS, Sir W. A. (1954), 'Economic Development with Unlimited Supplies of Labour', *The Manchester School*, 22: 139–91.

——(1955), *The Theory of Economic Growth* (Homewood, Illinois, R. D. Irwin).

LINDERT, Peter H. (1985), 'English Population, Wages, and Prices: 1541–1913', *Journal of Interdisciplinary History*, 15: 609–34.

——and WILLIAMSON, J. G. (1982), 'Revising England's Social Tables, 1688–1812', *Explorations in Economic History*, 19: 385–408.

——(1983), 'English Workers' Living Standards during the Industrial Revolution: A New Look', *Economic History Review*, 2nd series, 36: 1–25.

LIPSCOMB, George (1847), *The History and Antiquities of the County of Buckingham* (London, J. and W. Robins).

LIPSON, E. (1921), *The History of the Woollen and Worsted Industries* (London, A. and C. Black).

LLOYD, T. H. (1973), *The Movement of Wool Prices in Medieval England* (*Economic History Review*, Supplement No. 6).

LOUDON, J. C. (1831), *An Encyclopedia of Agriculture*. 2nd edn. (London, Longman, Rees, Orme, Brown, and Green).

LOW, David (1838), *Elements of Practical Agriculture* (London, Longman, Orme, Brown, Green, and Longman).

MACAULAY, Catherine (1766–83), *History of England* (London).

MACCULLOCH, Diarmaid (1988), 'Bondmen under the Tudors', in Claire Cross, David Loades, and J. J. Scarisbrick (eds.), *Law and Government under the Tudors* (Cambridge, Cambridge University Press), pp. 91–109.

MACDONALD, Stuart (1979), 'The Diffusion of Knowledge Among Northumberland Farmers, 1780–1815', *Agricultural History Review*, 27: 30–9.

——(1983), 'Agricultural Improvement and the Neglected Labourer', *Agricultural History Review*, 31: 81–90.

MACFARLANE, Alan (1978), *The Origins of English Individualism* (Oxford, Basil Blackwell).

MACPHERSON, C. B. (1962), *The Political Theory of Possessive Individualism* (Oxford, Clarendon Press).

MALTHUS, T. R. (1798), *An Essay on the Principle of Population*, ed. Philip Appleman (New York, W. W. Norton, 1976).

——(1836), *Principles of Political Economy*, (New York, A. M. Kelley, 2nd edn. with additions, 1951).

——(1872), *An Essay on the Principle of Population* (London, Reeves and Turner, 7th edn.).

MANTOUX, P. (1905), *La Révolution industrielle au xviiie siècle* (Paris).

MARCOMBE, David (1981), 'Church Leaseholders: the Decline and Fall of a Rural Elite', in Rosemary O'Day and Felicity Heal (eds.), *Princes and Paupers in the English Church, 1500–1800* (Leicester, Leicester University Press), pp. 255–75.

MARSHALL, William (1804), *On the Landed Property of England* (London, G. and W. Nicol).

——(1818), *The Review and Abstract of the County Reports of the Board of Agriculture*, 5 vols. (York, Thomas Wilson and Sons).

MARTIN, David E. (1976), 'The Rehabilitation of the Peasant Proprietor in Nineteenth-Century Economic Thought: A Comment', *History of Political Economy*, 8/2: 297–302.

MARTIN, John E. (1983), *Feudalism to Capitalism: Peasant and Landlord in English Agrarian Development* (London, Macmillan).

——(1988), 'Sheep and Enclosure in Sixteenth-Century Northamptonshire', *Agricultural History Review*, 36: 39–54.

MARTIN, J. M. (1966), 'Landownership and the Land Tax Returns', *Agricultural History Review*, 14: 96–103.

——(1967), 'The Cost of Parliamentary Enclosure in Warwickshire', in E. L. Jones (ed.), *Agriculture and Economic Growth in England, 1650–1815* (London, Methuen), pp. 128–151.

——(1982), 'Enclosure and the Inquisitions of 1607: An Examination of Dr Kerridge's Article "The Returns of the Inquisitions of Depopulation"', *Agricultural History Review*, 30: 41–8.

MARX, K. (1867), *Capital*, trans. Samuel Moore and Edward Aveling, 3 vols., 1867–94, vol. i (New York, The Modern Library, Random House, 1906).

MASTIN, John (1792), *History and Antiquities of Naseby* (Cambridge, E. Hodson).

MATHIAS, P. (1983), *The First Industrial Nation* (London, Methuen, 2nd edn.).

MAVOR, W. (1813), *General View of the Agriculture of Berkshire* (London).

McCLOSKEY, D. (1972), 'The Enclosure of Open Fields: Preface to a Study of Its Impact on the Efficiency of English Agriculture in the Eighteenth Century', *Journal of Economic History*, 32: 15–35.

——(1975), 'The Economics of Enclosure: A Market Analysis', in W. N. Parker and E. L. Jones (eds.), *European Peasants and Their Markets* (Princeton, Princeton University Press), pp. 123–60.

McCULLOCH, J. R. (1880), *A Dictionary of Commerce and Commercial Navigation*, ed. Hugh G. Reid (London, Longmans, Green and Co., new edn).

MEGARRY, Robert, and WADE, H. W. R. (1975), *The Law of Real Property* (London, Stevens and Sons, 4th edn).

MELTON, Frank T. (1986), *Sir Robert Clayton and the Origins of English Deposit Banking, 1658–1685* (Cambridge, Cambridge University Press).

MENDELS, Franklin (1984), 'Des industries rurales à la protoindustrialisation: historique d'un changement de perspective', *Annales: économies, sociétés, civilisations*, 39: 977–1008.

MERTENS, J. A. and VERHULST, A. E. (1966), 'Yield-Ratios in Flanders in the Fourteenth Century', *Economic History Review*, 2nd series, 19: 175–82.

MILL, John Stuart (1848), *Principles of Political Economy*, ed. W. J. Ashley (New York, Augustus M. Kelley, 1965).

MILLS, D. R. (1980), *Lord and Peasant in Nineteenth Century Britain* (London, Croom Helm).

MINGAY, G. E. (1962), 'The Size of Farms in the Eighteenth Century', *Economic History Review*, 2nd series, 14: 469–88.

——(1963), *English Landed Society in the Eighteenth Century* (London, Routledge and Kegan Paul).

——(1964), 'The Land Tax Assessments and the Small Landowner', *Economic History Review*, 2nd series, 17: 381–8.

——(1968), *Enclosure and the Small Farmer in the Age of the Industrial Revolution* (London, Macmillan).

——(ed.) (1975), *Arthur Young and His Times* (London, Macmillan).

——(1984), 'The East Midlands', in Joan Thirsk (ed.), *The Agrarian History of England and Wales*, v, pt I, *1640–1750: Regional Farming Systems* (Cambridge, Cambridge University Press), pp. 89–128.

MITCHELL, B. R. (1978), *European Historical Statistics, 1750–1970* (New York, Columbia University Press).

——and DEANE, Phyllis (1971), *Abstract of British Historical Statistics* (Cambridge, Cambridge University Press).

MITRANY, D. (1951), *Marx Against the Peasant* (Chapel Hill, University of North Carolina Press).

MOKYR, Joel (1985), *Why Ireland Starved: A Quantitative and Analytical History of the Irish Economy, 1800–1850* (London, George Allen and Unwin).

——(1988), 'Is There Still Life in the Pessimist Case? Consumption during the Industrial Revolution, 1790–1850', *Journal of Economic History*, 43: 69–92.

——and SAVIN, N. Eugene (1976), 'Stagflation in Historical Perspective: The Napoleonic Wars Revisited', *Research in Economic History*, 1: 198–259.

MONEY, W. (1899), 'A Religious Census of the County of Berks in 1676', *The Berks, Bucks and Oxon Archaeological Journal*, 4: 112–15.

MONK, John (1794), *General View of the Agriculture of the County of Leicester* (London, J. Nichols).

MORE, Thomas (1516), *Utopia*, trans. Paul Turner (Harmondsworth, Penguin Books, 1985).

MORGAN, David Hoseason (1982), *Harvesters and Harvesting, 1840–1900* (London, Croom Helm).

MORINEAU, Michel (1968), 'Y-a-t-il eu une révolution agricole en France au XVIIIe siècle?' *Revue historique,* 239: 299–326.

——(1970), *Les Faux-semblants d'un démarrage économique* (Paris, Librarie Armand Colin).

MORTON, J. C. (1855), *A Cyclopedia of Agriculture*, 2 vols. (Glasgow, Blackie and Son).

——(1868), *Handbook of Farm Labour* (London, Cassell, Petter, and Galpin).

MOYLE, W. (1699), *An Essay Upon the Constitution of the Roman Government*, in Caroline Robbins (ed.), *Two English Republican Tracts* (Cambridge, Cambridge University Press, 1969).

MUNN, John, (1738), *Observations on British Wool* (London, H. Kent).

MURRAY, A. (1813), *General View of the Agriculture of the County of Warwickshire* (London, B. McMillan).

NEESON, J. M. (1984), 'The Opponents of Enclosure in Eighteenth-Century Northamptonshire', *Past and Present*, No. 105: 114–39.

——(1989), 'Parliamentary Enclosure and the Disappearance of the English Peasant, Revisited', in George Grantham and Carol S. Leonard (eds.), *Agrarian Organization in the Century of Industrialization: Europe, Russia, and North America*, (*Research in Economic History*, Supplement 5), pp. 89–120.

NEVILLE, H. (1681), *Plato Redivivus*, in Caroline Robbins (ed.), *Two English Republican Tracts* (Cambridge, Cambridge University Press, 1969).

NICHOLAS, Stephen, and SHERGOLD, Peter R. (1985), 'Intercounty Labour Mobility during the Industrial Revolution', Australian National University, *Working Papers in Economic History*, No. 43.

NICHOLS, John (1795–1815), *History and Antiquities of the County of Leicester*, 4 vols. (London, J. Nichols).

O'BRIEN, Patrick K. (1977), 'Agriculture and the Industrial Revolution', *Economic History Review*, 2nd series, 30: 160–87.

——(1985), 'Agriculture and the Home Market for English Industry, 1660–1820', *English Historical Review*, 100: 773–800.

——and KEYDER, C. (1978), *Economic Growth in Britain and France, 1780–1914: Two Paths to the Twentieth Century* (London, George Allen and Unwin).

OFFER, Avner (1981), *Property and Politics, 1870–1914* (Cambridge, Cambridge University Press).

OGILVIE, William (1782), *An Essay on the Right of Property in Land* (London, J. Walter).

O GRADA, Cormac (1988), *Ireland Before and After the Famine: Explorations in Economic History* (Manchester, Manchester University Press).

OHKAWA, Kazushi, and ROSOVSKY, Henry (1963), 'The Significance of the Japanese Experience', in Takekazu Ogura (ed.), *Agricultural Development in Modern Japan* (Tokyo, Japan FAO Association), pp. 617–84.

ORWIN, C. S. and WHETHAM, E. H. (1964), *History of British Agriculture, 1846–1914* (London, Longmans).

OSCHINSKY, D. (1971), *Walter of Henley and Other Treatises on Estate Management and Accounting* (Oxford, Clarendon Press).

OSTERUD, Nancy Grey (1986), 'Gender Divisions and the Organization of Work in the Leicester Hosiery Industry', in Angela V. John (ed.), *Unequal Opportunities: Women's Employment in England, 1800–1918* (Oxford, Basil Blackwell), pp. 45–68.

OVERTON, M. (1977), 'Computer Analysis of an Inconsistent Data Source: the Case of Probate Inventories', *Journal of Historical Geography*, 3: 317–26.

——(1979), 'Estimating Crop Yields from Probate Inventories: An Example from East Anglia, 1585–1735', *Journal of Economic History*, 39: 363–78.

——(1984), 'Agricultural Productivity in Eighteenth-Century England: Some Further Speculations', *Economic History Review*, 2nd series, 37: 244–51.

——(1989), 'The Determinants of Crop Yields in Early Modern England', paper delivered at the conference Land and Labour Productivity in the European Past, Rockefeller Conference Centre, Bellagio, Italy.

PAINE, Thomas (1797), 'Agrarian Justice', in *The Writings of Thomas Paine*, ed. M. C. Conway (New York, G. P. Putnam's Sons, 1895), iii, pp. 322–44.

PALMER, W. M., and SAUNDERS, H. W. (n.d.), *Documents Relating to Cambridgeshire Villages* (Cambridge).

PARKER, L. A. (1947), 'The Depopulation Returns for Leicestershire in 1607', *Transactions of the Leicestershire Archaeological Society*, 23: 231–41, 290–3.

——(1948), '*Enclosure in Leicestershire, 1485–1607*', University of London, Ph.D. thesis.

——(1949), 'The Agrarian Revolution at Cotesbach, 1501–1612', in W. G. Hoskins (ed.), *Studies in Leicestershire Agrarian History* (Leicestershire Archaeological Society), xxiv, pp. 41–76.

PARKER, R. A. C. (1975), *Coke of Norfolk* (Oxford, Clarendon Press).

PARKER, W. N. and KLEIN, J. L. V. (1966), 'Productivity Growth in Grain Production in the United States, 1840–60 and 1900–10', in *Output, Employment, and Productivity in the United States after 1800* (New York, National Bureau of Economic Research, Studies in Income and Wealth), vol. 30, pp. 523–80.

PARKINSON, R. (1808), *A General View of the Agriculture of the County of Rutland* (London).

——(1810), *Treatise on the Breeding and Management of Livestock*, 2 vols. (London, Cadell, Davies and R. Sholey).

——(1811), *A General View of the Agriculture of the County of Huntingdon* (London, R. Phillips).

PATTEN, John (1976), 'Patterns of Migration and Movement of Labour to Three Pre-Industrial East Anglian Towns', *Journal of Historical Geography*, 2: 111–29.

PAYNE, E. O. (1946), *Property in Land in South Bedfordshire, 1750–1832* (Bedfordshire Historical Record Society, Publication 23).

PERKINS, Dwight H. (1969), *Agricultural Development in China, 1368–1968* (Chicago, Aldine).

PETTY, W. (1927), *The Petty Papers*, ed. Marquis of Lansdowne, 2 vols. (London, Constable).

PHELPS BROWN, E. H., and HOPKINS, Sheila V. (1955), 'Seven Centuries of Building Wages', *Economica*, NS, 22: 195–206.

——(1956), 'Seven Centuries of the Prices of Consumables, Compared with Builders' Wage-rates', *Economica*, NS, 23: 296–314.

PINCHBECK, I. (1930), *Women Workers and the Industrial Revolution* (London, Routledge).

PITT, W. (1809a), *A General View of the Agriculture of the County of Leicester* (London).

——(1809b), *A General View of the Agriculture of the County of Northampton* (London).

PLOT, R. (1677), *The Natural History of Oxfordshire* (Oxford).

PLUMMER, Alfred (1934), *The Witney Blanket Industry* (London, George Routledge and Sons).

——and EARLY, Richard E. (1969), *The Blanket Makers, 1669–1969* (London, Routledge and Kegan Paul).

POCOCK, J. G. A. (1975), *The Machiavellian Moment: Florentine Political Thought and the Atlantic Republican Tradition* (Princeton, Princeton University Press).

POLANYI, K. (1944), *Origins of Our Time, the Great Transformation* (New York, Farrar and Reinhart).

POLLARD, Sidney, (1965), *The Genesis of Modern Management* (Harmondsworth, Penguin Books, 1968).

——(1978), 'Labour in Great Britain', in Peter Mathias and M. M. Postan (eds.), *The Cambridge Economic History of Europe*, ii, *The Industrial Economies: Capital, Labour, and Enterprise*, pt I, *Britain, France, Germany, and Scandinavia* (Cambridge, Cambridge University Press), pp. 97–179.

——(1981), *Peaceful Conquest: The Industrialization of Europe, 1760–1970* (Oxford, Oxford University Press).

POLLOCK, F. and MAITLAND, F. W. (1968), *The History of English Law*, 2 vols. (Cambridge, Cambridge University Press, 2nd edn.).

POSTAN, M. M. (1950), 'Some Agrarian Evidence of Declining Population in the Later Middle Ages', *Economic History Review*, 2nd series, 2: 221–46.

——(1965), 'Agricultural Problems of Under-Developed Countries in the Light of European Agrarian History', *Deuxième conférence internationale d'histoire économique, Aix-en-Provence, 1962* (Paris, Mouton), pp. 9–24.

——(1975), *The Medieval Economy and Society.* (Harmondsworth, Penguin Books).

POSTGATE, M. R. (1973), 'Field Systems in East Anglia', in A. R. H. Baker and R. A. Butlin (eds), *Studies of Field Systems in the British Isles* (Cambridge, Cambridge University Press), pp. 281–324.

POUNDS, Norman J. G. (1985), *An Historical Geography of Europe, 1800–1914* (Cambridge, Cambridge University Press).

PRICE, R. (1792), *Observations on Reversionary Payments* (London, T. Cadell, 5th edn).

PRIEST, St J. (1813), *General View of the Agriculture of Buckingham* (London).

PRINCE, Hugh (1977), 'Regional Contrasts in Agrarian Structure', in Hugh D. Clout (ed.), *Themes in the Historical Geography of France* (London, Academic Press), pp. 129–84.

PURDY, Frederick (1861), 'On the Earnings of Agricultural Labourers in England and Wales, 1860', *Journal of the Royal Statistical Society*, 24: 328–73.

PUSEY, P. (1843–4), 'Evidence on the Antiquity, Cheapness, and Efficacy of Thorough-Draining, or Land-Ditching, as Practised through the Counties of Suffolk, Hertford, Essex, and Norfolk', *Journal of the Royal Agricultural Society of England*, 4: 23–49.

QUESNAY, F. (1756), 'Fermiers', in A. Oncken (ed.), *Oeuvres économiques et philosophiques* (Darmstadt, Scientia Verlag Aalen, 1965), pp. 159–92.

——(1757), 'Grains', in A. Oncken (ed.), *Oeuvres économiques et philosophiques* (Darmstadt, Scientia Verlag Aalen, 1965), pp. 193–249.

RAFTIS, J. Ambrose (1957), *The Estates of Ramsey Abbey* (Toronto, Pontifical Institute of Mediaeval Studies).

——(1964), *Tenure and Mobility* (Toronto, Pontifical Institute of Mediaeval Studies).

——(1974), *Warboys* (Toronto, Pontifical Institute of Mediaeval Studies).

RANDALL, H. A. (1970/1, 1971/2), 'The Kettering Worsted Industry of the Eighteenth Century', *Northamptonshire Past and Present*, 49: 1970/1, pp. 313–20; 1971/2, pp. 349–56.

RAO, C. H. Hanumantha (1975), *Technological Change and Distribution of Gains in Indian Agriculture* (Delhi, Macmillan Company of India).

REDFORD, Arthur (1926), *Labour Migration in England, 1800–1850*, ed. W. H. Chaloner, 2nd edn (Manchester, Manchester University Press, 1964).

RICARDO, David (1817), *The Principles of Political Economy and Taxation*, ed. William Fellner (Homewood, Illinois, R. D. Irwin, 1963).

——(1817), *On the Principles of Political Economy and Taxation*. 1st edn, in P. Sraffa (ed.), *The Works and Correspondence of David Ricardo* (Cambridge, Cambridge University Press, i, 1951).

RICHARDS, Alan (1982), *Egypt's Agricultural Development, 1800–1980: Technical and Social Change* (Boulder, Colorado, Westview Press).

RICHESON, A. W. (1966), *English Land Measuring to 1800: Instruments and Practices* (Cambridge, Mass., MIT Press).

*Road Atlas of Great Britain* (1980), (St Albans, Hertfordshire, Geographia Ltd, 11th revised edn).

ROBERTSON, J. (1983), 'The Scottish Enlightenment at the Limits of the Civic Tradition', in I. Hont and M. Ignatieff (eds.), *Wealth and Virtue: The Shaping of Political Economy in the Scottish Enlightenment* (Cambridge, Cambridge University Press), pp. 137–78.

ROBINSON, Joan (1933), *The Economics of Imperfect Competition* (London, Macmillan).

RODEN, David (1969), 'Enclosure in the Chilterns', *Geografiska Annaler*, 51: 115–26.

ROGERS, J. E. T. (1866–1892), *A History of Agriculture and Prices in England*, 7 vols. (Oxford, Clarendon Press).

ROUSSEAU, Jean-Jacques (1762), *Du contrat social*, in J. Ehrard (ed.), *Du contrat social et autres oeuvres politiques* (Paris, Editions Garnier Frères, 1975), pp. 235–336.

ROWNTREE, B. Seebohm (1911), *Land and Labour: Lessons from Belgium* (London, Macmillan).

ROYCE, D. (1879), 'Historical Notices of the Parish of Cropredy, Oxon,' *Transactions of the North Oxfordshire Archaeological Society*.

RUDRA, Ashok (1982), *Indian Agricultural Economics: Myths and Realities* (New Delhi, Allied Publishers Private Ltd.).

SEN, A. K. (1964), 'Size of Holdings and Productivity', *The Economic Weekly*, 16: 323–6.

SHERLOCK, R. L. (1960), *British Regional Geology: London and Thames Valley* (London, HMSO, 3rd edn.).

SHIRLEY, E. P. (1867), *English Deer Parks* (London, John Murray).

SIMMONS, Jack (1974), *Leicester Past and Present* (London, Eyre Methuen).

SIMON, Herbert A. (1982), *Models of Bounded Rationality* (Cambridge, Mass., MIT Press).

SIMPSON, A. W. B. (1961), *An Introduction to the History of the Land Law* (Oxford, Oxford University Press).

——(1986), *A History of the Land Law* (Oxford, Clarendon Press, 2nd edn).

SLICHER VAN BATH, B. H. (1963), *Yield Ratios, 810–1820, A.A.G. Bijdragen*, No. 10.

SMITH, A. (1776), *An Inquiry into the Nature and Causes of the Wealth of Nations*, ed. E. Cannan (New York, The Modern Library, Random House, 1937).

SNELL, K. D. M. (1985), *Annals of the Labouring Poor: Social Change and Agrarian England, 1660–1900* (Cambridge, Cambridge University Press).

SPENCELEY, G. F. R. (1973), 'The Origins of the English Pillow Lace Industry', *Agricultural History Review*, 21: 81–93.

SPRING, David (1963), *The English Landed Estate in the Nineteenth Century: Its Administration* (Baltimore, Johns Hopkins University Press).

SPUFFORD, Margaret (1974), *Contrasting Communities* (Cambridge, Cambridge University Press).

STAFFORD, William (1987), *Socialism, Radicalism, and Nostalgia: Social Criticism in Britain, 1775–1830* (Cambridge, Cambridge University Press).

STEUART, Sir James (1767), *An Inquiry into the Principles of Political Oeconomy*, ed. Andrew S. Skinner (Edinburgh, Oliver and Boyd, 1966).

STEWART, Dugald (1855), *Lectures on Political Economy*, vols. viii and ix of *The Collected Works of Dugald Stewart, Esq., F.R.S.S.*, ed. Sir William Hamilton, 11 vols., 1854–60 (Edinburgh, Thomas Constable).

STONE, Lawrence (1965), *The Crisis of the Aristocracy, 1558–1641* (Oxford, Clarendon Press, corrected edn, 1979).

——and STONE, Jeanne C. Fawtier (1984), *An Open Elite? England, 1540–1880* (Oxford, Clarendon Press).

STONE, Thomas (1785), *An Essay on Agriculture* (London, W. Whittingham).

——(1794), *General View of the Agriculture of the County of Bedford* (London, E. Hodson).

STURGESS, R. W. (1966), 'The Agricultural Revolution on the English Clays', *Agricultural History Review*, 14: 104–21.

TATE, W. E. (1939–42), 'Cambridgeshire Field Systems', *Proceedings of the Cambridge Antiquarian Society*, 40: 56–88.

——(1949), 'Enclosure Acts and Awards Relating to Warwickshire', *Transactions* (Birmingham Archaeological Society) 65: 45–104.

——(1946), *A Hand-List of Buckinghamshire Enclosure Acts and Awards* (Aylesbury, Buckinghamshire County Council).

——(1952), 'The Cost of Parliamentary Enclosure in England (With Special Reference to the County of Oxford)', *Economic History Review*, 2nd series, 5: 258–65.

——(1967), *The English Village Community and the Enclosure Movements* (London, Victor Gollancz).

——(1978), *A Domesday of English Enclosure Acts and Awards*, ed. M. E. Turner (unpublished MS, Reading, The Library, University of Reading).

TAWNEY, R. H. (1912), *The Agrarian Problem in the Sixteenth Century* (London, Longmans, Green).

——and POWER, E. (1924), *Tudor Economic Documents*. 3 vols. (London, Longmans, Green).

TAYLOR, E. G. R. (1930), *Tudor Geography, 1485–1583* (London, Methuen).

——(1934), *Late Tudor and Early Stuart Geography, 1583–1650* (London, Methuen).

TEMPLE, M. S. (1929), *A Survey of the Soils of Buckinghamshire* (University of Reading, Faculty of Agriculture and Horticulture, Department of Agricultural Chemistry, Bulletin No. 38).

THICK, Malcolm (1985), 'Market Gardening in England and Wales', in Joan Thirsk (ed.), *The Agrarian History of England and Wales*, v, pt II, 1640–1750: Agrarian Change (Cambridge, Cambridge University Press), pp. 503–32.

THIRSK, Joan (1952), 'The Sales of Royalist Land during the Interregnum', *Economic History Review*, 2nd series, 5/2: 188–207.

——(1954a), 'The Restoration Land Settlement', *Journal of Modern History*, 26/4: 315–28.

——(1954b), 'Agrarian History, 1540–1950', *Victoria County History of Leicestershire* (London, Oxford University Press), ii, pp. 199–264.

——(1961), 'Industries in the Countryside', in F. J. Fisher (ed.), *Essays in the Economic and Social History of Tudor and Stuart England* (Cambridge, Cambridge University Press), pp. 70–88.

——(1967a), 'The Farming Regions of England', in Joan Thirsk (ed.), *The Agrarian History of England and Wales, iv, 1500–1640* (Cambridge, Cambridge University Press), pp. 1–112.

——(1967b), 'Enclosing and Engrossing', in Joan Thirsk (ed.), *The Agrarian History of England and Wales, iv, 1500–1640*, pp. 200–55.

——(1973), 'The Fantastical Folly of Fashion: The English Stocking Knitting Industry, 1500–1700', in N. B. Harte and K. G. Ponting (eds.), *Textile History and Economic History* (Manchester, Manchester University Press), pp. 50–73.

——(1985), 'Agricultural Innovations and their Diffusion', in Joan Thirsk (ed.), *The Agrarian History of England and Wales, v, pt II, 1640–1750: Agrarian Change* (Cambridge, Cambridge University Press), pp. 533–89.

——(1987), *England's Agricultural Regions and Agrarian History, 1500–1750* (Houndmills, Basingstoke, Hants, Macmillan Educational).

THOMPSON, E. P. (1963), *The Making of the English Working Class* (New York, Vintage Books).

——(1971), 'The Moral Economy of the English Crowd in the Eighteenth Century', *Past and Present*, No. 50: 76–136.

THOMPSON, E. P. (1976), 'The Grid of Inheritance: A Comment', in Jack Goody, Joan Thirsk, and E. P. Thompson (eds.), *Family and Inheritance: Rural Society in Western Europe, 1200–1800* (Cambridge, Cambridge University Press), pp. 328–60.

THOMPSON, F. M. L. (1963), *English Landed Society in the Nineteenth Century* (London, Routledge and Kegan Paul).

——(1966), 'The Social Distribution of Landed Property in England since the Sixteenth Century', *Economic History Review*, 19: 505–17.

——(1968), *Chartered Surveyors: The Growth of a Profession* (London, Routledge and Kegan Paul).

——(1969), 'Landownership and Economic Growth in England in the Eighteenth Century', in E. L. Jones and S. J. Woolf (eds.), *Agrarian Change and Economic Development* (London, Methuen), pp. 41–60.

THORNTON, W. T. (1843), *A Plea for Peasant Proprietors* (London).

TILLY, C. (1981), *As Sociology Meets History* (New York, Academic Press).

TIMMER, C. P. (1969), 'The Turnip, the New Husbandry, and the English Agricultural Revolution', *Quarterly Journal of Economics*, 83: 375–95.

——FALCON, W. R., and PEARSON, S. R. (1983), *Food Policy Analysis* (Baltimore, Johns Hopkins University Press.)

TITOW, J. Z. (1969), *English Rural Society, 1200–1350* (London, George Allen and Unwin).

——(1972), *Winchester Yields: A Study in Medieval Agricultural Productivity* (Cambridge, Cambridge University Press).

TOCQUEVILLE, Alexis de (1833), *Journeys to England and Ireland*, trans. George Lawrence and K. P. Mayer, ed. J. P. Mayer (London, Faber and Faber, n.d.)

TOMALIN, G. H. J. (1975), *The Book of Henley-on-Thames* (Chesham, Buckinghamshire, Barracuda Books).

TOMSON, James (1847), 'Account of Hall Farm', *Journal of the Royal Agricultural Society of England*, 8: 33–46.

TOYNBEE, A. (1884), *Lectures on the Industrial Revolution in England* (Newton Abbot, David and Charles Reprints, 1969).

TRACY, Michael (1964), *Agriculture in Western Europe* (London, Jonathan Cape).

TROW-SMITH, Robert (1959), *A History of British Livestock Husbandry, 1700–1900* (London, Routledge and Kegan Paul).

TUCKER, R. S. (1936), 'Real Wages of Artisans in London, 1729–1935', *Journal of the American Statistical Association*, 31: 73–84.

TUCKETT, Philip D. (1863), 'On Land Valuing', *Journal of the Royal Agricultural Society of England*, 24: 1–7.

TURNER, Michael (1973a), 'The Cost of Parliamentary Enclosure in Buckinghamshire', *Agricultural History Review*, 21: 35–46.

——(1973b), *'Some Social and Economic Considerations of Parliamentary Enclosure in Buckinghamshire, 1738–1865'*, University of Sheffield, Ph.D. thesis.

——(1975), 'Parliamentary Enclosure and Landownership Change in Buckinghamshire', *Economic History Review*, 2nd series, 28: 565–81.

——(1980), *English Parliamentary Enclosures: Its Historical Geography and Economic History* (Folkestone, Wm. Dawson and Sons).

——(1981), 'Cost, Finance, and Parliamentary Enclosure', *Economic History Review*, 2nd series, 34: 236–48.

——(1982*a*), 'Agricultural Productivity in England in the Eighteenth Century: Evidence from Crop Yields', *Economic History Review*, 2nd series, 35: 489–510.

——(1982*b*), *Home Office Acreage Returns: Bedfordshire to Isle of Wight* (London, Public Record Office, List and Index Society, No. 189).

——(1982*c*), *Home Office Acreage Returns: Jersey to Somerset* (London, Public Record Office, List and Index Society, No. 190).

——(1983*a*), *Home Office Acreage Returns: Staffordshire to Yorkshire* (London, Public Record Office, List and Index Society, No. 195).

——(1983*b*), *Home Office Acreage Returns: Index* (London, Public Record Office, List and Index Society, No. 196).

——(1984*a*), *Enclosures in Britain, 1750–1830* (London, Macmillan).

——(1984*b*), 'Agricultural Productivity in Eighteenth Century England: Further Strains of Speculation', *Economic History Review*, 2nd series, 37: 252–7.

——(1986), 'English Open Fields and Enclosures: Retardation or Productivity Improvements', *Journal of Economic History*, 41: 669–92.

——and MILLS, Dennis (eds.) (1986), *Land and Property: The English Land Tax 1692–1832* (New York, St Martin's Press).

TURNER, R. W. (1931), *The Equity of Redemption* (Cambridge, Cambridge University Press).

VAGGI, Gianni (1987), *The Economics of François Quesnay* (London, Macmillan).

VANCOUVER, C. (1794), *General View of the Agriculture of the County of Cambridge* (London).

——(1795), *General View of the Agriculture of the County of Essex* (London).

VANDENBROEKE, Christian (1984), 'Le Cas flamand: évolution sociale et comportements démographiques aux XVIIe–XVIIIe siècles', *Annales: économies, sociétés, civilisations*, 39: 915–38.

VAN DER WEE, Herman (1963), *The Growth of the Antwerp Market and the European Economy*, 3 vols. (The Hague, Martinus Nijhoff).

VAN ZANDEN, J. J. (1985), *De economische ontwikkeling van de Nederlandse landbouw in de negentiende eeuw, 1800–1914*, A.A.G. Bijdragen, No. 25.

VCH (1900–), *The Victoria History of the Counties of England*.

WALLACE, Robert (1753), *A Discourse on the Numbers of Mankind* (Edinburgh, Archibald Constable, 2nd edn, 1809).

WALLERSTEIN, I. (1974), *The Modern World-System: Capitalist Agriculture and the Origins of the European World-Economy in the Sixteenth Century* (New York, Academic Press).

WARD, W. R. (1953), *The English Land Tax in the Eighteenth Century* (Oxford, Oxford University Press).

WAREING, John (1980), 'Changes in the Geographical Distribution of the Recruitment of Apprentices to the London Companies, 1486–1750', *Journal of Historical Geography*, 6/3: 241–9.

WAREING, John (1981), 'Migration to London and Transatlantic Emigration of Indentured Servants, 1683–1775', *Journal of Historical Geography*, 7/4: 336–78.

WARRINER, Doreen (1939), *Economics of Peasant Farming* (New York, Barnes and Noble, 2nd edn, 1965).

WATKINS, Charles (1816), *A Treatise on Copyholds* (London, W. Clarke and Sons, 2nd edn).

WEDGE, Thomas (1794), *General View of the Agriculture of the County Palatine of Chester* (London, Colin Macrae).

WELLS, F. A. (1935), *The British Hosiery and Knitwear Industry: Its History and Organization* (Newton Abbot, David and Charles, revised edn, 1972).

WESTERFIELD, Ray B. (1915), *Middlemen in English Business, 1660–1760* (New Haven, Yale University Press).

WILLIAMSON, Jeffrey G. (1985), *Did British Capitalism Breed Inequality?* (Boston, Allen and Unwin).

——(1987), 'Did English Factor Markets Fail During the Industrial Revolution?' *Oxford Economic Papers*, 39: 641–78.

WILSON, C. (1984), *England's Apprenticeship, 1603–1763* (London, Longman, 2nd edn).

WILSON, J. M. (1847), *The Rural Encyclopedia* (Edinburgh, A. Fullarton).

WILTSE, Charles M. (1935, 1960), *The Jefferson Tradition in American Democracy* (New York, Hill and Wong).

WORDIE, J. R. (1974), 'Social Change on the Leveson-Gower Estates', *Economic History Review*, 2nd series, 27: 593–609.

——(1981), 'Rent Movements and the English Tenant Farmer, 1700–1839', *Research in Economic History*, 6: 193–243.

——(1983), 'The Chronology of English Enclosure, 1500–1914', *Economic History Review*, 2nd series, 36: 483–505.

WRATISLAW, Charles (1861), 'The Amount of Capital Required for the Profitable Occupation of a Mixed Arable and Pasture Farm in a Midland County', *Journal of the Royal Agricultural Society of England*, 22: 167–89.

WRIGHT, J. F. (1955), 'A Note on Mr Bowden's Index of Wool Prices', *Yorkshire Bulletin of Economic and Social Research*, 7.

WRIGLEY, E. A. (1967), 'A Simple Model of London's Importance in Changing English Society and Economy, 1650–1750', *Past and Present*, 37: 44–70.

——(1985), 'Urban Growth and Agricultural Change: England and the Continent in the Early Modern Period', *Journal of Interdisciplinary History*, 15: 683–728.

——(1986), 'Men on the Land and Men in the Countryside: Employment in Agriculture in Early-Nineteenth-Century England', in Lloyd Bonfield, Richard M. Smith, and Keith Wrightson (eds.), *The World We Have Gained* (Oxford, Basil Blackwell), pp. 293–336.

——and SCHOFIELD, R. S. (1981), *The Population History of England, 1541–1871* (Cambridge, Mass., Harvard University Press).

YELLING, J. A. (1977), *Common Field and Enclosure in England, 1450–1850* (Hamden, Conn., Archon Books).

YOUNG, Arthur (1769), *A Six Weeks Tour Through the Southern Counties of England and Wales*, 2nd edn (London, W. Nicoll).

——(1770), *The Farmer's Guide in Hiring and Stocking Farms* (London, W. Strahan).

——(1771*a*), *A Six Month's Tour Through the North of England*, 2nd edn, 4 vols. (London, W. Strahan).

——(1771*b*), *The Farmer's Tour Through the East of England*, 4 vols. (London, W. Strahan).

——(1774), *Political Arithmetic* (London, W. Nicoll).

——(1785), 'A Tour to Shropshire', *Annals of Agriculture*, 4: 138–90.

——(1786), 'A Ten Days Tour to Mr Bakewell's', *Annals of Agriculture*, 6: 452–502.

——(1791), 'A Month's Tour to Northamptonshire, Leicestershire, &c.', *Annals of Agriculture*, 16: 480–607.

——(1792), 'Observations', *Annals of Agriculture*, 17: 45–7.

——(1794), *Travel During the Years 1787, 1788, 1789, Undertaken with a View of Ascertaining the Cultivation, Wealth, Resources, and National Prosperity of . . . France* (London, 2nd edn).

——(1804*a*), 'Minutes Concerning Parliamentary Inclosures in the County of Bedford', *Annals of Agriculture*, 42: 22–57.

——(1804*b*), 'Minutes Concerning Parliamentary Inclosures in the County of Cambridge', *Annals of Agriculture*, 42: 471–502.

——(1804*c*), *General View of the Agriculture of Hertfordshire* (London, R. Phillips).

——(1805), *The Farmer's Kalendar* (London, R. Phillips).

——(1807), *General View of the Agriculture of the County of Essex*, 2 vols. (London, R. Phillips).

——(1808), *General Report on Enclosures* (London).

——(1813), *General View of the Agriculture of Oxford* (London).

——(1815), 'An Inquiry into the Rise of Prices in Europe', *The Pamphleteer*, vi.

ZOHARY, Daniel (1969), 'The Progenitors of Wheat and Barley in Relation to Domestication and Agricultural Dispersal in the Old World', in Peter J. Ucko and G. W. Dimbleby (eds.), *The Domestication and Exploitation of Plants and Animals* (London, Gerald Duckworth) pp. 47–66.

ZUPKO, Ronald Edward (1968), *A Dictionary of English Weights and Measures From Anglo-Saxon Times to the Nineteenth Century* (Madison, University of Wisconsin Press).

——(1978), *French Weights and Measures Before the Revolution* (Bloomington, Indiana University Press).

ZYTKOWICZ, Leonid (1971), 'Grain Yields in Poland, Bohemia, Hungary, and Slovakia in the 16th to 18th Centuries', *Acta Poloniae Historica*, 24: 51–72.

# INDEX